ELEMENTS OF STATISTICAL MECHANICS

TO MY WIFE

ELEMENTS OF STATISTICAL MECHANICS

Third editon

D. ter Haar
Emeritus Fellow
Magdalen College
Oxford

Butterworth-Heinemann Ltd
Linacre House, Jordan Hill, Oxford OX2 8DP

℞ A member of the Reed Elsevier plc group

OXFORD LONDON BOSTON

MUNICH NEW DELHI SINGAPORE SYDNEY

TOKYO TORONTO WELLINGTON

First published by Rinehart 1954
Second edition 1966
Third edition published by Butterworth-Heinemann Ltd 1995

British Library Cataloguing in Publication Data

A catalogue record for this book is available from the British Library

ISBN 0 7506 2347 0

Library of Congress Cataloguing in Publication Data

A catalogue record for this book is available from the Library of Congress

Printed in Great Britain by Clays Ltd, St Ives plc

TABLE OF CONTENTS

PREFACE TO THE THIRD EDITION

It is nearly thirty years since the publication of the second edition and fully forty years after the publication of the first edition of this book. There have been many advances in statistical mechanics in that period, but it seems to me that the basic methods as described in the present book have been changed very little. This, combined with the urging of several friends and colleagues to make it again available, has persuaded me to present a third edition.

It will be seen that about two thirds of the book is essentially the same as it was in the first two editions. This refers especially to the first six chapters. The main change in those chapters since the second edition is that I have now included a discussion of the relation between spin and statistics in the main text and that I have mentioned anyons. My original intention was to devote more space to anyons and the rôle they play in the theory of the fractional quantum Hall effect and possibly in the theory of high-temperature superconductivity, but it soon became clear to me that in order to do this adequately I would have to go well beyond the general level of the book and also allot a disproportionate amount of space to the subject.

When I prepared the second edition it was my intention that it would be the first of a two-volume book, the second volume containing an update of those parts of the first edition which had not been covered in the first volume. This second volume was never written — to a large extent because developments had been so fast and extensive that I felt it was beyond my capability to do justice to the subject. However, this meant that such topics as the equation of state, phase transitions, the Ising model, and also modern ergodic theory, were not covered. This omission has now been made good to some extent in the present edition. Unfortunately the same reasons which made me omit a more extensive discussion of anyons also made me decide not to discuss in detail modern developments dealing with the foundations of statistical mechanics. However, I have once again a chapter on the equation of state, albeit not as extensive a one as in the first edition. Also, I have included again a chapter on phase transitions which follow to a large extent what was covered in the first edition, but with a more extensive discussion of Yang and Lee's theory of phase transitions — which appeared just when the manuscript for the first edition was completed — and a brief consideration of recent theories of critical phenomena, including the renormalisation group method.

The only real new topic in this edition is covered in Chapter 8 where I discuss the occupation number representation and the Green function method. I had originally intended to cover also the statistical mechanics of solitons, but there again this would have made it necessary to include a thorough discussion of the inverse scattering technique and would lead to a disproportionate amount of space being devoted to a subject which, though extremely interesting, is not really central to modern statistical mechanics. I also had hoped to discuss diagram techniques, but when looking at them I found also here that this would involve expanding the book much more than I wanted to do and, moreover, I found that the variety of diagram techniques in statistical mechanics is *legio*: they range from the statistical mechanics counterpart of the quantummechanical Feynman technique to various *ad hoc* techniques — all demonstrating that diagram techniques are, indeed, essentially bookkeeping methods in perturbation theory.

Due partly to a desire not too make too many changes where I, possibly mistakenly, thought that the treatment was satisfactory and partly to the fact that in my retirement I do no longer have easy access to extensive library facilities the coverage of the literature since 1966 is far from satisfactory and I offer my sincere apologies for this — if I had tried to remedy this, I fear that this edition would never have been finished.

In conclusion I would like to thank many friends and colleagues for help in pointing out shortcomings and in helping me with references to more modern sources. I would especially like to mention Ian Aitchison, Dusan Radojicic, and Stig Stenholm.

D.t.H.

Petworth, West Sussex
February, 1994

PREFACE TO THE SECOND EDITION

More than thirteen years have elapsed since the manuscript of the first edition was completed. For some time it has been clear that a second edition was needed. In the last fifteen years statistical mechanics has developed at an enormous pace, and a textbook on statistical mechanics should take these developments into account. When I started to contemplate what changes should be made, it soon became clear that, if the book were to cover even approximately the same ground as the first edition did, it would have to be greatly expanded. It was then decided to split the book into two parts, the first volume covering the basic theory; a second volume, advanced theory. As a result, the present volume is now probably more suitable as a textbook for advanced undergraduate courses than the first edition was. To improve its usefulness a variety of problems has been included. I am greatly indebted to R. Kubo and the North-Holland Publishing Company for permission to incorporate some of the problems from Kubo's *Statistical Mechanics* and to the Oxford University Press for permission to incorporate problems from Oxford University Examination Papers. To some extent I have adopted Kubo's philosophy in that I feel that, to get the greatest advantage from the present volume, the student should work through most of the problems and so learn the subject by practising it. A number of topics — especially in the theory of metals and semiconductors — which were treated in the text of the first edition, have now been relegated to the problems section, partly to give the students experience in using statistical methods and partly to make room for other topics.

I should like to express my thanks to the many readers, users, and reviewers who have given me the benefit of their comments and criticism. In many cases I have followed their advice, and I have found that the reactions of my own students especially have been invaluable in finding out obscure passages that needed clarification.

I have finally had the strength of my convictions and have used Kramers' terminology "Thermostatistics" in the title of the second edition.

A comparison between the first edition and the second edition will show that the present volume contains Chapters I to IV of the first edition as Chapters 1 to 4; Chapters V and VI of the first edition make up the present Chapter 5, while Chapter VII becomes Chapter 6. Parts of Chapter IX of the first edition are incorporated in Chapter 4 and in a problem at the end of Chapter 5; Chapters X and XI of the old edition have been nearly completely incorporated in various problems; and some topics of Chapter XIII have been put either in problems or in Chapters 5 and 6; Appendix II has become Chapter 8; Appendix III, Chapter 9; Appendix IV, Section 4.5; and Appendix VI, Chapter 7. The various parts of the mathematical appendix have been inserted at the appropriate spots in the earlier chapters. The main new topics covered in this book are the Saha equilibrium (Section 6.9), the Kramers-Casimir discussion of the third law (Section 9.2), fluctuations (Sections 7.1 and 7.2), and a more extensive discussion of the density matrix. A fair amount of new material is also to be found in the problems sections. Readers will probably also notice that the notation has been changed considerably to bring it in line with present-day practice.

Once again I appeal to readers and reviewers to let me have the benefit of their detailed criticism.

D.t.H.

Magdalen College, Oxford
May 1966

PREFACE TO THE FIRST EDITION

It is usual for an author to explain in a preface the reasons for writing the particular book which he is presenting to the public, to state who are the readers whom he has in mind for the book, to sketch the history of the writing of the book, and, last but not least, to express his thanks to all people who have been of assistance during the completion of the book.

The reason for writing another textbook on statistical mechanics was the feeling that there should be a textbook which combined in not too large a volume an outline of the main elements of statistical mechanics, starting from the Maxwell distribution and ending with quantum mechanical grand ensembles, with an account of a number of successful applications of these elements. Almost all existing textbooks stress only one or the other of these two aspects. It is hoped that readers will point out to me how far my own attempt has been successful, and I should at this point like to express the hope that reviewers will let me have the benefit of their detailed criticism.

The book is meant to be a textbook and is thus primarily intended for students. I have had in mind graduate students. This means that it should be used as a text for graduate lectures in the United States and for post-graduate lectures in the United Kingdom. It will probably be too advanced as a textbook for honours courses in British universities, although parts of it might be used as such and have been used as such by me. It is hoped that the book can also be used as a research tool and that it is possible to see from the applications how the theory might be applied to other subjects. For that reason I have tried to give as complete a bibliography as was feasible in the framework of a textbook. As the manuscript of the book was essentially completed at the beginning of 1952, references to papers published in 1951 or 1952 will not be complete.

A first rough outline of the book was sketched during the last war-winter in Leiden. A number of students were deprived of the regular courses because Leiden University was closed by the occupying authorities, and the outline of the present book served as a substitute for the regular statistical mechanics course. The manuscript then rested until I wrote the first draft of Parts A and B at Purdue University during 1947-1948. There was another interval until 1950, when I came to St. Andrews, where the manuscript in its present form was started and finished.

It is a pleasant task to thank the many physicists who have given me their advice on parts of the manuscript. My thanks are particularly due to Professors F.J.Belinfante, H.M.James, K.Lark-Horovitz, H.Margenau, R.E.Peierls, and F.E.Simon, who have helped me with their criticism and advice. If the approach is sometimes not very clear and if my English has sometimes a distinct foreign flavour, it cannot be blamed on Professor F.Y.Poynton, who has tried to make Parts A and B as far as possible easy reading for students, or on Professors E.S.Akeley and J.F.Allen, who have tried to weed out of the text all barbarisms. I should like to express to them my warmest gratitude. Finally I should like to express my great indebtedness to Professor H.A.Kramers. Anybody who is familiar with his lectures on statistical mechanics will immediately see how much this book owes to him. It is far from a platitude to say that it would never have been written but for Professor Kramers. Not only did he give me an outline of the contents of Parts A and B; in discussions and lectures he has taught me the fundamental ideas of the subject. I can therefore with some justification claim that the method of treatment in Parts A and B goes straight back to Boltzmann, via Kramers and Ehrenfest. Large parts of the book are, indeed, nearly wholly based on a series of lectures given by Professor Kramers in Leiden during 1944-1945.

In conclusion, I should like to express my thanks to Professor K.Lark-Horovitz and Miss A.Scudder for their help in editing the manuscript.

D.t.H.

Department of Natural Philosophy
St. Andrews
January, 1954

CHAPTER 1

THE MAXWELL DISTRIBUTION

1.1. The Maxwell Distribution

Equilibrium statistical mechanics, or "thermostatistics", to use a term coined by Kramers,[1] is that branch of physics which attempts to derive the equilibrium or thermal properties of matter in bulk, and of radiation, from the properties of the constituent particles. Such an atomistic interpretation of the thermal properties of matter was first attempted in the kinetic theory of gases, developed in the nineteenth century by Clausius, Maxwell, and Boltzmann. In this theory it was shown how such phenomenological concepts as temperature and entropy could be interpreted in terms of the *average* properties of the particles that were the constituent parts of the system under consideration. Kinetic theory could be applied as long as one could either completely neglect the interactions between the particles or could neglect them to a first approximation, taking them into account in a second approximation. Such an approach leads, for instance, to the van der Waals law of an imperfect gas, as we shall see in § 1.3 of this chapter. In most physical systems, however, this approach is much too simplified, and one needs have recourse to more sophisticated methods of averaging.

This more sophisticated approach is statistical mechanics proper, a term coined by J.W.Gibbs in 1901 for that branch of rational mechanics which deals statistically with systems consisting of large numbers of constituents. In the first four chapters we shall consider systems of independent particles, a subject which is really part of kinetic theory, while in later chapters we shall consider statistical mechanics proper. The reason for using statistical methods to treat physical systems is partly because they are so complicated that they present us with "well-nigh unsurmountable mathematical difficulties",[2] if we try to solve their equations of motion exactly; and partly because, even if we could solve these equations of motion exactly, we have only incomplete experimental data from which we can obtain the boundary conditions for the equations of motion.

In the present chapter we shall confine ourselves to the simplest possible

---

[1] H.A.Kramers, *Nuovo cimento* **6** Suppl., 158 (1949).

[2] H.A.Kramers, *ibid.*

system: a monatomic gas, that is, a system of point particles, enclosed in a vessel of volume V without any external forces acting upon the gas apart from the forces which the walls of the vessel will exert upon the gas and which, in fact, will keep the gas within the vessel. We introduce Cartesian coordinates x, y, and z to describe the system; the position of the i-th particle is thus determined by its three position coordinates x_i, y_i, and z_i, or by the vector r_i.[3] We denote by u, v, and w the x, y, and z components of a velocity c, and u_i, v_i, and w_i are thus the components of the velocity c_i of the i-th particle. Once the positions and velocities of all the atoms are given, the microscopic behaviour of the system is completely determined, provided the interatomic forces and the forces exerted by the walls on the atoms are known. If N is the number of atoms in the system, we need $6N$ quantities, for instance, x_i, y_i, z_i, u_i, v_i, w_i ($i = 1$ to N) to determine the microscopic behaviour. However, usually we are interested in only a few combinations of these $6N$ quantities which will determine the macroscopic behaviour of the system. We mentioned a moment ago that the exact knowledge of all $6N$ coordinates and velocities is outside the experimental possibilities, while the computation of their values from the equations of motion is outside our mathematical powers, as N is usually an extremely large number — of the order of 10^{25} for one m^3 of a gas at NTP. We can use thet fact that N is such a huge number to apply statistical methods in the safe knowledge that, because of the extremely large numbers of degrees of freedom, fluctuations will in general be small (compare §§ 5.4 and 5.10).

Let us for the moment neglect first of all the influence of the wall and secondly possible fluctuations. In that case the number of atoms in a unit volume of the gas will be independent of the position of that unit volume in the gas. If we denote the number of atoms per unit volume by n, we have

$$n = \frac{N}{V}. \tag{1.101}$$

Let us denote by

$$f(u, v, w) \, du \, dv \, dw$$

the number of atoms per unit volume, the velocity components of which lie in the specified intervals $(u, u + du)$, $(v, v + dv)$, and $(w, w + dw)$. The function $f(u, v, w)$ will be called the *distribution function*. It determines the fraction of atoms with velocities within given intervals. This fraction is obtained by dividing $f(u, v, w)$ by n.

We shall call the Cartesian three-dimensional space in which we can plot the x, y, and z components of the velocities *velocity space*, and the point u, v, w in velocity space will be called the *representative point* of an atom with velocity components u, v, and w.

From the definition of $f(u, v, w)$ it follows that it satisfies a *normalisation*

---

[3] Vectors are denoted by bold face italic type.

condition:

$$\int_{-\infty}^{+\infty} du \int_{-\infty}^{+\infty} dv \int_{-\infty}^{+\infty} dw \, f(u, v, w) \; = \; n. \qquad (1.102)$$

In the present chapter we shall assume that $f(u, v, w)$ is not only independent of x, y, and z, but also does not depend explicitly upon the time t. Let A be a quantity that is a function of the velocity components of an atom, but which does not depend explicitly on either x, y, and z or t. As an example we may give the kinetic energy of an atom. We can now ask for the *average value*, $\overline{A}$, of $A(u, v, w)$, where the average is taken over all atoms of the gas and where the average value is equivalent to the arithmetic mean, that is, defined by the equation

$$\overline{A} \; = \; \frac{1}{n} \int\!\!\!\int\!\!\!\int_{-\infty}^{+\infty} du \, dv \, dw \, A(u, v, w) \, f(u, v, w). \qquad (1.103)$$

Neither A nor f depends on x, y, z, or t so that $\overline{A}$ will also be independent of x, y, z, and t.

In the following sections of this chapter it will be proved that the distribution function of a gas in equilibrium at an absolute temperature T will be given by the equation

$$f(u, v, w) \; = \; n \left(\frac{m}{2\pi kT} \right)^{3/2} e^{-m(u^2+v^2+w^2)/2kT}, \qquad (1.104)$$

where m is the mass of one atom and k Boltzmann's constant.[4] The distribution given by Eq.(1.104) is called the *Maxwell distribution* and was first introduced by Maxwell in 1859. It can easily be shown that the $f(u, v, w)$ given by Eq.(1.104) satisfies Eq.(1.102). Equations (1.103) and (1.104) can now be used to calculate average values.

Denoting by c the absolute value of the velocity of an atom,

$$c^2 \; = \; u^2 + v^2 + w^2, \qquad (1.105)$$

and by $\mathcal{T}(= \frac{1}{2}mc^2)$ the kinetic energy of an atom, we have

$$\overline{c^2} \; = \; \frac{3kT}{m}, \qquad (1.106)$$

$$\overline{c} \; = \; \sqrt{\frac{8kT}{\pi m}}, \qquad (1.107)$$

$$\overline{u} \; = \; \overline{v} \; = \; \overline{w} \; = \; 0, \qquad (1.108)$$

$$\overline{u^2} \; = \; \overline{v^2} \; = \; \overline{w^2} \; = \; \frac{kT}{m}, \qquad (1.109)$$

$$\overline{\mathcal{T}} \; = \; \overline{\tfrac{1}{2}mc^2} \; = \; \tfrac{3}{2}kT. \qquad (1.110)$$

---

[4] Boltzmann never deduced the value of k; this was first done by Planck in 1900 in connection with his radiation law. Smekal (*Enzyklopädie der Mathematischen Wissenschaften*, Vol.V, Part 28, Leipzig-Berlin, 1926) and Meissner (*Science*, **113**, 78, 1951) suggest calling k the Boltzmann-Planck constant. It is interesting to note that Lorentz (see, for example, *Lectures in Theoretical Physics*, Vol.II, p.175, London 1927) calls k Planck's constant!

There are three points in connection with Eqs.(1.106) to (1.110) worth noting:

1. Compare Eqs.(1.106) and (1.107). We see that there exists a difference between the mean absolute velocity $\bar{c}$ and the root-mean-square velocity $\sqrt{\overline{c^2}}$:

$$\sqrt{\overline{c^2}} \;=\; \sqrt{\frac{3\pi}{8}}\,\bar{c} \;=\; 1.085\,\bar{c}. \tag{1.111}$$

2. Equation (1.110) shows that the mean kinetic energy is independent of the mass of the atoms. Thus we get the same result for any gas, provided the temperature is the same. We can therefore use Eq.(1.110) as a definition of the temperature of a gas.

3. From Eq.(1.109) it follows that the average kinetic energy pertaining to the x, y, or z direction is the same and is equal to one third of the average total kinetic energy. This is an example of the *equipartition of kinetic energy*.

1.2.   The Perfect Gas Law

In this section we shall assume that there is no interaction between the atoms of the gas which we are considering.

If we assume that the distribution function is given by Eq.(1.104), we can calculate the pressure of the gas in terms of n and T. Two derivations of the formula for the pressure will be given. The first derivation is due to Clausius and the second one to Lorentz.

In order to derive the formula in the way it was done by Clausius, we must remember that according to Newton's second law force is equivalent to the rate of change of momentum. The pressure of the gas is defined either as the force exerted by the gas on unit area of the wall, or as the total transfer of momentum per unit time from the gas to unit area of the wall. The Cartesian axes will be chosen in such a way that the negative x axis falls along the normal to the wall (see Fig.1.1). We shall again neglect any effect arising from a possible potential energy between the wall and the gas atoms, and we shall assume that the atoms will be perfectly reflected by the wall.

In the case of a perfect reflection, only the velocity component perpendicular to the wall will change during the collision of a gas atom with the wall. This component will change its sign, but the other components will remain unchanged. If an atom with velocity c strikes the wall at an angle θ to the normal, the momentum transferred during the collision will be given by the expression

$$2mc\cos\theta \;=\; 2mu. \tag{1.201}$$

In order to calculate the total transfer of momentum, we must next determine the number of collisions per unit time per unit area. For a given velocity c this number is zero, if u is negative. If u is positive, the number of collisions is given by the total number of atoms contained in a cylinder with a base of unit area and a slant height c, parallel to $\mathbf{c}$. The volume of this cylinder is $c \cos \theta$. If the number of atoms per unit volume with given velocity $\mathbf{c}$ is $n(\mathbf{c})$, the number of collisions N_{coll} per unit time per unit area will be given by the equation

$$N_{\text{coll}}(\mathbf{c}) \; = \; n(\mathbf{c}) \, c \cos \theta \; = \; n(\mathbf{c}) \, u. \tag{1.202}$$

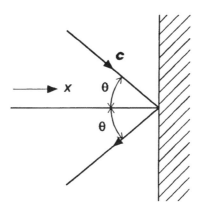

Fig.1.1. The elastic reflection of a point particle by a wall. The normal to the wall is along the negative x-axis, and a particle with velocity $\mathbf{c}$ strikes the wall at an angle θ to the normal.

The total momentum transferred to unit area of the wall per unit time by atoms of a given velocity $\mathbf{c}$ is found by combining Eqs.(1.201) and (1.202) and is equal to

$$2m \, n(\mathbf{c}) \, u^2. \tag{1.203}$$

The function $n(\mathbf{c})$ will be the distribution function $f(u, v, w)$, which is assumed to be given by Eq.(1.104). To obtain the pressure P we must integrate expression (1.203) over all velocities for which $u \geq 0$, so that we get

$$P \; = \; 2m \int_0^{+\infty} du \int_{-\infty}^{+\infty} dv \int_{-\infty}^{+\infty} dw \, f(u, v, w) \, u^2,$$

or,

$$P \; = \; nkT. \tag{1.204}$$

Equation (1.204) is called the *perfect gas law* or the Boyle–Gay-Lussac law. Introducing the absolute gas constant R by the equation

$$R \; = \; kN_{\text{A}}, \tag{1.205}$$

where N_A is the Avogadro number, that is the number of atoms in one mole, and introducing the molar volume V_m, we can write Eq.(1.204) in the form

$$PV_m = RT. \tag{1.206}$$

Clausius' derivation of the perfect gas law involves certain assumptions about the reflection of the atoms from the wall. These assumptions are unnecessary, if the pressure is defined in a slightly different manner. Consider for this purpose a surface of unit area somewhere in the gas. The pressure can now be defined as the total momentum in the direction of the normal to the surface transported per unit time through this surface in the direction of the positive normal.[5] Momentum transported in the opposite direction should be counted with a negative sign.

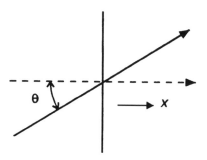

Fig.1.2. The transfer of momentum through a unit area somewhere in the gas.

The above definition is due to Lorentz, whose derivation of the perfect gas law proceeds as follows. Let the positive x axis be in the direction of the positive normal to the surface considered (see Fig.1.2). An atom that crosses the chosen surface in the direction of the positive normal with a velocity c will transport in the $+x$ direction an amount of momentum in the $+x$ direction equal to the x component of its own momentum ($= mu$). The total number of atoms crossing the unit area per unit time is again given by expression (1.202). In this way, the total momentum transported by atoms with a velocity c, the x component of which is positive, will be given by the expression

$$n(c)\, u\, mu, \qquad u > 0. \tag{1.207}$$

---

[5] Strictly speaking, one must still prove that the pressure defined in this way is identically equal to the force per unit area on the wall, since that force is the pressure which is measured. The proof follows, if one considers a cylinder with unit area cross-section and axis parallel to the normal to the wall, starting from the wall and extending into the gas. From the requirement that the gas inside the cylinder be in equilibrium, the equivalence of the two pressures follows.

The total momentum in $+x$ direction transported by atoms with velocities, the x component of which is negative, is also given by expression (1.207). The momentum is negative, but it is transported from the positive to the negative axis and should thus be subtracted from the total momentum transported through the surface.

The total pressure is thus obtained by integrating expression (1.207) over *all* possible c-values, or

$$ P \; = \; \overline{m\,n(c)\,u^2} \; = \; m \int\!\!\!\int\!\!\!\int\limits_{-\infty}^{+\infty} du\,dv\,dw\, f(u,v,w)\,u^2, \qquad (1.208) $$

which leads again to Eq.(1.204).

1.3.   The van der Waals Law

We mentioned at the beginning of this chapter that the simple approach of kinetic theory — with which we are concerned in this part of the book — entails that one either completely neglects all interactions between the atoms or at best takes them into account as a small perturbation which leads to slight modifications of the results obtained when neglecting them. Along these lines we shall in the present section investigate qualitatively how the perfect gas law must be modified when the particles in the gas, which we assume to be neutral atoms, exert forces upon one another. We shall assume that these forces are central forces and that they can be derived from a potential energy curve such as the one shown in Fig.1.3. From this figure we see that at large distances apart two atoms will attract one another, while at small distances apart they repel one another. The attraction is due to the polarisation of the electron cloud of the one atom by fluctuations in the charge distribution in the electron cloud of the other atom, while the repulsion occurs when the two electron clouds begin to overlap; the repulsion has the same effect, as if the atoms had finite dimensions.

Various equations have from time to time been proposed for the potential energy $U(r)$. We mention just three of those which are possibly the ones which are most widely used.

The Morse potential:[6]

$$ U(r) \; = \; D \left[e^{-2a(r-r_1)} - 2e^{-a(r-r_1)} \right], \quad D > 0, \quad a > 0, \quad r_1 > 0; \quad (1.301) $$

the Lennard-Jones potential:[7]

$$ U(r) \; = \; \frac{A}{r^n} - \frac{B}{r^m}, \quad n > m, \quad A > 0, \quad B > 0; \qquad (1.302) $$

[6] P.M.Morse, *Phys.Rev.* **34**, 57 (1929).

[7] J.E.Jones, *Proc.Roy.Soc.(London)* **A106**, 463 (1924).

and the so-called "exp-six" potential:[8]

$$U(r) = Ae^{-ar} - \frac{B}{r^6}, \quad A > 0, \quad B > 0, \quad a > 0. \tag{1.303}$$

The last potential approximates most closely the actual potential energies that can be derived from first principles, but the first two were suggested before the advent of powerful computers as they are more convenient to use for calculations.

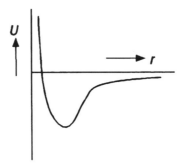

Fig.1.3. The interatomic potential, U, as a function of the distance apart, r, of the atoms.

As the repulsive forces between the atoms are equivalent to the atoms taking up a certain volume, this will lead to a smaller available volume for the gas. We must thus replace the factor V_m in Eq.(1.206) by a slightly smaller factor, say, $V_m - b$. If repulsive forces decrease the factor V_m, we should expect attractive forces to increase it; we must thus introduce a further correction and replace V_m by $V_m - b + a$, where we have now — completely artificially — separated the effects of the attractive and the repulsive forces. Instead of Eq.(1.206) we now have the following *equation of state*, as the relation between the pressure, the volume, and the temperature is called,

$$P(V_m - b + a) = RT. \tag{1.304}$$

It is essential for our argument that b and a are small corrections to V_m — a condition satisfied as long as the temperature or the density of the gas are sufficiently low. We can then rewrite Eq.(1.304) as follows

$$\left(P + \frac{a'}{V_m^2}\right)(V_m - b) = RT, \tag{1.305}$$

---

[8] R.A.Buckingham, *Proc.Roy.Soc.(London)* **A168**, 264 (1938).

where the left-hand sides of Eqs.(1.304) and (1.305) differ only by terms of second order in a/V_m or b/V_m, provided a' is given by the equation

$$a' = aRT.$$

The reason for rewriting Eq.(1.304) in the form (1.305) is purely a historical one. It is done in order to show that to a first approximation the changes introduced by considering the attractive and repulsive forces between the atoms in a gas will lead to the famous equation introduced by van der Waals in 1873 and hence called the *van der Waals law*. This equation was of the form (1.305). Its importance lies mainly in the very great number of successful empirical applications, but the reader may be reminded that neither Eq.(1.304) nor Eq.(1.305) gives us the correct equation of state.

It is instructive to write Eq.(1.304) in yet another way:

$$PV_m = RT \left[1 + \frac{b-a}{V_m} + \cdots \right]. \tag{1.306}$$

Equation (1.306) is the form of the equation of state which is usually employed; this gives us a power series in descending powers of V_m which can be written in the form

$$PV_m = RT \left[1 + \frac{B}{V_m} + \frac{C}{V_m^2} + \cdots \right]. \tag{1.307}$$

In Eq.(1.307) the quantities $B, C, \cdots$, which are functions of T are called the *virial coefficients*, B being called the second virial coefficient, C the third virial coefficient, and so on. Comparing Eqs.(1.306) and (1.307), we see that the second virial coefficient is related to the b and a from the van der Waals equation through the equation

$$B = b - a. \tag{1.308}$$

In our present approximation we cannot say anything about the higher virial coefficients, and this must be left to the discussion in Chapter 7 when we shall consider the equation of state of an imperfect gas in detail.

The term "virial" used in this connection arises from the function

$$\mathcal{V} = \sum_i (\boldsymbol{r}_i \cdot \boldsymbol{F}_i), \tag{1.309}$$

which was introduced into kinetic theory by Clausius and called by him the *virial*. In Eq.(1.309) $\boldsymbol{F}_i$ is the total force (hence the name "virial" from the Latin *vis* for force) acting upon the i-th atom. The summation in Eq.(1.309) is over all atoms in the gas.

Clausius established an important theorem relating $\mathcal{V}$ to the pressure as follows. From Newton's second law we have

$$m_i \frac{d^2 \boldsymbol{r}_i}{dt^2} = \boldsymbol{F}_i,$$

and hence

$$V = \sum_i m_i \left(r_i \cdot \frac{d^2 r_i}{dt^2} \right).$$

Taking the time average (denoted by a tilde) of the virial, we get

$$\tilde{V} = \frac{1}{\tau} \int_0^\tau V \, dt = \frac{1}{\tau} \int_0^\tau \sum_i m_i \left(r_i \cdot \frac{d^2 r_i}{dt^2} \right) dt$$

$$= -\frac{1}{\tau} \int_0^\tau \sum_i m_i \left\{ \frac{dr_i}{dt} \right\}^2 + \frac{1}{\tau} \left[\sum_i m_i \left(r_i \cdot \frac{dr_i}{dt} \right) \right]_0^\tau \quad (1.310)$$

As the gas is confined to a finite volume, all r_i and dr_i/dt are always finite and by letting τ tend to ∞ the integrated term can be made arbitrarily small compared to the first term on the right-hand side of Eq.(1.310). Thus we get

$$\tilde{V} = -2N\tilde{\tilde{\mathcal{J}}}, \quad (1.311)$$

where N is again the number of atoms in the gas and $\tilde{\tilde{\mathcal{J}}}$ the average kinetic energy per atom; this last average is now an average both over time and over the system of all the atoms in the gas.

We can obtain another equation for $\tilde{V}$ by dividing the virial into two parts, one part, $\tilde{W}$, due to the intermolecular forces and one part due to the pressure on the walls of the vessel (or the forces exerted by the wall on the gas):

$$\tilde{V} = \tilde{W} + \int_{\text{walls}} \widetilde{(r \cdot F_{\text{wall}})}, \quad (1.312)$$

where the integral is over all atoms in contact with the wall. As the time-averaged force exerted by the gas on an element of the wall (see Fig.1.4) is equal to $P d^2 S$, where $d^2 S$ is a vector along the normal to the wall with a length equal to the area of the element, we have

$$\tilde{V} = \tilde{W} - \oint_{\text{walls}} P(r \cdot d^2 S), \quad (1.313)$$

or, after the usual transformation from a surface to a volume integral,

$$\tilde{V} = \tilde{W} - 3PV, \quad (1.314)$$

where V is the volume occupied by the gas.

Combining Eqs.(1.311) and (1.314) and using Eq.(1.110) for the average value of the kinetic energy per atom, we obtain the equation

$$PV = NkT + \tfrac{1}{3}\tilde{W}. \quad (1.315)$$

If we compare Eqs.(1.307) and (1.315), it is clear why B, C, ... are called virial coefficients.

1.4.   Collisions

In this section we shall consider the influence of collisions between the atoms on their velocity distribution. If we maintained the assumption of § 1.2 that the atoms do not exert forces upon one another, there should be no collisions. On the other hand, if we introduce interactions between the atoms, we are no longer dealing with a system of independent particles. We resolve this apparent contradiction by assuming that we may treat the interactions as being so weak that to a very good approximation the atoms are still independent, but at the same time there are a sufficient number of collisions to set up an equilibrium distribution as far as their velocities are concerned. To simplify our discussions we assume that the atoms in the gas may be treated as elastic spheres — or, *hard spheres*. The interatomic potential energy will then be of the form shown in Fig.1.5.

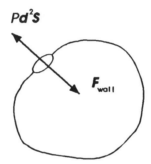

Pd^2S

F_{wall}

Fig.1.4. The contribution from the pressure to the virial. The force on the element of the wall is Pd^2S and is directed along the normal to the wall, while the force on the gas, F_{wall}, is equal to that force, but in the opposite direction.

If there are no external forces acting upon the system, it is plausible to assume that in any unit volume in the gas there will be $f(u, v, w)\, du\, dv\, dw$ atoms whose representative points in velocity space can be found within a given volume $du\, dv\, dw$ in velocity space; $f(u, v, w)$ is again the distribution function. If there are collisions in the gas, the velocities of the atoms will occasionally change and there will thus be atoms that leave the volume[9] $du\, dv\, dw$ and atoms that enter that volume. If the number of atoms per unit volume leaving $du\, dv\, dw$ per unit time is denoted by A and the number of atoms per unit volume entering $du\, dv\, dw$ per unit time by B, we have

---

[9] In order to avoid cumbersome sentences, we shall use the expression "atoms leaving the volume $du\, dv\, dw$" when we want to say "atoms whose representative points are leaving the volume $du\, dv\, dw$".

the following equation for the change in f per unit time:

$$\frac{\partial f}{\partial t}\, du\, dv\, dw \;=\; -A + B, \tag{1.401}$$

where the partial derivative indicates that for the moment we are interested only in the change in f due to collisions.

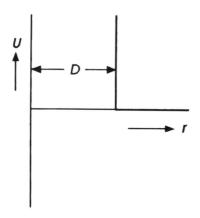

Fig.1.5. The potential U as a function of the distance r between the centres of two hard spheres of diameter D.

In order to calculate A and B we have to consider in detail collisions between two atoms (see Figs.1.6 and 1.7). We denote by c_1 and c_2 the velocities of the two atoms before the collision and by c_1' and c_2' the velocities after the collision. The velocity of the centre of mass will be denoted by w and is given by the equation

$$(m_1 + m_2)w \;=\; m_1 c_1 + m_2 c_2. \tag{1.402}$$

The velocities c_1' and c_2' are not completely determined by c_1 and c_2, since there are only four equations from which the six components have to be determined. These four equations are

$$\tfrac{1}{2}m_1 c_1^2 + \tfrac{1}{2}m_2 c_2^2 \;=\; \tfrac{1}{2}m_1 {c_1'}^2 + \tfrac{1}{2}m_2 {c_2'}^2 \;\text{(conservation of energy)} \tag{1.403}$$

and

$$m_1 c_1 + m_2 c_2 \;=\; m_1 c_1' + m_2 c_2'. \quad \text{(conservation of momentum)} \tag{1.404}$$

If we give the direction of the *line of centres*, that is, the line connecting the centre of atom 1 with the centre of atom 2, c_1' and c_2' are completely determined. Denoting by ω the unit vector in the direction of the line of centres, we have (compare Figs.1.6 and 1.7)

$$\omega = \frac{c_1 - c_1'}{|c_1 - c_1'|},\qquad(1.405)$$

where $|c|$ denotes the absolute magnitude of the vector c and where we have assumed that the two atoms collide like smooth spheres so that there is no change in the components of the momentum which are perpendicular to the line of centres.

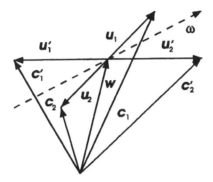

Fig.1.6. The velocities of two colliding hard spheres before and after their collisions: c_1 and c_2 are the velocities before, and c_1' and c_2' those after, the collision; w is the centre of mass velocity; u_1 and u_2 are the velocities in the centre of mass system before the collision, and u_1' and u_2' the velocities in the centre of mass system after the collision; finally ω is the unit vector in the direction of the line of centres.

If we introduce the centre of mass system, that is, the coordinate system in which the centre of mass is at rest ($w = 0$), the mathematical description of the collision is much simpler. Denoting the corresponding velocities by u_1, u_2, u_1', and u_2',[10] we have

$$u_1 = c_1 - w, \quad u_2 = c_2 - w, \quad u_1' = c_1' - w, \quad u_2' = c_2' - w, \quad(1.406)$$

$$m_1 u_1 + m_2 u_2 = m_1 u_1' + m_2 u_2' = 0,\qquad(1.407)$$

$$\tfrac{1}{2} m_1 u_1^2 + \tfrac{1}{2} m_2 u_2^2 = \tfrac{1}{2} m_1 {u_1'}^2 + \tfrac{1}{2} m_2 {u_2'}^2,\qquad(1.408)$$

---

[10]The reader should note the difference between the vector u (velocity in the centre of mass system) and the scalar u (x component of c) or between w and w.

$$\omega = \frac{u_1 - u'_1}{|u_1 - u'_1|}.$$

(1.409)

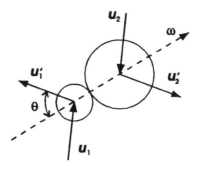

Fig.1.7. The collisions of two hard spheres in the centre of mass system.

Figures 1.6 and 1.7 show the velocities of the two atoms before and after the collision in the system in which the centre of mass has a velocity w and in the centre of mass system.

The number of collisions per unit volume in a time interval dt between atoms with velocities between c_1 and $c_1 + dc_1$,[11] on the one hand, and atoms with velocities between c_2 and $c_2 + dc_2$, on the other hand, while the line of centres lies within a solid angle $d^2\omega$,[12] is given by the expression

$$a_{12 \to 1'2'} \, f(u_1, v_1, w_1) f(u_2, v_2, w_2) \, du_1 \, dv_1 \, dw_1 \, du_2 \, dv_2 \, dw_2 \, d^2\omega \, dt.$$

(1.410)

In the case of our spherical atoms $a_{12 \to 1'2'}$ can easily be calculated. It follows from the consideration that expression (1.410) is equal to the number of atoms per unit volume with velocities between c_1 and $c_1 + dc_1$, which is $f(u_1, v_1, w_1) \, du_1 \, dv_1 \, dw_1$, multiplied by the number of atoms with velocities between $c_2 + dc_2$ that collide with one atom in a time interval dt in such a way that the line of centres has the direction ω within the element of solid angle $d^2\omega$. If c_{rel} is the relative velocity of atom 1 with respect to atom 2 ($c_{rel} = c_1 - c_2$) and if we denote its absolute value by c_{rel}, it can be seen from Fig.1.8 that the number of atoms which collide in a time interval dt with one atom is equal to the number of atoms which is contained in

---

[11] We shall use this short-hand notation to indicate velocities with an x component between u and $u + du$, a y component between v and $v + dv$, and a z component between w and $w + dw$.

[12] We shall denote an element of solid angle by $d^2\omega$. If ω is characterized by two angles ϑ and φ, ϑ being the angle between the $+z$ axis and ω, and φ being the angle between the $+x$ axis and the projection of ω on the x, y plane, $d^2\omega$ will be equal to $\sin \vartheta \, d\vartheta \, d\varphi$.

a cylinder with base $D^2 d^2\omega$ (D is the diameter of an atom;[13] compare Fig.1.5) and slant height $c_{\text{rel}} dt$.

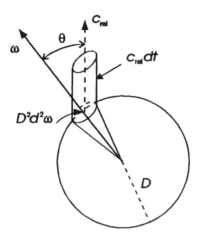

Fig.1.8. Collisions between hard sphere atoms. All atoms lying in a cylinder of base $D^2 d^2\omega$ and slant height $c_{\text{rel}} dt$ will undergo a collision with a given atom during a time interval dt.

Since the number of atoms per unit volume with velocities between c_2 and $c_2 + dc_2$ is equal to $f(u_2, v_2, w_2) \, du_2 \, dv_2 \, dw_2$, we find for $a_{12 \to 1'2'}$

$$
\left.
\begin{aligned}
a_{12 \to 1'2'} &= D^2 \, c_{\text{rel}} \cos\theta, & \cos\theta &> 0, \\
a_{12 \to 1'2'} &= 0, & \cos\theta &\leqslant 0,
\end{aligned}
\right\} \tag{1.411}
$$

where θ is the angle between ω and c_{rel}.

If the atoms do not behave like hard spheres, $a_{12 \to 1'2'}$ will be much more complicated, but will still depend only on θ and c_{rel}.

The number A of atoms leaving the volume element $du_1 \, dv_1 \, dw_1$ in velocity space per unit time will now be obtained by integrating expression (1.410) over all possible values of u_2, v_2, w_2 and over all allowed values of ω, and dividing by dt,

$$
\begin{aligned}
A = f(u_1, v_1, w_1) \, du_1 \, dv_1 \, dw_1 \int f(u_2, v_2, w_2) \, du_2 \, dv_2 \, dw_2 \\
\times \int a_{12 \to 1'2'} \, d^2\omega.
\end{aligned} \tag{1.412}
$$

We must remark here that in deriving Eq.(1.412) we have tacitly assumed — as did Clausius in 1858 — that there is no correlation whatever between

---

[13]If the radii of the two atoms are unequal, D will be equal to $r_1 + r_2$.

velocities and positions of different atoms. More specifically, whereas $f(u_2,$ $v_2, w_2)\, du_2\, dv_2\, dw_2 \cdot dV$ is the number of atoms with velocities between c_2 and $c_2 + dc_2$ that will be found in a volume element dV selected at random, it does not follow that we may use the same expression for a particular volume element which is not selected at random, such as the volume of the cylinder which we just considered. We must introduce as a basic assumption that we may still use the same expression. This assumption is called the *Stosszahlansatz*. It is also often called the assumption of *molecular chaos*. However, with Boltzmann and the Ehrenfests we shall distinguish between these two assumptions and use the German expression for the assumption introduced at this point, and use the English expression only in connection with the discussion of fluctuations in the number of collisions.

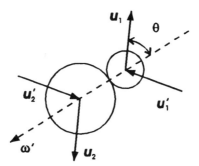

Fig.1.9.   The inverse collision in the centre of mass system. This collision is the inverse of the one shown in Fig.1.7 in that the velocities which are now the velocities before the collision were those after the collision in Fig.1.7, and *vice versa*.

We must now calculate the number of collisions that will bring atoms back into the volume element $du_1\, dv_1\, dw_1$. These are collisions where the velocity of one of the two colliding atoms lies between c_1 and $c_1 + dc_1$ *after* the collision. We have just considered a collision where the velocities before the collision were c_1 and c_2, the line of centres was determined by ω, and the velocities after the collision were c_1' and c_2'. Let us now consider a collision where the velocities before the collision were c_1' and c_2' and after the collision c_1 and c_2. It then follows from Eq.(1.405) that the new line of centres (ω') is determined by $-\omega$ instead of ω (see Fig.1.9, where this inverse collision is pictured in the centre of mass system). The second collision will be called the *inverse* collision, corresponding to the original collision picured in Fig.1.7. It can easily be verified that there is a one-to-one correspondence between the original and the inverse collisions. This is a consequence of the fact that c_1, c_2, and ω determine c_1', c_2', and ω' $(= -\omega)$ completely.

The total number of inverse collisions per unit volume and unit time where the velocities before the collision lie between c_1' and $c_1' + dc_1'$, and c_2' and $c_2' + dc_2'$, while ω' lies within a solid angle $d^2\omega'$, is given by (compare expression (1.410))

$$a_{1'2'\to12} f(u_1', v_1', w_1')\, du_1'\, dv_1'\, dw_1'\, f(u_2', v_2', w_2')\, du_2'\, dv_2'\, dw_2'\, d^2\omega'. \quad (1.413)$$

In order to get B we must integrate expression (1.413) over all allowed values of c_1', c_2', and ω'. Only those values of c_1', c_2', and ω' are allowed for which one of the velocities after the collision lies between c_1 and $c_1 + dc_1$. If we indicate this restriction by priming the integral sign, we have

$$B = \int' f(u_1', v_1', w_1')\, du_1'\, dv_1'\, dw_1'\, f(u_2', v_2', w_2')\, du_2'\, dv_2'\, dw_2'$$

$$\times \int a_{1'2'\to12}\, d^2\omega'. \quad (1.414)$$

It is possible to simplify the expression for B, since c_1', c_2', and ω' are completely determined by c_1, c_2, and ω through Eqs.(1.403) to (1.405) and the equation $\omega' = -\omega$ which is a consequence of Eq.(1.405). If we change variables, using $u_1, v_1, w_1, u_2, v_2, w_2$, and ω instead of the primed quantities, we can write Eq.(1.414) in the form

$$B = du_1\, dv_1\, dw_1 \int f(u_1', v_1', w_1') f(u_2', v_2', w_2')\, du_2\, dv_2\, dw_2$$

$$\times \int J a_{1'2'\to12}\, d^2\omega. \quad (1.415)$$

The arguments of the distribution functions in Eq.(1.415) are functions of the unprimed quantities through Eqs.(1.403) to (1.405), and J is the Jacobian of the transformation:

$$J = \frac{\partial(u_1', v_1', w_1', u_2', v_2', w_2', \omega')}{\partial(u_1, v_1, w_1, u_2, v_2, w_2, \omega)}$$

$$= \begin{vmatrix} \dfrac{\partial u_1'}{\partial u_1} & \dfrac{\partial v_1'}{\partial u_1} & \dfrac{\partial w_1'}{\partial u_1} & \cdots & \dfrac{\partial w_2'}{\partial u_1} \\ \vdots & \vdots & \vdots & \ddots & \vdots \\ \dfrac{\partial u_1'}{\partial w_2} & \dfrac{\partial v_1'}{\partial w_2} & \cdots & \cdots & \dfrac{\partial w_2'}{\partial w_2} \end{vmatrix} \cdot \frac{\partial \omega'}{\partial \omega}. \quad (1.416)$$

It can be proved that J is always equal to unity.[14]

We shall prove that $J = 1$ for the case of an elastic collision between two particles with masses m_1 and m_2. We choose the $+x$-axis parallel to ω. Since the collision is elastic, we have from Eqs.(1.403) to (1.405)

$$\left. \begin{array}{llll} v_1 = v_1' & w_1 = w_1'; & v_2 = v_2' & w_2 = w_2'; \\[2mm] u_1' = \dfrac{(m_1 - m_2)u_1 + 2m_2 u_2}{m_1 + m_2}, & u_2' = \dfrac{(m_2 - m_1)u_2 + 2m_1 u_1}{m_1 + m_2}. \end{array} \right\} \quad (1.417)$$

---

[14]This is a consequence of Liouville's theorem.

Equations (1.417) are most easily obtained by remembering that in the centre of mass system, the y- and z-components of the velocities are unchanged while the x-components only change sign. The Jacobian now reduces to

$$J = \begin{vmatrix} \dfrac{\partial u_1'}{\partial u_1} & \dfrac{\partial u_2'}{\partial u_1} \\[2mm] \dfrac{\partial u_1'}{\partial u_2} & \dfrac{\partial u_2'}{\partial u_2} \end{vmatrix} \cdot \frac{\partial \omega'}{\partial \omega} = -\frac{\partial \omega'}{\partial \omega} = 1. \qquad (1.418)$$

Since $a_{12 \to 1'2'}$ or $a_{1'2' \to 12}$ depend only on c_{rel} and θ, neither of which changes by going over to the inverse collision, $a_{12 \to 1'2'} = a_{1'2' \to 12}$, and we get from Eqs.(1.401), (1.412), and (1.415), dropping the subscripts of the a's, putting $J = 1$, and dividing by $du_1 \, dv_1 \, dw_1$,

$$\frac{\partial f_1}{\partial t} = -\int (f_1 f_2 - f_1' f_2') \, a \, du_2 \, dv_2 \, dw_2 \, d^2\omega. \qquad (1.419)$$

In Eq.(1.419) we used the abbreviated notation

$$f_i \equiv f(u_i, v_i, w_i), \qquad f_i' \equiv f(u_i', v_i', w_i'). \qquad (1.420)$$

If the distribution function is given by Eq.(1.104), the conservation of energy (Eq.(1.403)) will mean that

$$f_1 f_2 = f_1' f_2'. \qquad (1.421)$$

For a Maxwell distribution we find therefore that

$$\frac{\partial f}{\partial t} = 0. \qquad (1.422)$$

This means that the Maxwell distribution is an equilibrium distribution: once it has been established, collisions will not alter it.

The distribution function can be generalised as follows:

$$\widetilde{f} = K \exp \left[\frac{-m(c - c_0)^2}{2kT} \right], \qquad (1.423)$$

where K is a normalising constant and c_0 a given constant vector. This distribution corresponds to a case where the system as a whole is moving with a constant velocity c_0. This can be proved by performing a transformation to a coordinate system moving with a velocity c_0 with respect to our original system of reference; the distribution function $\widetilde{f}$ is then transformed into a distribution $\widetilde{f}'$, which is just the Maxwell distribution (1.104). By evaluating the average velocity $\bar{c}$ of the atoms in the system, one finds — as one should expect — that

$$\bar{c} = c_0. \qquad (1.424)$$

The distribution $\tilde{f}$, like the Maxwell distribution, is a stationary distribution, because by virtue of the conservation of energy (Eq.(1.403)) and the conservation of momentum (Eq.(1.404)) we have

$$\tilde{f}_1 \tilde{f}_2 = \tilde{f}'_1 \tilde{f}'_2 \quad \text{or} \quad \frac{\partial \tilde{f}}{\partial t} = 0. \tag{1.425}$$

In conclusion we draw attention to the fact that Eq.(1.421) means that there are as many inverse collisions per unit time as there are original ones; this is but one example of the far more general *principle of detailed balancing*.

1.5.   The H-Theorem

In 1872 Boltzmann introduced his H-function which is defined by the equation

$$H = n \overline{\ln f} = \int f \ln f \, du \, dv \, dw, \tag{1.501}$$

where f is any distribution function defined as in § 1.1, and not necessarily the equilibrium distribution function given by Eq.(1.104) or Eq. (1.423). The integration in Eq.(1.501) is over the whole of velocity space and H will thus be a function of the time only. For its derivative with respect to the time we find[15]

$$\frac{dH}{dt} = \int \frac{\partial f}{\partial t} (\ln f + 1) \, d^3c. \tag{1.502}$$

If we use Eq.(1.419) for $\partial f / \partial t$—which was derived without any restrictions on f—we get from Eq.(1.502):

$$\frac{dH}{dt} = \int (f'f'_1 - ff_1)(\ln f + 1) \, a \, d^3c \, d^3c_1 \, d^2\omega. \tag{1.503a}$$

In Eq.(1.503a) c and c_1 are integration variables and both integrations are over the whole of velocity space; we can thus interchange them without affecting the result of the integration. As $a_{12\to 1'2'}$ depends on c_{rel} and θ only, it will not change when this interchange is performed; thus we get

$$\frac{dH}{dt} = \int (f'f'_1 - ff_1)(\ln f_1 + 1) \, a \, d^3c \, d^3c_1 \, d^2\omega. \tag{1.503b}$$

If in Eqs.(1.503a) and (1.503b) we denote the velocities before the collision by c' and c'_1 and those after the collision by c and c_1, and if we write ω' for the unit vector along the line of centres and a' for $a_{1'2'\to 12}$, we have the equations

$$\frac{dH}{dt} = \int (ff_1 - f'f'_1)(\ln f' + 1) \, a' \, d^3c' \, d^3c'_1 \, d^2\omega', \tag{1.503c}$$

---

[15]We use the abbreviation d^3c for $du \, dv \, dw$. It must be noted that d^3c is not a vector, but a symbolical notation for a volume element in velocity space; it should not be confused with the vector dc which is an increment in c.

$$\frac{dH}{dt} = \int (ff_1 - f'f_1')(\ln f_1' + 1)\, a'\, d^3c'\, d^3c_1'\, d^2\omega'. \tag{1.503d}$$

We saw in the preceding section that the quantities c, c_1, and ω ($= (c_1 - c_1')/|c_1 - c_1'|$) are completely determined by c', c_1', and ω' ($= (c_1' - c_1)/|c_1' - c_1| = -\omega$) and that the Jacobian of the transformation from the former to the latter is equal to unity. If we perform the transformation from the primed to the unprimed variables in Eqs.(1.503c) and (1.503d), we get

$$\frac{dH}{dt} = \int (ff_1 - f'f_1')(\ln f' + 1)\, a\, d^3c\, d^3c_1\, d^2\omega, \tag{1.503e}$$

$$\frac{dH}{dt} = \int (ff_1 - f'f_1')(\ln f_1' + 1)\, a\, d^3c\, d^3c_1\, d^2\omega, \tag{1.503f}$$

where the primed variables which still occur under the integral signs must be regarded as functions of the unprimed variables (see the remark after Eq.(1.415)).

Equations (1.503a), (1.503b), (1.503e), and (1.503f) are four equivalent equations for dH/dt. Adding these four equations and dividing by four, we get for dH/dt

$$\frac{dH}{dt} = -\frac{1}{4} \int (ff_1 - f'f_1') \ln \frac{f'f_1'}{ff_1}\, a\, d^3c\, d^3c_1\, d^2\omega. \tag{1.504}$$

Since a is a positive quantity and since $(p - q)\ln(p/q)$ is positive, except when $p = q$, in which case it is zero, it follows from Eq.(1.504) that

$$\frac{dH}{dt} \leqslant 0. \tag{1.505}$$

The equality in Eq.(1.505) is valid only, if for *every* two velocities c and c_1 we have $ff_1 = f'f_1'$. Equation (1.505), which expresses that H can never increase, is known as Boltzmann's *H-theorem*.

One can prove that H is a bounded function, and Eq.(1.505) thus entails that H will decrease as long as the distribution function does not satisfy Eq.(1.421). Hence, we see that Eq.(1.421) is not only a sufficient condition for an equilibrium distribution — as was shown in the previous section — but also a necessary condition.

We shall now prove that the generalised Maxwell distribution given by Eq. (1.423) is the only possible equilibrium distribution in the case of a gas which is not in an external field of force. We first of all note that we have just seen that for equilibrium the following relation must hold:

$$\ln f_1 + \ln f_2 = \text{the same before and after the collision.} \tag{1.506}$$

If $\psi(u, v, w)$ is a function of the velocity components of an atom such that $\psi(u_1, v_1, w_1) + \psi(u_2, v_2, w_2)$ remains unchanged when two atoms with velocities c_1 and c_2 collide, $\ln f = \psi$ will be a solution of Eq. (1.506). If there are several such functions, $\psi_1, \psi_2, \psi_3, \cdots$, the general solution of Eq.(1.506) will be

$$\ln f = \alpha_0 + \alpha_1 \psi_1 + \alpha_2 \psi_2 + \alpha_2 \psi_3 + \cdots , \tag{1.507}$$

where the α_i are arbitrary constants. We know four ψ's having this property: the energy and the three components of the momentum. As the conservation laws for energy and momentum, together with the direction of the line of centres, completely determine the six velocity components after the collision, once the six velocity components before the collision are given, it follows that there can be no other linearly independent ψ's but only those four. From Eq.(1.507) it then follows that

$$\ln f = \alpha_0 + \alpha_1 \cdot \tfrac{1}{2}m(u^2 + v^2 + w^2) + \alpha_2 \cdot mu + \alpha_3 \cdot mv + \alpha_4 \cdot mw,$$

or

$$f = K e^{-a(c - c_0)^2}, \tag{1.508}$$

where K is again a normalising constant.

In order to prove that the distribution (1.508) is the same as (1.423) we must show that

$$a = \frac{m}{2kT}. \tag{1.509}$$

This can be done in different ways, depending on the way the temperature is introduced. If the temperature is introduced through Eq.(1.110) for the average value of the kinetic energy (or rather in the present case of a moving distribution, the average value of the random or *thermal* kinetic energy $\tfrac{1}{2}m(c - c_0)^2$) or through the equation of state of a perfect gas (1.204), Eq.(1.509) follows immediately, since Eq.(1.508) leads to

$$\overline{(c - c_0)^2} = \frac{3}{2a}. \tag{1.510}$$

If a distribution differs appreciably from the equilibrium distribution, the return to equilibrium is usually very rapid. The relaxation time, that is, the time necessary to return to equilibrium, is, for instance, in the case of a gas under normal conditions ($T = 300°$ K, $P = 1$ atm) of the order of 10^{-9} s.

Another way to arrive at the generalised Maxwell distribution (1.423) is by showing that it is that distribution which makes H a minimum under the restricting conditions that the total number of atoms N in the gas, their total energy E, and their total momentum P are given. We thus take as the equilibrium condition the equation

$$H = \text{a minimum.} \tag{1.511}$$

If $f(u, v, w)$ is an equilibrium distribution, *any* small change δf (which itself will be a function of u, v, and w) will have to satisfy the following conditions:

$$\delta H = \int (\ln f + 1)\, \delta f\, d^3 c = 0 \quad \text{(equilibrium condition)}, \quad (1.512a)$$

$$\delta E = \int (\tfrac{1}{2} mc^2)\, \delta f\, d^3 c = 0 \quad \text{(total energy given)}, \quad (1.512b)$$

$$\delta P = \int mc\, \delta f\, d^3 c = 0 \quad \text{(total momentum given)}, \quad (1.512c)$$

$$\delta N = V \int \delta f\, d^3 c = 0 \quad \text{(total number of atoms given)}, \quad (1.512d)$$

where V is the volume of the gas.

Equations (1.512) express the fact that the equilibrium distribution f must satisfy Eq.(1.512a) for any small change $\delta f(u, v, w)$ in the distribution function which satisfies Eqs.(1.512b), (1.512c), and (1.512d). These restrictions on δf can be removed by the method of the *Lagrangian multipliers*. In using this method we multiply each of the four Eqs.(1.512) by a constant and add them. If we take

$$\delta H + \beta\, \delta E + (\boldsymbol{\lambda} \cdot \delta P) - \frac{\alpha + 1}{V}\, \delta N,$$

which must be zero according to Eqs.(1.512), we get

$$\int \delta f \left[\ln f + \beta \cdot \tfrac{1}{2} mc^2 + (\boldsymbol{\lambda} \cdot mc) - \alpha \right] d^3 c = 0, \quad (1.513)$$

and we can now treat δf, as if it were an arbitrary change in f without any restrictions.

The justification of the method of the Lagrangian or undetermined multipliers is the following one.

Consider a function $f(x_1, x_2, \cdots, x_n)$ of n variables. Let us ask for which values of the x_i this function reaches an extremum under the following p restraining conditions:

$$g_j(x_1, x_2, \cdots, x_n) = 0; \quad j = 1, 2, \cdots, p. \quad (1.514)$$

The extremum of f can be found by requiring that the variation of f will be zero for any possible variation of the x_i from their values corresponding to the extremum, or

$$\delta f = \sum_i \frac{\partial f}{\partial x_i}\, \delta x_i = 0. \quad (1.515)$$

If there were no restraining conditions, the x_i could be varied independently, and Eq.(1.515) would lead to the well known conditions for an extremum:

$$\frac{\partial f}{\partial x_i} = 0, \quad i = 1, \cdots, n. \quad (1.516)$$

However, if Eqs.(1.514) must be satisfied, the δx_i cannot be chosen arbitrarily, but must satisfy the p equations

$$\delta g_j = 0 = \sum_i \frac{\partial g_j}{\partial x_i} \delta x_i, \qquad j = 1, \cdots, p. \tag{1.517}$$

From Eqs.(1.517) we can find p of the δx_i in terms of the other $n - p$ variations and substitute for these p variations into Eq.(1.515). The remaining $n - p$ variations are then arbitrary and their coefficients must vanish identically in the x_i. We can also eliminate p of the δx_i, say δx_1, δx_2, $\cdots$, δx_p, by multiplying the p equations (1.517) with constants $\lambda_1, \lambda_2, \cdots, \lambda_p$ and adding them to Eq(1.515). The result will be

$$\sum_i \left[\frac{\partial f}{\partial x_i} + \lambda_1 \frac{\partial g_1}{\partial x_i} + \lambda_2 \frac{\partial g_2}{\partial x_i} + \cdots + \lambda_p \frac{\partial g_p}{\partial x_i} \right] \delta x_i = 0. \tag{1.518}$$

We eliminate δx_1 to δx_p by determining the λ_j in such a way that

$$\frac{\partial f}{\partial x_i} + \lambda_1 \frac{\partial g_1}{\partial x_i} + \lambda_2 \frac{\partial g_2}{\partial x_i} + \cdots + \lambda_p \frac{\partial g_p}{\partial x_i} = 0, \qquad i = 1, \cdots, p. \tag{1.519}$$

Once δx_1 to δx_p are eliminated, the remaining δx_i can be chosen arbitrarily, and Eq.(1.518) can only be satisfied for all choices of these variations, if their coefficients vanish, or

$$\frac{\partial f}{\partial x_i} + \lambda_1 \frac{\partial g_1}{\partial x_i} + \lambda_2 \frac{\partial g_2}{\partial x_i} + \cdots + \lambda_p \frac{\partial g_p}{\partial x_i} = 0, \qquad i = n - p, \cdots, n. \tag{1.520}$$

Combining Eqs.(1.519) and (1.520), we see that we could have obtained the same result by assuming the δx_i in Eq.(1.518) to be independent. Equations (1.519) and (1.520) give us the values of the x_i for which f is an extremum; these values are functions of the undetermined multipliers λ_j which can be eliminated by substituting the x_i into Eqs.(1.514) and solving for the λ_j.

In the case discussed earlier, f depended on a set of continuous variables. If we group the values of these variables in finite intervals, we have the case for which we have just justified the method, and after that the transition can be made to the continuous case.

Since δf in Eq.(1.513) can be treated as an arbitrary change, this equation can be satisfied only, if

$$\ln f = -\beta \cdot \tfrac{1}{2} mc^2 - (\boldsymbol{\lambda} \cdot mc) + \alpha, \tag{1.521}$$

or

$$f = e^\alpha \, e^{-\frac{1}{2} \beta mc^2 - m(\boldsymbol{\lambda} \cdot c)}, \tag{1.522}$$

which is the same as expressions (1.423) or (1.508) with a slightly different notation. The Lagrangian multipliers α, β, and $\boldsymbol{\lambda}$ must be determined from the total number of particles, that is, the normalisation condition (1.102), the total energy E of the system, and the total momentum $\boldsymbol{P}$ of the system. Comparing Eqs.(1.522) and (1.508), we see that $\boldsymbol{\lambda}$ will be proportional to the velocity c_0 of the moving Maxwell distribution, and that

$$\beta = \frac{1}{kT}. \tag{1.523}$$

where we have used Eq.(1.509). For the case of a gas at rest we get for α the equation

$$e^\alpha = \left(\frac{\beta m}{2\pi}\right)^{3/2} n. \tag{1.524}$$

1.6.   The Connection between H and Entropy

In this section, for the special case of a gaseous system of non-interacting particles, we shall discuss the connection between Boltzmann's H-function and the entropy.

From Eqs.(1.521) and (1.524) we get for the case of a gas at rest

$$\ln f = \ln n + \tfrac{3}{2} \ln \beta - \tfrac{1}{2}\beta mc^2 + \text{constant}. \tag{1.601}$$

Introducing the volume v per unit mass,

$$v = \frac{1}{nm}, \tag{1.602}$$

we can write Eq.(1.601) in the form

$$\ln f = -\ln v + \tfrac{3}{2} \ln \beta - \tfrac{1}{2}\beta mc^2 + \text{constant}. \tag{1.603}$$

It now follows from the definition (1.501) of H that

$$\frac{H}{n} = -\ln v + \tfrac{3}{2} \ln \beta - \tfrac{1}{2}\beta \,\overline{mc^2} + \text{constant},$$

or, after evaluating $\overline{c^2}$,

$$\frac{H}{n} = -\ln v + \tfrac{3}{2} \ln \beta + \text{constant}. \tag{1.604}$$

If we are dealing with a perfect gas, β will be equal to $1/kT$ (compare Eq.(1.523)) and we have from Eq.(1.604)

$$\frac{H}{n} = -\ln v - \tfrac{3}{2} \ln T + \text{constant}. \tag{1.605}$$

If the entropy and the internal energy per unit mass are denoted by s and u, respectively, the second law of thermodynamics gives us the following relation between the changes in s, u, and the work done by the pressure,[16]

$$ds = \frac{dq}{T} = \frac{du + Pdv}{T}, \tag{1.606}$$

where dq is the heat increase per unit mass.[17]

The perfect gas law can be written in the form

$$Pv = \frac{R}{M} T, \tag{1.607}$$

where M is the molecular weight of the gas. Since the internal energy of a perfect gas consists of the kinetic energy only, we have

$$du = c_v dT = \frac{3}{2} \frac{R}{M} dT, \tag{1.608}$$

where c_v is the specific heat per unit mass at constant volume.

From Eqs.(1.606), (1.607), and (1.608) it follows that

$$ds = c_v \frac{dT}{T} + \frac{R}{M} \frac{dv}{v}, \tag{1.609}$$

or

$$s = \frac{R}{M} \left(\tfrac{3}{2} \ln T + \ln v \right) + \text{constant}. \tag{1.610}$$

Using Eq.(1.205), we get for the entropy per unit volume

$$S_v = \frac{Mn}{N_A} s = nk \left(\tfrac{3}{2} \ln T + \ln v \right) + \text{constant}. \tag{1.611}$$

If we compare Eqs.(1.605) and (1.611), we see that, apart possibly from an additive constant, there exists the following relation between H and the entropy per unit volume:

$$S_v = -kH. \tag{1.612}$$

The fact that H never increases thus corresponds to the well known statement of the second law of thermodynamics that the entropy will never decrease, if the volume and internal energy are kept constant.[18]

It may be remarked here that, although in thermodynamics the entropy is, strictly speaking, defined only in the case of equilibrium, H is also

---

[16]In this section, quantities referring to the whole system will be denoted by capital letters (S, V, U, Q, and so forth); quantities referring to unit mass, by lower case letters (s, v, u, q, and so forth); and quantities referring to unit volume by a subscript v (S_v, and so forth).

[17]The bar through the symbols d and δ will indicate that we are not dealing with a total differential or variation.

[18]See, for example, D.ter Haar and H.Wergeland, *Elements of Thermodynamics*, Addison-Wesley, Reading, Mass., 1966, Chap.3.

defined for non-equilibrium situations and one can use Eq.(1.612) as a definition of the entropy in such non-equilibrium situations.

1.7.   The Connection between H and Probability

To conclude this chapter, we shall for the special case of a perfect gas discuss the connection between Boltzmann's H and the probability that a given distribution is realised.

We must first of all define the probability of a distribution. To do this we divide velocity space into a number of cells Z_k, each cell being a volume element $du\,dv\,dw$, and we shall distribute the representative points of the N atoms in the system over velocity space. If this distribution results in N_1 points falling into Z_1, N_2 into Z_2, $\cdots$, we call this distribution an N_i-distribution. The probability $W(N_1,\cdots,N_k,\cdots)$ or $W(N_i)$ is defined as the fraction of all possible arrangements where this specially chosen distrbution is realised. If we make the plausible assumption that the a priori probability for a representative point to fall into a cell Z_k will be proportional to its size Z_k, we find for $W(N_i)$ the equation

$$W(N_1, N_2, \cdots, N_k, \cdots) \;=\; C\,N!\prod_i \frac{Z_i^{N_i}}{N_i!}, \qquad (1.701)$$

where C is a normalisation constant.

In order to derive Eq(1.701) we have to remember that the number of possible ways of choosing A objects out of B is given by the binomial coefficient

$$\binom{A}{B} \;=\; \frac{B!}{A!(B-A)!}. \qquad (1.702)$$

In our case we have to choose first N_1 objects out of N, then N_2 objects out of $N-N_1$, N_3 out of $N-N_1-N_2$, and so on; for the total number of possible ways we therefore get

$$\binom{N}{N_1}\binom{N-N_1}{N_2}\binom{N-N_1-N_2}{N_3}\cdots$$
$$= \frac{N!}{N_1!(N-N_1)!}\cdot\frac{(N-N_1)!}{N_2!(N-N_1-N_2)!}\cdots \;=\; \frac{N!}{N_1!N_2!\cdots}. \qquad (1.703)$$

Multiplying this *multinomial* coefficient with the a priori probability, we get expression (1.701).

The name "multinomial coefficient" arises from the fact that these coefficients appear in the multinomial expansion

$$(a_1 + a_2 + \cdots)^N \;=\; \sum \frac{N!}{N_1!N_2!\cdots}a_1^{N_1}a_2^{N_2}\cdots, \qquad (1.704)$$

where the summation extends over all non-negative values of the N_i in such a way that their sum equals N.

We now define the equilibrium distribution as that distribution which makes $W(N_i)$ a maximum under the restricting conditions that

$$\sum N_i = N \tag{1.705}$$

and

$$\sum N_i \varepsilon_i = E, \tag{1.706}$$

where ε_i is the energy of an atom in cell Z_i. Conditions (1.705) and (1.706) fix the total number of particles in the gas and the total energy of the system.

In order to find the distribution which we are looking for, we shall require $\ln W$ to be a maximum instead of W. We have from Eq.(1.701)

$$\ln W = \ln N! + \sum_i [N_i \ln Z_i - \ln N_i!] + \text{constant}. \tag{1.707}$$

If we may assume that all of the N_i are large compared to unity, we can use for the factorials the Stirling formula,

$$x! \approx x^x\, e^{-x}. \tag{1.708}$$

The Stirling formula in its form (1.708) can be derived as follows.[19] Since $\ln x$ is a monotonically increasing function of x, we have

$$\sum_{k=1}^{N} \ln k < \int_0^N \ln x\, dx < \sum_{k=1}^{N} \ln(k+1),$$

or

$$\int_0^{N-1} \ln x\, dx < \sum_{k=1}^{N} \ln k < \int_0^N \ln x\, dx,$$

and hence

$$(N-1)[\ln(N-1)-1] < \ln N! < N(\ln N - 1);$$

since $(N-1)[\ln(N-1)-1]$ and $N(\ln N - 1)$ differ relatively less the larger N is, we have asymptotically for large N

$$\ln N! \approx N \ln N - N,$$

which is the same as Eq.(1.708).

Using Eqs.(1.708) and (1.705), we get from Eq.(1.707)

$$\ln W = -\sum_i N_i \ln \frac{N_i}{N Z_i} + \text{constant}. \tag{1.709}$$

---

[19]See, for example, P.Jordan, *Statistische Mechanik auf Quantentheoretischer Grundlage*, Vieweg, Braunschweig, 1933; J.Satterly, *Nature* **111**, 220 (1923).

Looking for the maximum of $\ln W$ under the conditions (1.705) and (1.706), we can again use Lagrange's method of undetermined multipliers. For a variation δN_i of the N_i, we have

$$\delta \ln W \;=\; 0 \;=\; - \sum_i \delta N_i \left[\ln \frac{N_i}{N Z_i} + 1 \right], \tag{1.710a}$$

$$\delta N \;=\; 0 \;=\; \sum_i \delta N_i, \tag{1.710b}$$

$$\delta E \;=\; 0 \;=\; \sum_i \varepsilon_i \, \delta N_i. \tag{1.710c}$$

Next we multiply these equations by 1, $\alpha + 1$, and $-\beta$, respectively, add them, and obtain

$$\sum_i \delta N_i \left[-\ln \frac{N_i}{N Z_i} + \alpha - \beta \varepsilon_i \right] \;=\; 0. \tag{1.711}$$

Since in Eq.(1.711) the δN_i can be taken to be arbitrary, it follows that

$$\ln \frac{N_i}{N Z_i} \;=\; \alpha - \beta \varepsilon_i,$$

or

$$N_i \;=\; N Z_i \, e^{\alpha - \beta \varepsilon_i}. \tag{1.712}$$

The quantities α and β can be determined from conditions (1.705) and (1.706).

We can now apply these formulæ to the case of a perfect gas. The cells Z_i are then to be thought of as "elements of volume" in velocity space, or

$$Z_i \;=\; du \, dv \, dw. \tag{1.713}$$

As N is the total number of particles, N_i is the total number of particles with velocities between c and $c + dc$. Thus

$$\frac{n_i}{N} \;=\; \frac{1}{n} f(u, v, w) \, du \, dv \, dw, \tag{1.714}$$

while the energies ε_i of the atoms are given by

$$\varepsilon_i \;=\; \tfrac{1}{2} m c^2. \tag{1.715}$$

The transition to infinitesimal cells is trivial, and the distribution (1.712) goes over into the Maxwell distribution (1.104).

From Eqs.(1.709), (1.713), and (1.714) we then get

$$\ln W = - \int f \ln f \, du \, dv \, dw + \text{constant}. \tag{1.716}$$

Comparing Eqs.(1.501) and (1.716), we see that, apart from an additional constant, $\ln W$ and $-H$ are the same. The tendency for a decrease of H (the H-theorem) or an increase of entropy, corresponds thus to a tendency for an increase in probability.

Combining Eqs.(1.612), (1.501), and (1.716) and omitting for the time being possible additive constants, we have

$$-kH = S_v = k \ln W, \tag{1.717}$$

which gives us the well known relation between entropy and probability.

Problems[20]

1. Assume that the atoms in a gas satisfy the Maxwell distribution; evaluate the number of atoms that strike a unit area in unit time.

2. Use the result of the preceding problem to find
   (a) the velocity distribution of the atoms that escape through an (infinitesimal) hole of area A in the wall of the vessel, inside of which the velocity distribution is Maxwellian;
   (b) the root mean square speed of the escaping molecules;
   (c) the rate of loss of energy in the vessel;
   (d) the rate at which the temperature of the residual gas is changing.

3. Prove Dalton's law for a mixture of non-interacting gases consisting of independent atoms.

4. For a gas of identical atoms in thermal equilibrium evaluate the number of collisions per unit time and unit volume between atoms (taken to be perfectly elastic spheres) with relative velocities in the range c_{rel} to $c_{rel} + dc_{rel}$.
   Determine the average length and duration of a free path.
   Find an expression for the probability of an atom's describing a free path of a given length.

5. Find an expression for the mean free path of those atoms that cross a given plane with speed c, and explain why this differs from the mean free path of such atoms.

6. Explain the difference between the "Maxwell free path" (that is, the average free path per unit time) and the "Tait free path" (that is, the average free path of a given particle).

---

[20] All problems, except the last one, are concerned with kinetic theory; they take the development of kinetic theory one stage further than was done in the text.

7. Consider a monatomic gas in a steady state in a vessel closed by a piston; assume that the piston moves infinitesimally slowly, with speed U. Use kinetic theory to find a relation between the pressure P and the volume V during this quasistatic adiabatic process. Assume that through collisions the change in energy upon hitting the moving piston is distributed evenly over all degrees of freedom.

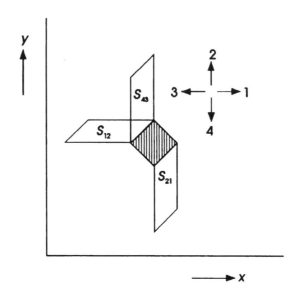

Fig.1.10. The Ehrenfests' wind-wood model.

8. Consider the following simplified model of a gas, due to the Ehrenfests. In the plane of the paper there may be a large number of point particles, N per unit area, which we shall call the P-molecules. They do not interact with each other, but collide elastically with another set of entities which we shall call the Q-molecules. These Q-molecules are squares with edgelength a, distributed at random over the plane (see Fig.1.10). They are fixed in their positions in such a way that their diagonals are exactly parallel to the x- and y-axes. Their average surface density is n, and their mean distance apart is large compared to a.

Suppose that at a certain moment all the P-molecules have velocities which have the same absolute magnitude, c, and which are limited in direction to (1) the positive x-axis, (2) the positive y-axis, (3) the negative x-axis, and (4) the negative y-axis (Fig.1.10). Due to the nature of our system and the interactions allowed in it, the situation at any other time will be similar to the one just described, the only possible difference being in the values of the numbers of P-molecules moving in the four possible directions. Let f_1, f_2, f_3, and f_4 be the numbers of P-molecules per unit area moving in those four directions. These numbers will be functions of t and together will take the place of the distribution function $f(u, v, w)$. The equilibrium distribution will be given by the equation

$$f_1 = f_2 = f_3 = f_4 = \tfrac{1}{4}N.$$

Prove that, if we assume that the number $N_{ij} \Delta t$ of P-molecules per unit area changing direction from i to j during a time interval Δt is given by the Stosszahlansatz expression

$$N_{ij} \Delta t = f_i S_{ij} n,$$

where S_{ij} is the area of a parallelogram of length $c \Delta t$ on that edge of one of the squares which is in the $-i, j$ quadrant (see Fig.1.10), the f_i will approach their equilibrium values.

9. Consider the Lorentz model of a metal in which the conduction electrons, which are assumed to be non-interacting, are scattered in such a way that the number per unit volume $N_{\omega, \omega'} \, dt \, d^2\omega \, d^2\omega'$ which during a time interval dt change their direction from within an element of solid angle $d^2\omega$ to within an element of solid angle $d^2\omega'$ is given by the equation

$$N_{\omega, \omega'} \, dt \, d^2\omega \, d^2\omega' = A f(\vartheta, \varphi) \frac{d^2\omega}{4\pi} \frac{d^2\omega'}{4\pi} \, dt,$$

where $f(\vartheta, \varphi) \sin \vartheta \, d\vartheta \, d\varphi / 4\pi$ is the number of electrons per unit volume with velocities within the solid angle $d^2\omega$. It is assumed in this model that all electrons are moving with the same speed.

Find the equilibrium distribution of the electrons and show that the scattering mechanism will produce an exponential approach to this equilibrium distribution.

10. A vessel of volume V contains N atoms in thermal equilibrium. If n is the number of atoms in a subvolume v, and if the probability of finding any atom in v is equal to v/V, find the probability $f(n)$ for v containing exactly n atoms. Calculate the mean value $\bar{n}$ and also its dispersion, that is, the quantity $\overline{(n - \bar{n})^2}$, where the averages are all taken with $f(n)$ as the distribution function.

Prove that $f(n)$ is approximately Gaussian when both N and n are large. Use Stirling's formula in the form (1.708).

Show that in the limit as $v/V \to 0$, $N \to \infty$, and $V \to \infty$, while N/V is kept constant, $f(n)$ tends to the Poisson distribution

$$f(n) = e^{\bar{n}} \frac{\bar{n}^n}{n!}.$$

Bibliographical Notes

General references dealing with the material of this chapter are

1. L.Boltzmann, *Vorlesungen über Gastheorie*, 2 vols., Leipzig, 1896–1898; *Lectures on Gas Theory*, Berkeley, California, 1964.
2. P. and T. Ehrenfest, *Enzyklopädie der Mathematischen Wissenschaften*, Vol. IV, Part 32, Leipzig-Berlin, 1911; *The Conceptual Foundations of the Statistical Approach in Mechanics*, Ithaca, N.Y., 1959.
3. L.Boltzmann and J.Nabl, *Enzyklopädie der Mathematischen Wissenschaften*, Vol.V, Part 8, Leipzig, Berlin, 1905.

For a more modern review of the historical development of kinetic theory, leading to statistical mechanics, we refer to the work of Brush.

4. S.G.Brush, *Annals of Sc.* **13**, 188, 273 (1957); **14**, 185, 243 (1958); *Am.J.Phys.* **29**, 593 (1961); *Am.Scientist* **49**, 202 (1961).
5. S.G.Brush, *Kinetic Theory*, 3 vols, Oxford, 1965–1966.

In these papers and books, Brush carefully traces the development of the subject. Many of the classical papers, for instance, Refs. 6, 16, 17, 23, and 24 of the present bibliographical notes, are reprinted in his books.

Section 1.1. The Maxwell distribution was introduced by Maxwell in 1859:

6. J.C.Maxwell, *Phil.Mag.* **19**, 19 (1960).

This paper was read before the British Association in Aberdeen on 1859, September 21. In the paper, Maxwell derives his distribution from the following assumptions:

(a) The distribution depends only on the absolute magnitude of the velocity.
(b) The probability that the x-component of the velocity lies within a certain range is independent of the values of the y- or z-components.

From these two assumptions the Maxwell distribution follows in a straightforward manner. The derivation which is given in Section 1.5, and which does not depend on the above assumptions, was given by Maxwell in 1867:

7. J.C.Maxwell, *Trans.Roy.Soc.* **157**, 49 (1867) (= *Phil.Mag.* **35**, 129, 185,1868).
   See also
8. L.Boltzmann, *Wien.Ber.* **58**, 517 (1868).[21]

Experimental verifications of the Maxwell distribution were obtained by Ornstein and van Wijk, who studied the shape of spectral lines, and by Stern and collaborators and by Kofsky and Levinstein, who used the molecular beam method:

9. L.S.Ornstein and W.R.van Wijk, *Z.Physik* **78**, 734 (1932).
10. I.Estermann, O.C.Simpson, O.Stern, *Phys.Rev.* **71**, 238 (1947).
11. I.L.Kofsky and H.Levinstein, *Phys.Rev.* **74**, 500 (1948).

Equations (1.106) to (1.111) can be found in Maxwell's papers (References 6 and 7).

---

[21]Wien.Ber. ≡ Sitzungsberichte der kaiserlichen Akademie der Wissenschaften in Wien, Klasse IIa.

Section 1.2. The perfect gas law as a formula describing experimental data is due to the combined efforts of many scientists, such as Boyle, Mariotte, and Gay-Lussac. In Russian literature (e.g., E.M.Lifshitz and L.P.Pitaevskii, *Statistical Physics*, Part 1, p.122, Oxford, 1980; A.I.Khinchin, *Statistical Mechanics*, p.121, New York, 1949) the perfect gas law is often called Clapeyron's law, probably because Clapeyron was the first to write the law in the form $pV = RT$:

12. E.Clapeyron, *J.d'école polytech.* 14, 153 (1834).

Clapeyron, however, explicitly mentions Mariotte and Gay-Lussac as the ones who discovered and stated the gas law.

The idea that the pressure of a gas is due to the movement of its constituent particles is very old. Both Maxwell (Reference 7) and Clausius (*Ann.Physik* 115, 22 (1862)) give long lists of early papers discussing this idea.

Joule was the first to show that the pressure is proportional to the square of the velocity:

13. J.P.Joule, *Mem. and Proc. Manchester Lit. and Phil. Soc.* 9, 107 (1851) (= *Phil. Mag.* 14, 211, 1857).

By assuming that all the atoms have the same speed c and that one sixth of all the atoms were flying in the $+x$-direction, one sixth in the $-x$-direction, and so on, Krönig derived the equation $pV = \frac{1}{3}nmc^2$, and from this equation deduced that mc^2 should be proportional to T:

14. A.Krönig, it Ann.Physik 99, 315 (1856).

Krönig can probably be considered to be the father of kinetic gas theory. In the introduction to his paper he writes: "$\cdots$ die Bahn jedes Gasatoms muss deshalb eine so unregelmässige seyn, dass sie sich der Berechnung entzieht. Nach den Gesetzen der Wahrscheinlichkeitsrechnung wird man jedoch statt dieser vollkommenen Unregelmässigkeit eine vollkommene Regelmässigkeit annehmen dürfen." [22]

Clausius derives the perfect gas law under slightly less stringent assumptions:

15. R.Clausius, *Phil.Mag.*14, 108 (1857).

In this paper, Clausius actually gives a definition of a perfect gas. There are three conditions that the gas must fulfill:

(a) Its own volume must be small compared to the total volume of the system.

(b) The period during which two atoms are actually colliding must be small compared to the period between two collisions (or, the molecular dimensions must be small compared to the mean free path).

(c) There may only be a weak interaction between the atoms.

More precisely and in more modern language we can say: A perfect gas is a gas where the total potential energy of the system at any time is negligibly small compared to the total kinetic energy of the system.

In all papers mentioned so far, the perfect gas law is derived by considering the transfer of momentum from the gas to the wall. In

---

[22] $\cdots$ the orbit of each gas atom must thus be so irregular that it cannot be calculated. However, according to the laws of probability theory one is allowed to assume a total regularity rather than this complete irregularity.

describing pressure and tension in liquids, Maxwell (Reference 7) uses the idea of transfer of momentum through an imaginary unit area. The use of this definition in deriving the gas law is usually attributed to Lorentz; see, for instance, his paper in *Collected Papers*, Vol.VI, p.143, The Hague, 1938.

Section 1.3 The idea that the volume taken up by the atoms would change the gas law is very old. Bernoulli derived in 1738 that b should simply be equal to the volume ω of the atoms:

16. D.Bernoulli, *Hydrodynamica*, Chap.X, Argentorati, 1738.

Van der Waals showed, however, that as long as $b \ll V$, $b = 4\omega$:

17. J.D.van der Waals, *Over de Continuïteit van den Gas- en Vloeistof-toestand*, Thesis, Leiden, 1873; *Studies in Statistical Mechanics*, Vol. 14, p.121, Amsterdam, 1988.

For an account of van der Waals' thesis see:

18. J.C.Maxwell, *Nature* **10**, 477 (1874).

19. J.S.Rowlinson, *Studies in Statistical Mechanics*, Vol.14, p.1, Amsterdam, 1988.

For extensive bibliographies about the van der Waals law, see References 3 and 19, and

20. H.Kamerlingh Onnes and W.H.Keesom, *Enzyklopädie der mathematischen Wisenschaften*, Vol.V, Part 10, Leipzig-Berlin, 1912 (= Comm.Leiden, Suppl. 23).

21. J.R.Partington, *Physical Chemistry*, Vol.I, Section VIIC, London, 1949.

The term "virial coefficient" was first introduced by Onnes:

22. H.Kamerlingh Onnes, *Proc. Kon. Ned. Akad. Wet.* **4**, 125 (1902) (= Comm. Leiden 71).

The use of the virial for calculating the pressure of a gas and the term "virial" are, however, due to Clausius:

23. R.Clausius, *Phil.Mag.* **40**, 122 (1870).

For a discussion of the virial theorem see also

24. M.Toda in *Selected Papers of Morikazu Toda* (Ed. M. Wadati), World Scientific (1993) p.7.

Section 1.4. The derivation of the Maxwell distribution by considering collisions between elastic spheres was given by Maxwell (Reference 7) and Boltzmann (Reference 8); see also Reference 1.

The Stosszahlansatz was implied in many early calculations. For instance, Clausius used it in his calculations of the mean free path in the form: number of collisions per atom = (volume swept through by the atom) × (number of atoms per cm^3):

25. R.Clausius, *Phil.Mag.* **17**, 81 (1859).

For a detailed discussion and bibliography we refer to

25. D.ter Haar, *Elements of Statistical Mechanics*, 1st Edition, Appendix I, New York, 1954.

27. D.ter Haar, *Rev.Mod.Phys.* **27**, 289 (1955).

28. I.E.Farquhar, *Ergodic Theory in Statistical Mechanics*, New York, 1964.

29. R.Jancel, *Foundations of Classical and Quantum Statistical Mechanics*, Oxford, 1969.

Section 1.5. The function H and the H-theorem are both due to Boltzmann:

30. L.Boltzmann, *Wien.Ber.* **66**, 275 (1872).

In this paper Boltzmann still uses the letter E (entropy) for H. See also References 26 and 27.

Sections 1.6 and 1.7. The considerations of these two sections can, for instance, be found in Boltzmann's papers (see also Reference 1).

CHAPTER 2

THE MAXWELL-BOLTZMANN DISTRIBUTION

2.1.   The Barometer Formula

In the preceding chapter we restricted our discussion to the case of a perfect gas which, moreover, was not under the influence of any external field of force. In the present chapter we want to lift some of these restrictions, and we shall consider the case where external fields of force are present. Also, we shall consider molecular gases. We shall, however, restrict ourselves to conservative systems where the force on each atom (or molecule) can be derived from a potential energy.

Before discussing the general case in detail we shall first consider in a rough, empirical manner the case of a perfect monatomic gas in a uniform gravitational field.

Consider a disk of cross-section A and thickness dz (see Fig.2.1). If this disk is in equilibrium, the following relation must hold:

$$(P + dP)A = PA - g\varrho A\,dz, \qquad (2.101)$$

where g is the gravitational acceleration, ϱ the density of the gas, and P and $P + dP$ the pressures in the gas at heights z and $z + dz$. If we assume that the pressure at any point in the gas is related to the density by the perfect gas law (1.204), which we can write in the form

$$P = \frac{\varrho}{m}\,kT, \qquad (2.102)$$

where m is the mass of one atom, we find from Eq.(2.101) that

$$\varrho = \varrho_0\,e^{-mgz/kT}. \qquad (2.103)$$

Equation (2.103) is known as the *barometer formula*.

We can interpret this formula in the following way. Let us introduce a generalised distribution function $f(x, y, z; u, v, w)$. This function is defined such that $f(x, y, z; u, v, w)\,dx\,dy\,dz\,du\,dv\,dw$ is the number of atoms with positions between r and $r + dr$ and velocities between c and $c + dc$.[1]

---

[1] The vectors r and $r + dr$ have the components x, y, z and $x + dx$, $y + dy$, $z + dz$. A volume element $dx\,dy\,dz$ will often symbolically be denoted by d^3r (not to be confused with dr).

The density $\varrho(x, y, z)$ can be obtained from $f(x, y, z; u, v, w)$ by integration over the whole of velocity space:

$$\varrho(x, y, z) = m \int\!\!\!\int\limits_{-\infty}^{+\infty}\!\!\!\int f(x, y, z; u, v, w)\, du\, dv\, dw. \tag{2.104}$$

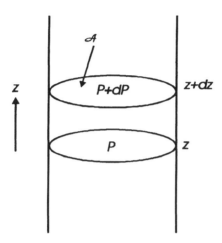

Fig.2.1. Equilibrium of a gas in a uniform gravitational field; $\mathcal{A}$ is the cross-sectional area of a disk of gas between heights z and $z + dz$; P and $P + dP$ are the pressures at these heights.

If we compare Eqs.(2.103) and (2.104) and make the reasonable assumptions that the velocity distribution is independent of the coordinate distribution and that the velocity distribution is again a Maxwell distribution, the function $f(x, y, z; u, v, w)$ will contain a factor

$$e^{-mc^2/2kT}; \tag{2.105}$$

If we combine this result with Eq.(2.103), we find the distribution function

$$f(x, y, z; u, v, w) = C \exp\left[-\frac{mc^2}{2kT} - \frac{mgz}{kT}\right], \tag{2.106}$$

where C is a normalising constant.

Expression (2.106) can be written in the form

$$f(x, y, z; u, v, w) = C\,e^{-\varepsilon/kT}, \tag{2.107}$$

where ε is the total energy of one atom in the gravitational field,

$$\varepsilon = \mathcal{T} + U = \tfrac{1}{2}mc^2 + mgz, \tag{2.109}$$

if $\mathcal{T}$ and U are the kinetic and potential energies of one atom.

2.2.   The μ-Space

We arrived at the distribution function (2.107) by a rather heuristic procedure. Before proving in the next section that the distribution (2.107) — the so-called *Maxwell-Boltzmann distribution* — is, indeed, the equilibrium distribution, we shall generalise it in such a way that ε is the energy of a particle in terms of the so-called generalised coordinates and momenta that are used in Hamiltonian mechanics.[2]

If the particles in our system have s degrees of freedom, the condition of one of the particles will be completely determined by the values of s suitably chosen parameters q_i $(i = 1, 2, \cdots, s)$ and the values of their time derivatives $\dot{q}_i$. The q_i are called *generalised coordinates* and their time derivatives *generalised velocities*.

Let us consider a few examples. One atom will have three degrees of freedom, if we may disregard its internal degrees of freedom, that is, if we may forget that an atom consists of a complicated nucleus with accompanying electrons.[3] In this case we can use for the q_i the Cartesian coordinates of its centre of mass, x, y, and z, and for the generalised velocities the linear velocities of the centre of mass, u, v, and w.

A more complicated case is the so-called dumbbell molecule, a diatomic molecule where the two atoms are a fixed distance, d, apart. The number of degrees of freedom is here five. It is possible to use the Cartesian coordinates of the centres of mass of the two atoms, x_1, y_1, z_1, x_2, y_2, and z_2 as the coordinates with the restriction, or constraint,

$$(x_1 - x_2)^2 + (y_1 - y_2)^2 + (z_1 - z_2)^2 = d^2. \qquad (2.201)$$

However, it is much more convenient to introduce as the q's the three coordinates of the centre of mass, X, Y, and Z and the two angles θ and φ, defining the direction of the axis of the molecule. The generalized velocities are then $\dot{X}$, $\dot{Y}$, $\dot{Z}$, $\dot{\theta}$, and $\dot{\varphi}$. We may draw attention to the fact that, as one can see from this example, generalized coordinates have not necessarily the dimensions of a length.

The total energy of a particle consists of two parts: the kinetic energy $\mathcal{T}$ and the potential energy U. It can be shown that the kinetic energy $\mathcal{T}$ is a quadratic function of the $\dot{q}_i$, where the coefficients in the quadratic

---

[2] For an account of Hamiltonian mechanics see, for instance, D.ter Haar, *Elements of Hamiltonian Mechanics*, Pergamon, Oxford, 1971. Here we shall briefly summarise those results which we need for our present discussion.

[3] The disregard of internal degrees of freedom, such as the degrees of freedom corresponding to the electron motion in an atom or the vibrational degree of freedom of a diatomic molecule, can only be justified by quantum statistics where one shows that these degrees of freedom are *frozen in* (compare the discussion in Chapter 3); for a discussion of the freezing-in of degrees of freedom leading to constraints in classical mechanics, see D.ter Haar, *Elements of Hamiltonian Mechanics*, Pergamon, Oxford, 1971, Chap.2.

expression may still depend on the q_i:

$$\mathcal{T} = \mathcal{T}(q_i, \dot{q}_i) = \tfrac{1}{2} \sum_{k,l=1}^{s} a_{kl}(q_i)\, \dot{q}_k \dot{q}_l. \tag{2.202}$$

We shall not at this point consider velocity-dependent potential energies; the potential energy will then be a function of the q_i only:

$$U = U(q_i). \tag{2.203}$$

The total energy ε is given by the equation

$$\varepsilon = \varepsilon(q_i, \dot{q}_i) = \mathcal{T}(q_i, \dot{q}_i) + U(q_i). \tag{2.204}$$

It is more convenient to introduce instead of the generalised velocities $\dot{q}_i$ a new set of variables, the *generalised momenta*, p_i, which are defined by the equations

$$p_i = \frac{\partial \mathcal{T}}{\partial \dot{q}_i} = \sum_j a_{ij} \dot{q}_j, \tag{2.205}$$

where we have used Eq.(2.202) for $\mathcal{T}$. If q_k is a Cartesian coordinate, p_k will be the linear momentum (= mass times velocity). Transforming from the $\dot{q}_i$ to the p_i we have, instead of Eq.(2.204),

$$\mathcal{H}(p_i, q_i) = \mathcal{T}(p_i, q_i) + U(q_i), \tag{2.206}$$

where $\mathcal{T}$ is now a quadratic function of the p_i. In Eq.(2.206) the total energy is denoted by $\mathcal{H}$. If the energy is expressed in terms of the p_i and the q_i, it is called the *Hamiltonian* of the system, whence the notation $\mathcal{H}$. The $2s$ coordinates $p_1, p_2, \cdots, p_s, q_1, q_2, \cdots, q_s$ are called *canonical coordinates*. In theoretical mechanics it is shown that, expressed in the canonical coordinates, the equations of motion take on the canonical or Hamiltonian form, or

$$\dot{q}_i = \frac{\partial \mathcal{H}}{\partial p_i}, \qquad \dot{p}_i = -\frac{\partial \mathcal{H}}{\partial q_i}. \tag{2.207}$$

The s p_i and s q_i together form the *phase* of the molecule. Once they are given at one time, they follow at any other time from the equations of motion. The behaviour of the molecule can be pictured by an orbit in the $2s$-dimensional space which is spanned by the $2s$ coordinates $p_1, \cdots, p_s$, $q_1, \cdots, q_s$. This space is called the *phase space* of the molecules or, with the Ehrenfests, μ-space, where μ stands for the first letter of *m*olecule.

A volume element in μ-space will be denoted by $d\omega$,

$$d\omega = \prod_{i=1}^{s} dp_i\, dq_i. \tag{2.208}$$

If $f(p, q)\, d\omega$ is the number of molecules, the p_i and q_i of which lie in the intervals $(p_i, p_i + dp_i)$ and $(q_i, q_i + dq_i)$, then in the case of thermal equilibrium the distribution function $f(p_i, q_i)$ will be given by

$$f(p_i, q_i) = C\, e^{-\mathcal{H}(p_i, q_i)/kT}, \qquad (2.209)$$

where C is again a normalising constant. This form of the distribution function was introduced by Boltzmann in 1871. In 1868 Boltzmann had generalised the distribution function (1.104) to include a potential energy, and in 1871 he extended it further to include the case of polyatomic molecules. The distribution of Eq.(2.209) is called the *Maxwell-Boltzmann distribution*.

The average value $\overline{G}$ of a function $G(p_i, q_i)$ of the p_i and q_i is now defined as

$$\overline{G} = \frac{\int G(p_i, q_i)\, f(p_i, q_i)\, d\omega}{\int f(p_i, q_i)\, d\omega}, \qquad (2.210)$$

where the integration extends over the whole of μ-space.

We can apply Eq.(2.210) to calculate the average value of the kinetic energy. Since $\mathcal{T}$ is a homogeneous, quadratic polynomial in the p_i, we have from Euler's theorem on homogeneous functions

$$\mathcal{T} = \tfrac{1}{2} \sum_i \frac{\partial \mathcal{T}}{\partial p_i}\, p_i. \qquad (2.211)$$

For the average value of $\mathcal{T}$ we then find

$$\overline{\mathcal{T}} = s \cdot \tfrac{1}{2} kT. \qquad (2.212)$$

The result (of Eq.(2.212) is obtained as follows. If the expression (2.211) for $\mathcal{T}$ is substituted into Eq.(2.210), there will result s integrals of the type

$$\frac{1}{2} \int \frac{\partial \mathcal{T}}{\partial p_k}\, p_k\, e^{-\mathcal{H}/kT}\, dp_k\, d\omega', \qquad (2.213)$$

where $d\omega'$ stands for the product of all the dp_i and dq_i, except dp_k.

Integrating by parts, we obtain

$$\tfrac{1}{2} kT \int f(p_i, q_i)\, d\omega, \qquad (2.214)$$

since the integrated parts will vanish. Equation (2.212) then follows immediately.

Expression (2.212) means that for every degree of freedom there is a contribution $\tfrac{1}{2}kT$ to the average kinetic energy; this is called the *equipartition* of the kinetic energy. We met with an example of this equipartition in § 1.1 of the previous chapter when we considered a monatomic gas and found that the average kinetic energy pertaining to either the x-, or the y-, or the z-direction was the same and equal to $\frac{1}{2}kT$. Another case would be the dumbbell molecule, where the number of degrees of freedom is five and where we find for the average kinetic energy $\frac{5}{2}kT$.

2.3.   The H-Theorem; H and Probability

In order to prove that the distribution function given by Eq.(2.209) is the equilibrium distribution for a system of independent particles, Boltzmann considered a function H,

$$H \;=\; \overline{\ln f} \;=\; \int f \ln f \, d\omega, \tag{2.301}$$

where the integration extends over the whole of μ-space. As long as the distribution function is not given by Eq.(2.209), dH/dt will be negative, provided a slightly modified Stosszahlansatz is valid. Furthermore, dH/dt will vanish, if f is given by Eq.(2.209). This is a consequence of the fact that during any collision the total energy will be conserved. Although we shall not give a proof of these properties of H, it is along the same lines as that given in § 1.5, albeit much more complicated. We may mention here that an essential point in the proof is that the integration in Eq.(2.301) is over μ-space and not over a $q_i, \dot{q}_i$-space.

The function H is again related to the entropy S_v per unit volume through Eq.(1.612) — if we forget about a possible additive constant. It is also possible to relate H again to the probability in phase space. The argument is the same as in § 1.7, but now we replace the cells Z_i by volume elements in μ-space,

$$Z_i \;\rightarrow\; d\omega, \tag{2.302}$$

the N_i by the total number of particles with p_k and q_k in the intervals $(p_k, p_k + dp_k)$ and $(q_k, q_k + dq_k)$,

$$N_i \;\rightarrow\; f(p_k, q_k)\, d\omega, \tag{2.303}$$

and the ε_i by the total energy of a molecule,

$$\varepsilon_i \;\rightarrow\; \mathcal{H}(p_k, q_k). \tag{2.304}$$

2.4.   Applications of the Maxwell-Boltzmann Formula

In many cases the potential energy may contain a term that depends only on the position of the centre of mass of the molecule. This will, for instance, be the case, if we are considering a system in a gravitational field, or inside a vessel the walls of which exert forces which can be derived from a potential energy such as shown in Fig.2.2. In such cases we can take as three of the q_i (say, q_1, q_2, and q_3) the x-, y-, and z-coordinates of the centre of mass. Integrating the distribution function over all the other q_i and over all the p_i, we get the density of the gas as a function of the position in space,

$$\varrho(x, y, z) = C \int dp_1 \cdots dp_s \, dq_4 \cdots dq_s \, e^{-\mathcal{H}/kT}, \qquad (2.401)$$

or

$$\varrho(x, y, z) = C' \, e^{-U(x,y,z)/kT}, \qquad (2.402)$$

where we have put

$$\mathcal{H} = \mathcal{T}(p_i, q_4, \cdots, q_s) + U(q_1, q_2, q_3) + U'(q_4, \cdots, q_s). \qquad (2.403)$$

In writing down Eq.(2.403) we have assumed that the potential energy can be split into two parts, one depending on the centre of mass coordinates only, and one not depending on those coordinates. It can be shown that the kinetic energy does not depend on the x-, y-, or z-coordinate of the centre of mass.

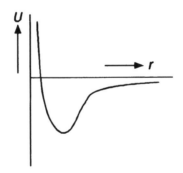

Fig.2.2. The potential energy U of a particle near a wall as function of its distance r from the wall.

One instance of Eq.(2.402) is the barometer formula (2.103). In general, we see that the density will be highest at places where the potential energy is lowest. In the case where the forces exerted by a wall are derived from a potential energy such as the one of Fig.2.2, a certain amount of adsorption on the wall will take place.

Let us now consider the case where the potential is a quadratic function of the q_i,

$$U = \tfrac{1}{2} \sum_{k,l=1}^{s} b_k \, q_k q_l, \qquad (2.404)$$

and where the kinetic energy $\mathcal{T}$ does not contain the q_i. (For an example where the potential energy is of the form (2.404), but the kinetic energy contains the q_i, see Problem 1 at the end of this chapter.) By a suitable choice of linear combinations of the q_i we can reduce U to a sum of t squares, with $t \leqslant s$,

$$U = \tfrac{1}{2} \sum_{k=1}^{t} b_k q_k'^{\,2}, \tag{2.405}$$

where the q_i' are the linear combinations just mentioned, which we use as new coordinates.

One finds the average value of the potential energy in much the same way as was used in § 2.2 to calculate the average value of the kinetic energy. We have thus

$$\overline{U} = t \cdot \tfrac{1}{2}kT. \tag{2.406}$$

The average value of the total energy is now obtained by combining Eqs.(2.212) and (2.406):

$$\bar{\varepsilon} = \overline{\mathcal{T}} + \overline{U} = (s+t) \cdot \tfrac{1}{2}kT. \tag{2.407}$$

Equation (2.407) expresses a more general equipartition of energy: each degree of freedom contributing to the kinetic energy and each degree of freedom contributing a quadratic term to the potential energy will give rise to a term $\tfrac{1}{2}kT$ in the average total energy, provided the kinetic energy does not contain the coordinates.

The specific heat per mole, C_v, will now be given by the equation

$$C_v = N_A \frac{\partial \bar{\varepsilon}}{\partial T} = \tfrac{1}{2}(s+t)R. \tag{2.408}$$

As an example we can calculate the average energy of an isotropic three-dimensional harmonic oscillator. The energy is given by the equation

$$\varepsilon = \frac{p_x^2 + p_y^2 + p_z^2}{2m} + \tfrac{1}{2}\alpha \left(x^2 + y^2 + z^2\right), \tag{2.409}$$

and its average value is given by the equation

$$\bar{\varepsilon} = 3kT, \tag{2.410}$$

leading to a specific heat of $3R$ per mole.

Let us now consider the case of a gas of dumbbell molecules, for instance, HCl. If we treat the molecules as real dumbbells, that is, molecules with a fixed distance apart between the hydrogen atom and the chlorine atom, we have $s = 5$, $t = 0$, and the specific heat per mole will be $\tfrac{5}{2}R$, which is observed experimentally, at any rate at not too high temperatures. If, however, we bear in mind that the idea of a completely fixed distance apart of the two atoms is an idealisation and that there will always be the possibility of vibrations along the molecular axis, we use $s = 6$ and $t = 1$, leading to a specific heat per mole of $\frac{7}{2}R$, which is not found experimentally. The solution of this apparent contradiction — which upset Gibbs so much that he resolved to treat statistical mechanics as a branch of rational mechanics rather than as a branch of physics — can only be given by quantum mechanics and is related to the problem of the freezing-in of degrees of freedom, mentioned briefly at the beginning of § 2.2.

A similar problem arises in the case of non-linear polyatomic molecules. If all the distances in the molecule are completely fixed, it will behave as a solid body, and since it then would have three rotational and three translational degrees of freedom, we should expect a specific heat per mole of $3R$ ($s = 6$, $t = 0$), which has been found experimentally. However, if there are N atoms in the molecules, and if we take into account all possible vibrations, we find $3N$ degrees of freedom of which $3N - 6$ contribute to the potential energy: only the translational and the rotational degrees of freedom do not make a contribution to the potential energy. If all $3N - 6$ terms in the potential energy are quadratic, we should expect a specific heat per mole of $3(N-1)R$, which is not found experimentally. The situation becomes even more complicated and would lead to even larger specific heats,[4] if we take the degrees of freedom corresponding to the electronic and nuclear motions into account.

Let us now consider the case of a dumbbell molecule that possesses a magnetic (or electric) dipole moment, and let us calculate its average moment in a uniform field. For the sake of simplicity we shall choose as our q_i the three coordinates of the centre of mass, X, Y, and Z, the angle θ between the magnetic moment μ and the magnetic field B, and the angle φ between the projection of μ on a plane perpendicular to B and a fixed axis in that plane. The corresponding momenta will be denoted by p_X, p_Y, p_Z, p_θ, and p_φ. If the absolute values of the permanent magnetic moment and of the uniform magnetic field are denoted by μ and B, the potential energy will be given by the expression

$$U = -\mu B \cos \theta. \tag{2.411}$$

For the average magnetic moment in the direction of the field we get

$$\frac{\overline{\mu}}{\mu} = \frac{\int \cos \theta \, e^{-(\Im+U)/kT} \, d\omega}{\int e^{-(\Im+U)/kT} \, d\omega} = \mathcal{L}\left(\frac{\mu B}{kT}\right), \tag{2.412}$$

where $\mathcal{L}(x)$ is the Langevin function,

$$\mathcal{L}(x) = \coth x - \frac{1}{x}. \tag{2.413}$$

---

[4] We must warn our readers here that, in the discussion of the paradox of the specific heats, we have presented the situation as being simpler than it, in fact, is. The reader can, however, easily verify the real state of affairs, once the discussion of the next chapter is understood. It is then also possible to see for what temperatures our present considerations will hold.

The behaviour of $\mathcal{L}(x)$ is shown in Fig.2.3.

In order to derive Eq.(2.412) we have to remember that the kinetic energy $\mathcal{T}$ in terms of the momenta is

$$\mathcal{T} = \frac{p_X^2 + p_Y^2 + p_Z^2}{2M} + \frac{p_\theta^2}{2A} + \frac{p_\varphi^2}{2A \sin^2\theta}, \qquad (2.414)$$

where M is the total mass of the molecule and where A is the moment of inertia of the molecule with respect to an axis through the centre of mass, at right angles to the axis of the molecule.

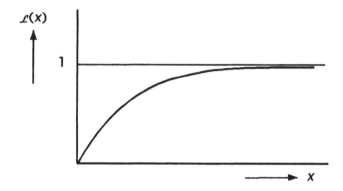

Fig.2.3. The Langevin function $\mathcal{L}(x) = \coth x - \frac{1}{x}$.

Integration over p_φ produces, apart from constants which cancel in numerator and denominator, a factor $\sin\theta$. In this way we must finally evaluate the expression

$$\int_0^\pi \cos\theta \; e^{\mu B \cos\theta/kT} \sin\theta \, d\theta \Big/ \int_0^\pi \cos\theta \; e^{\mu B \cos\theta/kT} \sin\theta \, d\theta, \qquad (2.415)$$

which leads to Eq.(2.412).[5]

Equation (2.412) was derived by Langevin in 1905, and in 1912 Debye derived a similar formula for a system of electric dipoles. We see from Fig.2.3 that $\bar{\mu}$ approaches μ for low temperatures (or large fields), that is, saturation sets in at low temperatures, while at higher temperatures thermal agitation produces a smaller mean dipole moment.

---

[5] We note that Eq.(2.412) will also hold for the more general case where the particle is a symmetric top with its dipole moment μ along the axis of symmetry; see Problem 8 at the end of this chapter.

2.5.   The Boltzmann Transport Equation

We saw in § 1.5 that collisions change the distribution function in such a way as to decrease the value of H, and we mentioned in § 2.3 that this remains true for the case where the distribution function depends on the coordinates as well as on the velocities. In that discussion we were not interested in the actual change in the distribution function itself; nevertheless, it is sometimes important to evaluate this change in f, for instance, if we wish to know how soon equilibrium will be reached.

In order to find the general equation that must be satisfied by the distribution function, we must consider not only the change of this distribution function by collisions, but also the change due to the continuous motion of the particles. For the sake of simplicity we shall consider only the case of systems of point particles; the distribution function will then be a function of the coordinates x, y, and z and of the velocity components u, v, and w.

We shall consider those atoms that have coordinates between r and $r+dr$ and velocities between c and $c + dc$. At time t there will be[6]

$$f(r, c, t)\, d^3r\, d^3c \tag{2.501}$$

such atoms, and at time $t + dt$ there will be

$$f(r, c, t + dt)\, d^3r\, d^3c \tag{2.502}$$

such atoms. The difference between these two numbers (2.501) and (2.502), which is equal to

$$\frac{\partial f}{\partial t}\, dt\, d^3r\, d^3c, \tag{2.503}$$

is due to two causes: (1) there may not be a balance between atoms entering the volume element $d^3r\, d^3c$ during the time interval $t, t + dt$ because of collisions and those atoms leaving this volume element during the given time interval as the result of collisions (compare Eq.(1.401)), and (2) the number of atoms that have acquired positions and velocities during $t, t + dt$ such that their representative points fall at the end of the time interval inside $d^3r\, d^3c$ may not be equal to the number of atoms that had positions and velocities at the beginning of the time interval within the volume element $d^3r\, d^3c$ but the representative points of which have left that volume element during the time interval.

We can express this fact by writing

$$\frac{\partial f}{\partial t} = \left(\frac{\partial f}{\partial t}\right)_{\text{drift}} + \left(\frac{\partial f}{\partial t}\right)_{\text{coll}}. \tag{2.504}$$

For $(\partial f/\partial t)_{\text{drift}}$ we can find the following equation:

$$-\left(\frac{\partial f}{\partial t}\right)_{\text{drift}} = \frac{\partial f}{\partial x}\, u + \frac{\partial f}{\partial y}\, v + \frac{\partial f}{\partial z}\, w + \frac{\partial f}{\partial u}\, \frac{F_x}{m} + \frac{\partial f}{\partial v}\, \frac{F_y}{m} + \frac{\partial f}{\partial w}\, \frac{F_z}{m}, \tag{2.505}$$

---

[6] We use the abbreviated expression $f(r, c, t)$ for $f(x, y, z; u, v, w; t)$.

where F is the external force on the particles and m their mass.

We can prove Eq.(2.505) by considering the flow of representative points out of and into the volume element $d^3r\,d^3c$ in the six-dimensional space of which x, y, z, u, v, and w are the coordinates. The total number of representative points that cross the "face" which is "orthogonal" to the x-axis during a time interval dt is equal to the product of the area of that face, $dy\,dz\,du\,dv\,dw$, the distance travelled in the x-direction by the representative point during dt, $u\,dt$, and the number of representative points per unit volume of c-space, f. The net decrease of representative points in $d^3r\,d^3c$ during that time interval is the difference between the number crossing the face at $x + dx$ and the number crossing the "opposite" face at x; that is, it is equal to $\left[f(x+dx, y, z; u, v, w; t) - f(x, y, z; u, v, w; t) \right] u\,dy\,dz\,du\,dv\,dw\,dt$, or $(\partial f/\partial t)\,u\,dx\,dy\,dz\,du\,dv\,dw\,dt$. From the flow through the faces orthogonal to the y-, the z-, the u-, the v-, and the w-axes we get similar contributions to the net decrease in the number of representative points in $d^3r\,d^3c$, that is, to $(\partial f/\partial t)_{\text{drift}}\,d^3r\,d^3c\,dt$; these contributions are equal to $(\partial f/\partial y)\,v\,d^3r\,d^3c\,dt$, $\cdots$, $(\partial f/\partial w)\,a_x\,d^3r\,d^3c\,dt$, where a_z is the z-component of the acceleration which by Newton's second law is equal to F_z/m.

If we use shorthand notation and introduce the vector operators ∇ (which is the gradient vector operator) and ∇_c with components $\partial/\partial x$, $\partial/\partial y$, $\partial/\partial z$ and $\partial/\partial u$, $\partial/\partial v$, $\partial/\partial w$, respectively, we can write Eq.(2.505) in the form

$$- \left(\frac{\partial f}{\partial t} \right)_{\text{drift}} = (c \cdot \nabla)f + \frac{1}{m}(F \cdot \nabla_c)f. \qquad (2.506)$$

We can use Eq.(1.419) for $(\partial f/\partial t)_{\text{coll}}$, since we may assume that in practically all cases of interest the collision mechanism is independent of where the collisions take place. The only difference between the present situation and the one leading to Eq.(1.419) is that the distribution function is now a function of r, as well as of c, which means that, for instance, f_1' stands for $f(r, c_1', t)$, and so on. Combining Eqs.(2.504), (2.506), and (1.419) we get the following integro-differential equation for the distribution function:

$$\frac{\partial f}{\partial t} + (c \cdot \nabla)f + \frac{1}{m}(F \cdot \nabla_c)f + \int (ff_1 - f'f_1')\,a\,d^3c_1\,d^2\omega = 0. \quad (2.507)$$

Equation (2.507) is known as *Boltzmann's transport equation* or, more simply, as *Boltzmann's equation*.

If there are no external forces, the Maxwell distribution will be an (equilibrium) solution of Eq.(2.507). Since f does not depend on r, $\nabla f = 0$, and we have also $F = 0$, while we saw in §1.4 that for the Maxwell distribution $(\partial f/\partial t)_{\text{coll}} = 0$, so that altogether we find that $\partial f/\partial t = 0$.

If the external forces can be derived from a potential energy function $U(r)$, the Maxwell-Boltzmann distribution (2.107) will be an equilibrium solution. We have in that case

$$F = -\nabla U, \qquad (2.508)$$

$$\nabla f = -\frac{1}{kT} f \nabla U, \tag{2.509}$$

$$\nabla_c f = \frac{1}{kT} f \nabla_c \mathcal{T} = -\frac{mc}{kT} f, \tag{2.510}$$

while

$$f f_1 = f' f_1', \tag{2.511}$$

because of the conservation of energy in a collision. From Eqs.(2.507) to (2.511) it then follows again that $\partial f / \partial t = 0$.

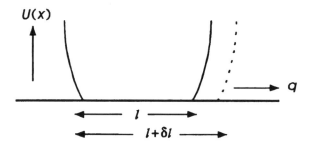

Fig.2.4. External parameters. The potential energy $U(q)$ of a particle in a one-dimensional "box" of "volume" l. The dashed curve indicates the potential energy when the "volume" is increased to $l + \delta l$.

Equation (2.507) can, for instance, be used to estimate the time necessary to re-establish equilibrium. It is in a modified form extensively used in solid state theory and in kinetic theory its most important applications have been in the discussion of steady non-equilibrium states. In 1916 and 1917 Chapman and Enskog, working independently and using powerful approximation methods, applied it to derive theoretical expressions for the viscosity coefficient, the thermal conductivity, and other transport coefficients.

2.6.   External Parameters

We met in § 2.4 a case where the energy of a molecule depended not only on the p_i and the q_i, but also on the magnitude of the magnetic (or electric) field. In general, it is possible for the energy of a molecule to depend on a number of *external parameters*, a_i. Examples of such parameters are the volume of a gas, the electric field strength, and so on. We have in that case

$$\varepsilon = \varepsilon(p_1, \cdots, p_s, q_1, \cdots, q_s; a_1, a_2, \cdots). \tag{2.601}$$

Let us first consider the case where a is the "volume" of a one-dimensional gas (see Fig.2.4). If $U(q)$ is the potential energy of the molecules, then $-\partial U/\partial q$ will be the force exerted by the wall on the molecule. Apart from on q, U will also depend on the "volume" l of the one-dimensional "box": $U = U(q, l)$. Let us now displace the wall a distance δl. If we displace the molecule at the same time over a distance $\delta q = \delta l$, there will be no change in potential energy, or

$$\delta U \; = \; 0 \; = \; \frac{\partial U}{\partial q} \, \delta q + \frac{\partial U}{\partial l} \, \delta l, \qquad \text{provided } \delta q \; = \; \delta l, \qquad (2.602)$$

or

$$\frac{\partial U}{\partial l} \; = \; - \, \frac{\partial U}{\partial q}. \qquad (2.603)$$

Since $-\partial U/\partial q$ is the force exerted by the "volume" on the molecule, $-\partial U/\partial l$ will be the force exerted by the molecule "on the volume". In general, we can introduce quantities A_i defined by the equation

$$A_i \; = \; - \, \frac{\partial \varepsilon}{\partial a_i}. \qquad (2.604)$$

These A_i can be called the *generalised forces* (per molecule) exerted by the system tending to increase its parameters. If N is the total number of particles in the system, the total force exerted by the system tending to increase the parameter a_i will be equal to $N\overline{A}_i$.

We see that the total work done by the molecule "on the parameters" which is equal to minus the change in the energy ε of the molecule, when the parameters change from a_i to $a_i + \delta a_i$, is given by the equation

$$-\varepsilon \; = \; \sum_i A_i \, \delta a_i. \qquad (2.605)$$

In the case considered in § 2.4 the parameter was the magnetic field and the generalised force was there $-\partial \varepsilon/\partial B = \mu \cos \theta$, that is, the magnetic moment in the direction of the magnetic field. In that case $-\delta \varepsilon$ was equal to $\mu \cos \theta \, \delta B$.

2.7.   The Phase Integral; Connection with Thermodynamics

Let us consider the following quantity:

$$Z_\mu \; = \; \mathrm{e}^{-\beta F} \; = \; K \int \mathrm{e}^{-\beta \varepsilon} \, d\omega, \qquad (2.701)$$

where the suffix μ indicates that the integration is over the whole of μ-space, and where now and henceforth β is related to the absolute temperature T through the equation

$$\beta \; = \; \frac{1}{kT}. \qquad (2.702)$$

The quantity Z_μ is called the *phase integral* or *partition function* of one molecule. Both Z_μ and F, the physical meaning of which we shall see in a moment, are functions of the temperature (or of β) and of the external parameters. The constant K, which we shall leave undetermined for the moment, enters in order that Z_μ may be made a dimensionless quantity.

We can write Eq.(2.701) in the form

$$\frac{1}{K} = \int e^{\beta(F-\epsilon)} \, d\omega. \tag{2.703}$$

If we vary the temperature and the external parameters, we get from Eq.(2.703)

$$\delta \frac{1}{K} = 0 = \int \delta(\beta F) \, e^{\beta(F-\epsilon)} \, d\omega - \int \epsilon \, \delta\beta \, e^{\beta(F-\epsilon)} \, d\omega - \int \beta \, \delta\epsilon \, e^{\beta(F-\epsilon)} \, d\omega. \tag{2.704}$$

Since

$$\delta\epsilon = \sum_i \frac{\partial \epsilon}{\partial a_i} \, \delta a_i = -\sum_i A_i \, \delta a_i, \tag{2.705}$$

Eq.(2.704) leads to

$$\delta(\beta F) = \bar{\epsilon} \, \delta\beta - \beta \sum_i \overline{A}_i \, \delta a_i, \tag{2.706}$$

where we have used Eqs.(2.701), (2.703), and the definition (2.210) of the average value of a physical quantity.

To find the physical meaning of F we shall express the right-hand side of Eq.(2.706) in terms of the internal energy U and the entropy S per molecule.[7] This can be done by considering the second law of thermodynamics in the form

$$\delta S = \frac{\delta Q}{T} = \frac{\delta U + \delta A}{T}, \tag{2.707}$$

where A is the work done by the system.

For one molecule we have the relations

$$U = \bar{\epsilon} \quad \text{and} \quad \delta A = \sum_i \overline{A}_i \, \delta a_i, \tag{2.708}$$

where the bar across the δ indicates that $\delta A \; (= -\overline{\delta\epsilon})$ is not necessarily a total variation.

From Eqs.(2.707) and (2.708) we find that the right-hand side of Eq. (2.706) can be written in the form

$$\bar{\epsilon} \, \delta\beta - \beta \sum_i \overline{A}_i \, \delta a_i = \delta \left(\frac{U}{kT} - \frac{S}{k} \right), \tag{2.709}$$

---

[7] In §1.6 S and U referred to the entropy and the internal energy of the whole system, but here they refer to one particle only.

and it follows that F/T, apart from a possible additive constant, satisfies the equation

$$\frac{F}{T} = \frac{U}{T} - S, \quad \text{or} \quad F = U - TS. \tag{2.710}$$

Thus, F corresponds to the (Helmholtz) free energy per molecule — apart possibly from an additive multiple of the temperature.

In the case where the volume V is the only external parameter Eqs.(2.706) and (2.707) can be written in the form (compare Eq.(1.606))

$$\delta \frac{S}{k} = \beta \, \delta \bar{\varepsilon} + \frac{P}{N} \, \beta \, \delta V \tag{2.711}$$

and

$$\delta(\beta F) = \bar{\varepsilon} \delta \beta - \frac{P}{N} \, \beta \, \delta V, \tag{2.712}$$

where all quantities refer to one molecule, bar the pressure whence the factor N.

The temperature (or β) and the external parameters determine the *thermodynamic state* of the system. Once they are given, one can calculate the phase integral Z_μ and thus the free energy F. All thermodynamic quantities can then be derived from Z_μ or F. Since $F = -kT \ln Z_\mu$, we shall give here only the equations involving F. From Eq.(2.706) it follows that

$$\bar{\varepsilon} = \frac{\partial(\beta F)}{\partial \beta} = -T^2 \frac{\partial(F/T)}{\partial T}, \tag{2.713}$$

$$\overline{A_i} = -\frac{1}{\beta} \frac{\partial(\beta F)}{\partial a_i} = -T \frac{\partial(F/T)}{\partial a_i}, \tag{2.714}$$

while we get from Eq.(2.713) for the specific heat per molecule, c_v,

$$c_v = \frac{\partial \bar{\varepsilon}}{\partial T} = -k\beta^2 \frac{\partial \bar{\varepsilon}}{\partial \beta} = -k\beta^2 \frac{\partial^2(\beta F)}{\partial \beta^2}$$
$$= -2T \frac{\partial(F/T)}{\partial T} - T^2 \frac{\partial^2(F/T)}{\partial T^2}. \tag{2.715}$$

If one of the a_i is the volume, we get for the presure P the equation

$$P = -\frac{N}{\beta} \frac{\partial(\beta F)}{\partial V}. \tag{2.716}$$

We can use Eq.(2.716) to obtain the equation of state. The equation of state can also be obtained from the equation

$$P = NT \frac{\partial S}{\partial V}, \tag{2.717}$$

once the entropy is known as a function of the volume; Eq.(2.717) follows from Eq.(2.711).

We note that the integrand in Eq.(2.701) is proportional to the distribution function f given by Eq.(2.107). As it follows from the definition of f that $\int f \, d\omega$, integrated over the whole of μ-space, must be equal to N, we find from Eqs.(2.701) and (2.107) that

$$f = KN e^{\beta(F-\epsilon)}, \tag{2.718}$$

and we get for Boltzmann's H-function

$$H = \int f \ln f \, d\omega = \beta(F - \bar{\epsilon}) + \text{const} = -\frac{S}{k} + \text{const}. \tag{2.719}$$

This shows once again that kH, apart possibly from an additive constant, is equal to $-S$.

In conclusion, let us apply our formulæ to the case of a monatomic perfect gas. The energy of one atom is given by the equation

$$\epsilon = \frac{p_x^2 + p_y^2 + p_z^2}{2m} + U(x, y, z), \tag{2.720}$$

where the potential energy is such that the gas stays inside the volume V. This can be realised, if U is given by the equations

$$\left.\begin{array}{ll} U(x, y, z) = 0 & \text{inside } V; \\ U(x, y, z) = \infty & \text{outside } V. \end{array}\right\} \tag{2.721}$$

Introducing expression (2.720) into Eq.(2.701) for Z_μ we get, after integrating over x, y, z, p_x, p_y, and p_z,

$$Z_\mu = KV(2\pi mkT)^{3/2}. \tag{2.722}$$

From Eqs.(2.722), (2.701), (2.713), and (2.715) we get for the energy and specific heat per atom:

$$\bar{\epsilon} = \tfrac{3}{2}kT, \tag{2.723}$$

$$c_v = \tfrac{3}{2}k. \tag{2.724}$$

Finally, we get from Eq.(2.716) for the equation of state the perfect gas law (1.204).

Problems

1. Consider a two-dimensional harmonic oscillator with energy

$$\epsilon = \frac{p_x^2 + p_y^2}{2m} + \tfrac{1}{2}\alpha(x^2 + y^2).$$

According to Eq.(2.407) the average energy will be equal to $2kT$. Introducing polar coordinates, $x = r \cos \varphi$, $y = r \sin \varphi$, the energy can be expressed in the form

$$\varepsilon = \frac{1}{2m} \left[p_r^2 + \frac{p_\varphi^2}{r^2} \right] + \tfrac{1}{2} \alpha r^2.$$

There are now two kinetic energy terms and one, quadratic, potential energy term. Verify by direct calculation, using polar coordinates, that just the same the average energy equals $2kT$, and comment on the result.

2. From the example of the preceding problem it is clear that the equipartition theorem must be formulated with caution. Prove the following general statement, which is due to Tolman,[8] stating clearly the conditions under which it is valid:

$$\overline{q_i \frac{\partial \mathcal{H}}{\partial q_j}} = kT\, \delta_{ij}, \qquad \overline{p_i \frac{\partial \mathcal{H}}{\partial p_j}} = kT\, \delta_{ij},$$

where δ_{ij} is Kronecker's symbol, which is equal to 1, if $i = j$, and equal to 0, if $i \neq j$.

3. Show that the following distribution function satisfies the transport equation (2.507) for the case of point particles moving either in force-free space or in a central field of force:

$$f(\boldsymbol{r}, \boldsymbol{c}, t) = A\, e^{-\alpha c^2 + \gamma(uy - vx)}.$$

Discuss the physical meaning of this distribution.

4. Consider the Lorentz model[9] of a metal in which the transport properties are ascribed to classical, charged particles ("electrons") that do not interact with one another, but only with the ions in the lattice. The electron-lattice collisions are supposed to be elastic and isotropic. This means that we assume that, if $\Theta(\boldsymbol{c}, \boldsymbol{c}')\, d^3 c'$ denotes the probability that per unit time an electron will change its velocity from $\boldsymbol{c}$ to a velocity in the range $\boldsymbol{c}', \boldsymbol{c}' + d\boldsymbol{c}'$, we can write $\Theta(\boldsymbol{c}, \boldsymbol{c}')$ in the form

$$\Theta(\boldsymbol{c}, \boldsymbol{c}') = \frac{\eta \delta(c - c')}{c^2},$$

where we assume η to be constant and where $\delta(c - c')$ is Dirac's δ-function.[10]

Consider the transport equation (2.504) with the first term on the right-hand side given by Eq.(2.505) and the second term by Eq.(1.401). Let

[8] R.C.Tolman, *Phys. Rev.* **11**, 261 (1918).

[9] H.A.Lorentz, *The Theory of Electrons*, pp. 47–50, Teubner, Leipzig, 1909.

[10] Dirac's δ-function is defined by the equations (see, for example, P.A.M.Dirac,

the system be subject to an electromagnetic field so that the force F acting on the electrons is the Lorentz force,

$$F \ = \ -e\left\{E + [c \times B]\right\},$$

where $-e$ is the charge of an electron. Find a stationary solution of the transport equation, assuming (1) that the electric field E is uniform and lies in the xy-plane, (2) that the magnetic field B is uniform and along the z-axis, and (3) that we may write the distribution function f in the form

$$f \ = \ f_0 + u\chi_1(c) + v\chi_2(c),$$

where f_0 is the Maxwell distribution and where we assume the terms with the χ_i to be small compared to similar terms involving f_0, so that we may neglect them when compared with the latter. Using the relation that exists between η and the mean free path λ, express the solution in terms of f_0, λ, E_x, E_y, and the ratio s of λ and the radius of curvature of an electron moving with speed c in the magnetic field B.

5. Use the solution obtained in the previous problem to evaluate the electrical conductivity of a (classical) metal. In this case we put $E_y = 0$, $B = 0$, and find the conductivity σ from Ohm's law, $j_x = \sigma E_x$, where

$$j_x \ = \ -e \int u f \, d^3 c.$$

6. Use the solution obtained in Problem 4 to evaluate the isothermal Hall coefficient R, defined by the relation

$$E_y \ = \ R j_x B_z,$$

Quantum Mechanics, Second Edition, §§ 20, 21, Clarendon Press, Oxford, 1935)

$$\delta(x) \ = \ 0, \qquad x \ \neq \ 0; \qquad \int_{-\infty}^{+\infty} \delta(x)\,dx \ = \ 1. \tag{A}$$

From Eqs.(A) it follows that for any function $f(x)$ we have

$$\left. \begin{array}{l} \displaystyle\int_{-\infty}^{+\infty} f(x)\,\delta(x)\,dx \ = \ f(0); \qquad \int_{-\infty}^{+\infty} f(x)\,\delta(x-a)\,dx \ = \ f(a); \\[3mm] \displaystyle\int_{a}^{b} f(x)\,\delta(x-c)\,dx \ = \ f(c), \quad \text{if } a < c < b; \\[3mm] \displaystyle\int_{a}^{b} f(x)\,\delta(x-c)\,dx \ = \ 0, \qquad \text{if either } b < c \text{ or } c < a. \end{array} \right\} \tag{B}$$

For a discussion of the mathematical aspects of the δ-function see, for instance, J.S.R.Chisholm and R.M.Morris, *Mathematical Methods in Physics*, § 18.1, North-Holland, Amsterdam, 1966; or K.F.Riley, *Mathematical Methods for the Physical Sciences*, § 8.8, Cambridge University Press, 1974.

for the case where a current j_x is flowing in the x-direction, where there is a magnetic field B_z in the z-direction, and where E_y is the electromotive force in the y-direction needed to prevent current flowing in that direction.

7. Consider the transport equation for the electrons in a metal for the case where $B = 0$, $E_y = 0$, but where there is a temperature gradient dT/dx in the x-direction so that the equilibrium distribution f_0 now depends on x. Solve the transport equation under the same conditions as those used in Problem 4. Find the coefficient of thermal conductivity κ, defined by the equation

$$w_x = -\kappa \frac{dT}{dx},$$

where

$$w_x = \int \tfrac{1}{2} m c^2 f \, d^3 c$$

is the thermal current in the metal. Bear in mind that we have to find w_x under the condition that there is no electrical current flowing in the metal ($j_x = 0$).

Having found κ, combine the result with that of Problem 5 to find the *Wiedemann-Franz ratio*[11] $\kappa/\sigma T$.

8. Evaluate the phase integral for the following systems:
(a) a perfect gas of particles for which the energy ε is proportional to the absolute magnitude of their momentum ($\varepsilon = sp$);
(b) a perfect gas of point particles in a spherical container of volume V, using spherical polars (compare your result with Eq.(2.722));
(c) a perfect gas of magnetic dipoles in an external magnetic field $\boldsymbol{B}$;
(d) a perfect gas of symmetric top molecules, the kinetic energy of which is

$$\mathcal{T} = \frac{1}{2m}\left(p_x^2 + p_y^2 + p_z^2\right) + \frac{1}{2}\left[\frac{p_\vartheta^2}{A} + \frac{\left(p_\varphi - p_\psi \cos \vartheta\right)^2}{A \sin^2 \vartheta} + \frac{p_\psi^2}{C}\right],$$

where m is the mass of the top, A, A, and C are the principal moments of inertia about the principal axes through the centre of mass, the Cartesian coordinates of which are x, y, and z, while ϑ, φ, and ψ are Euler angles;
(e) a perfect gas of asymmetric top molecules, the kinetic energy of which is

$$\mathcal{T} = \frac{1}{2m}\left(p_x^2 + p_y^2 + p_z^2\right)$$
$$+ \frac{1}{2}\left\{\frac{\left[\left(p_\varphi - p_\psi \cos \vartheta\right) \cos \psi - p_\vartheta \sin \vartheta \sin \psi\right]^2}{A \sin^2 \vartheta}\right.$$
$$\left. + \frac{\left[\left(p_\varphi - p_\psi \cos \vartheta\right) \sin \psi + p_\vartheta \sin \vartheta \cos \psi\right]^2}{B \sin^2 \vartheta} + \frac{p_\psi^2}{C}\right\},$$

---

[11]G.Wiedemann and R.Franz, *Ann.Physik* **89**, 497 (1853).

where A, B, and C are the principal moments of inertia, while the rest of the notation is the same as under (d);

(f) a perfect gas of point particles moving in a one-dimensional potential energy field given by the equation

$$U(x) = \frac{1}{2}\alpha x^2 - \gamma x^3 - \delta x^4,$$

assuming that the phase integral can be written as a power series in the supposedly small quantities $\gamma^2/\beta\alpha^3$ and $\delta/\beta\alpha^2$, and that only the first non-trivial term need be retained;

(g) a perfect gas of point particles in an infinitely high cylindrical vessel in a uniform gravitational field along the axis of the cylinder;

(h) a perfect gas of symmetric top molecules, each of which carries along its axis of symmetry a permanent dipole moment μ, in a uniform magnetic field $\boldsymbol{B}$.

9. Use the results of the preceding problem
   (a) to calculate the specific heat for cases (a), (f), and (g);
   (b) to calculate the magnetisation in case (h). Find also the asymptotic high-temperature expression for the magnetisation.

10. A point particle of mass m is fixed to the middle of a thin, inextensible string of length l (Fig.2.5). The mass is rotated about the x-axis connecting the two end points of the string, and the system is in thermal contact with its surroundings at a temperature T. Calculate the force needed to keep the end points at a distance x apart.

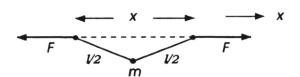

Fig.2.5. The mass m is fixed at the middle of the string and is rotated around the x-axis. The length of the string is l and the force needed to keep the ends at a distance x apart is F.

11. Consider a system of two rigid point dipoles at a fixed distance apart. Assuming these dipoles to be in thermal equilibrium as far as their orientation is concerned, find an expression (in the form of an integral) for the average force acting between them and evaluate the high-temperature limit of that expression.

The force between two dipoles μ and μ' at a distance R apart follows from the potential energy

$$U = -\frac{\mu_0\mu\mu'}{4\pi R^3} \left[2 \cos\vartheta \cos\vartheta' - \sin\vartheta \sin\vartheta' \cos(\varphi - \varphi')\right],$$

where ϑ, φ (ϑ', φ') are the angles defining the direction of $\boldsymbol{\mu}$ ($\boldsymbol{\mu}'$).

12. Prove the Bohr-van Leeuwen theorem, which states that the magnetic susceptibility of a system of charged point particles governed by classical mechanics and obeying classical statistics vanishes rigorously. Introduce the magnetic field by means of a vector potential, so that the magnetic field dependence of the energy is through the kinetic energy which will be of the form[12]

$$\mathcal{T} = \frac{(\boldsymbol{p} + e\boldsymbol{A})^2}{2m}.$$

Bibliographical Notes

General references for this chapter are

1. L. Boltzmann, *Vorlesungen über Gastheorie*, 2 vols., Leipzig, 1896–1898; *Lectures on Gas Theory*, Berkeley, California, 1964.
2. S. Chapman and T.G.Cowling, *The Mathematical Theory of Non-Uniform Gases*, Cambridge, 1939.
3. J. W. Gibbs, *Elementary Principles in Statistical Mechanics* (Vol. II of his Collected Works), New Haven, 1948.
4. S. G. Brush, *Kinetic Theory*, 3 vols, Oxford, 1965–1966.

Section 2.1. A rather crude attempt to derive the barometer formula can be found in Krönig's paper on kinetic theory:
5. A. Krönig, *Ann. Physik* **99**, 315 (1856).
The exact barometer formula was first given by Boltzmann:
6. L.Boltzmann, *Wien. Ber.* **78**, 7 (1879).

Section 2.2. The term "μ-space" was introduced by the Ehrenfests:
7. P. and T. Ehrenfest, *Enzyklopädie der Mathematischen Wissenschaften*, Vol. IV, Part 32, Leipzig-Berlin, 1911, p.36.
The Maxwell-Boltzmann distribution was derived by Boltzmann in 1871:
8. L. Boltzmann, *Wien. Ber.* **58**, 517 (1868).
9. L. Boltzmann, *Wien. Ber.* **63**, 397 (1871).
In Ref.8 Boltzmann gives the distribution in the case of an external field of force, while in Ref.9 he also takes into account the internal degrees of freedom of the molecules. The name "Maxwell-Boltzmann distribution" was suggested by G.H. Bryan in a paper read before the British Association at Oxford in 1894.
The equipartition of kinetic energy over the different degrees of freedom was proved by Boltzmann:
10. L. Boltzmann, *Wien. Ber.* **63**, 712 (1871).
Compare also:
11. R. C. Tolman, *Phys. Rev.* **11**, 261 (1918).

Section 2.3. The generalized H-theorem is also due to Boltzmann:
12. L. Boltzmann, *Wien. Ber.* **72**, 427 (1875).

---

[12]See, for instance, D.ter Haar, *Elements of Hamiltonian Mechanics*, Pergamon, Oxford, 1971, p.113.

Section 2.4. Equation (2.412) was first derived by Langevin for the case of molecules possessing magnetic dipole moments;

13. P. Langevin, *J. phys.* **4**, 678 (1905).
14. P. Langevin, *Ann. chim. et phys.* **5**, 70 (1905).
   The case of electric dipoles was studied by Debye, see, for example,,
15. P. Debye, *Physik. Zs.* **13**, 97 (1912).

Section 2.5. The transport equation is due to Boltzmann:
16. L. Boltzmann, *Wien. Ber.* **74**, 503 (1876).
   Probably the most important papers on the solution of the equation for the case of gases are those by Chapman and Enskog:[13]
17. S. Chapman, *Trans. Roy. Soc.* (*London*) **A216**, 279 (1916).
18. S. Chapman, *Trans. Roy. Soc.* (*London*) **A217**, 115 (1917).
19. D. Enskog, *Dissertation*, Uppsala, 1917.
   For a thorough discussion of the whole subject we refer to Ref.2; see also:
20. R.Balian, *Statistical Mechanics*, Springer, Berlin, 1991, § 15.3.

Section 2.6. The introduction of external parameters into the statistical discussion was made by Gibbs, Ref.3, p.4.

Section 2.7. The phase integral and its relation to the free energy were introduced and used by Gibbs, Ref.3, Chaps. IV *et seq.*.
   Darwin and Fowler were the first to introduce the term "partition function":
21. C. G. Darwin and R. H. Fowler, *Phil. Mag.* **44**, 450 (1922)
   Tolman uses in his textbook the term "sum over states".

---

[13]Born *Nuovo Cimento* **6**, Suppl. 296 (1949)) has drawn attention to the fact that Hilbert had developed methods similar to those of Chapman and Enskog for solving the Boltzmann equation. Hilbert's results were given in a series of lectures at Göttingen but only a short account of them was published (*Math. Ann.* **72**, 562 (1912)).

CHAPTER 3

THE PARTITION FUNCTION

3.1. The Partition Function

So far we have used classical mechanics to describe the systems in which we were interested. We know, however, that in many instances classical mechanics is inadequate to describe the behaviour of physical systems, and that only a quantum-mechanical treatment can give correct results. We should thus build statistical mechanics upon a quantum-mechanical basis and afterwards show that, indeed, the classical formulæ that we derived in the preceding chapters are limiting cases of the quantum-mechanical formulæ. We shall see in the next chapter that the statistical methods used are the same for all cases and that the term "quantum statistics" is thus a misnomer.

We remind ourselves that there are two aspects in which quantum mechanics differs from classical mechanics: "diffraction" effects and "symmetry" effects. The first arise because of the wave nature of matter; they lead to the Schrödinger wave equation and thus to the existence of energy levels for the systems considered. The second occur because the wavefunction of a system of identical particles must be totally symmetric or totally antisymmetric, if we are dealing with bosons or fermions. In the present chapter we shall be concerned with the diffraction effects only, postponing until the next chapter the discussion of symmetry effects as well as a brief discussion of to what particles these consideration apply.

Let us consider a system of particles, each of which has a discrete energy spectrum. Let the energy levels be ε_n, all of which we shall for the moment assume to be non-degenerate.[1] We can now ask the question: "How many particles will be in each of the various levels, if we know that the total energy of the system consisting of N particles is E?" This problem is

---

[1] We shall not consider here the complications of degeneracy or of a continuous spectrum; we leave the generalisation to the case where there is a continuous spectrum to the reader. The case of degenerate levels will be considered later on in this chapter. We refer to quantum-mechanics textbooks (for example, A.S.Davydov, *Quantum Mechanics*, Pergamon, Oxford, 1990) for details about how one obtains the energy spectrum of a particle moving in a given potential energy.

similar to the one treated in § 1.7. The cells Z_i introduced there correspond now to the discrete energy levels. Assuming that the *a priori* probability for a particle to be in a particular energy level is the same for all levels, we may put all Z_i equal to one another, and we find for the probability W that there are N_1 particles in the energy level ε_1, N_2 in the level ε_2, and so on (compare Eq.(1.701)),

$$W = C \frac{N!}{\prod_i N_i!}. \tag{3.101}$$

The assumption of equal *a priori* probabilities for the occupation of different non-degenerate energy levels is one of the basic assumptions in quantum-statistical mechanics. To a large extent its justification lies in the fact that the results of quantum-statistical mechanics are in accordance with experimental data. We shall discuss this basic assumption in some detail in § 5.13. At this point we may draw attention to the fact that its classsical counterpart lies in the assumption that the *a priori* probability for a representative point to fall into a cell Z_i will be proportional to its size. In § 3.4 we shall, indeed, see the connection between these two assumptions.

Having obtained an expression for W, we can use the same arguments as in § 1.7; we find for the equilibrium value of N_i the expression

$$N_i = e^{\alpha - \beta \varepsilon_i}, \tag{3.102}$$

where we have again identified the most probable distribution with the equilibrium distribution, and where we have imposed the constraints of a constant total energy and a constant total number of particles,

$$\sum_i N_i = N, \qquad \sum_i n_i \varepsilon_i = E. \tag{3.103}$$

The quantities α and β can be evaluated as functions of N and E by substituting Eq.(3.102) into Eq.(3.103).

The average energy $\bar\varepsilon$ per particle can be obtained from the equation

$$\bar\varepsilon = \frac{\sum_i N_i \varepsilon_i}{\sum_i N_i} = \frac{\sum_i \varepsilon_i e^{\alpha - \beta \varepsilon_i}}{\sum_i e^{\alpha - \beta \varepsilon_i}} = -\frac{\partial \ln Z_\mu^{\mathrm{qu}}}{\partial \beta}, \tag{3.104}$$

where the *partition function* Z_μ^{qu} is defined by the equation

$$Z_\mu^{\mathrm{qu}} = K' \sum_i e^{\alpha - \beta \varepsilon_i}, \tag{3.105}$$

where K' is an as yet unspecified constant.

If we introduce a function F by the relation

$$e^{-\beta F} = Z_\mu^{\mathrm{qu}}, \tag{3.106}$$

we can write Eq.(3.104) in the form (compare Eq.(2.713)):

$$\bar{\varepsilon} = \frac{\partial \beta F}{\partial \beta}, \tag{3.107}$$

and we see that, provided F is the Helmholtz free energy and β satisfies the equation

$$\beta = \frac{1}{kT}, \tag{3.108}$$

this equation is just the Gibbs-Helmholtz equation of thermodynamics. At this point we shall use this correspondence as a sufficient justification for the identification of F with the Helmholtz free energy and of β with $1/kT$; we shall discuss this point in more detail in the next chapter.

Since the Helmholtz free energy as a function of the temperature (or β) and of the external parameters, which enter into F through the dependence of the ε_i on them, completely determines the thermodynamic state of the system, we should be able to derive all thermodynamic quantities from F. Indeed, Eqs.(2.713) to (2.716) are again valid, and it follows that, once we have evaluated the partition function Z_μ^{qu}, we can find the average energy, the specific heat, the generalised forces such as the magnetisation or the pressure, and so on.

3.2.   The Harmonic Oscillator

It is of historical interest to consider for a moment the old quantum theory or model theory. A discussion of the connection between the old quantum theory and statistical mechanics is particularly appropriate because it was considerations of a statistical nature that led Planck to introduce his quantum of action, h, thus starting quantum theory.

We shall first of all consider a one-dimensional oscillator. It is defined by a quadratic potential energy function $U(q)$,

$$U(q) = \tfrac{1}{2}\alpha q^2, \tag{3.201}$$

where α is a constant and q the coordinate describing the position of the oscillator. If the mass of the oscillator is m, its frequency ω will be related to α by the equation

$$\omega = \sqrt{\frac{\alpha}{m}}. \tag{3.202}$$

Its total energy can then be written in the form

$$\varepsilon(p,q) = \frac{p^2}{2m} + \frac{1}{2}m\omega^2 q^2. \tag{3.203}$$

In this case μ-space, which is now the p, q plane, has only two dimensions. We can draw curves of constant energy in this phase plane. These curves will be ellipses, a few of which are drawn in Fig.3.1. As long as the oscillator is not disturbed from the outside, it will move in such a way that its representative point will stay on one ellipse.

The area enclosed by such an ellipse is given by the following equation:[2]

$$I = \oint p\,dq, \tag{3.204}$$

where the integral extends over one period of the oscillator.

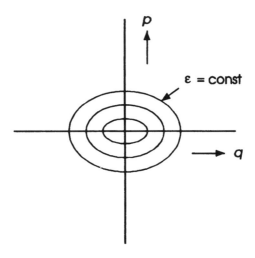

Fig.3.1. Constant energy curves in the μ-space (the pq-plane) of a one-dimensional harmonic oscillator.

Since p and q satisfy the equations

$$q = q_0 \sin \omega t, \qquad p = m\dot{q} = m\omega q_0 \cos \omega t, \tag{3.205}$$

we find for I

$$I = \int_0^{2\pi/\omega} p\dot{q}\,dt = \pi m\omega q_0^2 = \frac{\varepsilon}{\nu}, \qquad \nu = \frac{\omega}{2\pi}. \tag{3.206}$$

Although in classical mechanics any ellipse in the phase plane could represent a state of the oscillator, this will no longer be true in quantum theory. In order to find out how to select the possible energy levels we shall briefly recall how Planck was led to the introduction of his quantum of action.

---

[2] This integral is sometimes called the *phase integral*. Since we have already used this term in a different connection, we shall not use it for I to avoid confusion.

As Kirchhoff had shown that the energy density in the radiation field in thermal equilibrium was independent of the material of the surrounding walls, Planck considered a system of one-dimensional harmonic oscillators in equilibrium with the radiation field. To find the thermodynamic properties of the radiation field, Planck considered its entropy and evaluated it using Boltzmann's formula (1.717); to do this he needed to find the probability function W. He then divided the phase plane into sections of area h, allotting to each cell a representative energy (compare the discussion in § 2.3). He originally intended to take the limit as $h \to 0$, but found — rather to his surprise — that the radiation density calculated in that way corresponded to the experimentally observed one, provided h was not allowed to tend to zero, but was put equal to 6.6×10^{-34} J s.

Some time later, in 1915, Sommerfeld and Wilson suggested that this division of the phase plane into cells of area h was the correct way to derive the possible values which the energy of a one-dimensional periodic system could assume. In fact, the Sommerfeld-Wilson quantisation rule is

$$I = nh, \tag{3.207}$$

where n is a nonnegative integer.

If one uses the quantisation rule (3.207), the energy corresponding to the n-th cell is determined by Eq.(3.206) and given by

$$\varepsilon_n = n\hbar\omega, \qquad n = 0, 1, 2, \ldots , \tag{3.208}$$

where $\hbar \, (= h/2\pi)$ is Dirac's constant.

In this way we have allotted to each energy level a cell of equal area (equal to h) in μ-space. The energy corresponding to a cell is that corresponding to the inner boundary (ellipse) of the cell, if we call ε_n the $n + 1$-st level, n indicating not the the actual number of the level, but the value of the corresponding *quantum number*.

In order to get a better agreement with the experimental evidence regarding the three-dimensional rotator (compare Eq.(3.504)), Sommerfeld later introduced instead of Eq.(3.207) the so-called *"halbzahlige"* quantisation rule,

$$I = (n + \tfrac{1}{2})h. \tag{3.209}$$

For the energy levels we then get the same values as those following from the Schrödinger equation,

$$\varepsilon_n = (n + \tfrac{1}{2})\hbar\omega, \tag{3.210}$$

corresponding to a value inside the original cell. We see that the lowest energy available to a one-dimensional oscillator is not zero, but $\tfrac{1}{2}\hbar\omega$, the so-called *zero-point energy*.

That such a zero-point energy is not just a special way of choosing the energy origin, but a physical reality can, for instance, be seen in the case of liquid helium where the zero-point energy is sufficiently large to prevent the liquid from solidifying under its own vapour pressure.

Using Eq.(3.105), we can evaluate the partition function; the result is

$$Z_\mu^{\text{qu}} = K' \sum_{n=0}^{\infty} e^{-\beta n \hbar \omega} \left[e^{-\frac{1}{2}\beta \hbar \omega} \right], \qquad (3.211)$$

where the factor in the square brackets occurs if we use Eq.(3.210), and does not occur if we use Eq.(3.208).

From Eq.(3.104) we get for the average energy per oscillator

$$\bar{\varepsilon} = \frac{\hbar \omega}{e^{\beta \hbar \omega} - 1} \left[+ \tfrac{1}{2}\hbar \omega \right]. \qquad (3.212)$$

Equation (3.212) can be written as a series expansion, both at high temperatures ($\beta \hbar \omega \ll 1$) and at low temperatures ($\beta \hbar \omega \gg 1$), as follows,

$$\hbar \omega \ll kT : \quad \varepsilon = kT - \tfrac{1}{2}\hbar \omega \left[+ \tfrac{1}{2}\hbar \omega \right] + \text{series in } \frac{\hbar \omega}{kT}; \qquad (3.213)$$

$$\hbar \omega \gg kT : \quad \varepsilon = \left[\tfrac{1}{2}\hbar \omega \right] + \hbar \omega \cdot e^{-\beta \hbar \omega} + \ldots , \qquad (3.214)$$

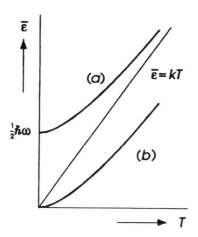

Fig.3.2. The average energy of a one-dimensional harmonic oscillator: (a) with zero-point energy; (b) without zero-point energy.

Figure 3.2 shows the behaviour of $\bar{\varepsilon}$ as a function of temperature. From the considerations of the previous chapter we might have expected $\bar{\varepsilon} = kT$ (Eq.(2.407) with $s = t = 1$). We see that this behaviour is realised at high temperatures. This is not surprising, as we should expect classical behaviour in the limit as $h \to 0$, and this means $\beta\hbar\omega \ll 1$, a condition also realised at high temperatures for the given, fixed value of h. We may remark here that it also gives us an example of the *correspondence principle*, enunciated by Bohr in 1918. This principle states that in the limit of large quantum numbers, the results obtained from quantum mechanics will be the same as those obtained from classical mechanics. That the requirement of large quantum numbers will lead in the present case to the same result as taking the limit $h \to 0$ can be seen from Eq.(3.208) where the condition $n \to \infty$ would entail $h \to 0$ for a given energy.

We shall now consider the case of the two-dimensional oscillator. Its Hamiltonian is given by the equation

$$\varepsilon = \frac{1}{2m}(p_1^2 + p_2^2) + \tfrac{1}{2}m\omega_1^2 q_1^2 + \tfrac{1}{2}m\omega_2^2 q_2^2, \tag{3.215}$$

and the energy levels of the stationary states, which follow from the appropriate Schrödinger equation, are given by

$$\varepsilon_{n_1 n_2} = (n_1 + \tfrac{1}{2})\hbar\omega_1 + (n_2 + \tfrac{1}{2})\hbar\omega_2. \tag{3.216}$$

For the partition function we have from Eq.(3.105)

$$Z_\mu^{\text{qu}} = K' \sum_{n_1, n_2} e^{-\beta\varepsilon_{n_1 n_2}}, \tag{3.217}$$

which leads to the expression

$$Z_\mu^{\text{qu}} = K' \frac{e^{-\frac{1}{2}\beta(\omega_1+\omega_2)}}{\left(1 - e^{-\beta\hbar\omega_1}\right)\left(1 - e^{-\beta\hbar\omega_2}\right)}. \tag{3.218}$$

For a three-dimensional oscillator the corresponding equations will be

$$\varepsilon = \frac{1}{2m}(p_1^2 + p_2^2 + p_3^2) + \tfrac{1}{2}m[\omega_1^2 q_1^2 + \omega_2^2 q_2^2 + \omega_3^2 q_3^2], \tag{3.219}$$

$$\varepsilon_{n_1 n_2 n_3} = (n_1 + \tfrac{1}{2})\hbar\omega_1 + (n_2 + \tfrac{1}{2})\hbar\omega_2 + (n_3 + \tfrac{1}{2})\hbar\omega_3. \tag{3.220}$$

$$Z_\mu^{\text{qu}} = K' \sum_{n_1, n_2, n_3} e^{-\beta\varepsilon_{n_1 n_2 n_3}}, \tag{3.221}$$

or

$$Z_\mu^{\text{qu}} = K' \frac{e^{-\frac{1}{2}\beta(\omega_1+\omega_2+\omega_3)}}{\left(1 - e^{-\beta\hbar\omega_1}\right)\left(1 - e^{-\beta\hbar\omega_2}\right)\left(1 - e^{-\beta\hbar\omega_3}\right)}. \tag{3.222}$$

These formulæ become interesting only if we consider isotropic oscillators. For the two-dimensional case, Eqs.(3.216) to (3.218) then becomes

$$\varepsilon_n = (n+1)\hbar\omega, \qquad (\omega_1 = \omega_2 = \omega), \tag{3.223}$$

$$Z_\mu^{\mathrm{qu}} = K' \sum_n (n+1)\,e^{-\beta\varepsilon_n}, \tag{3.224}$$

or

$$Z_\mu^{\mathrm{qu}} = K' \frac{e^{-\beta\hbar\omega}}{\left(1 - e^{-\beta\hbar\omega}\right)^2}. \tag{3.225}$$

Equation (3.224) follows from the fact that there are $n+1$ different ways of writing n as the sum of two non-negative integers n_1 and n_2. Equation (3.225) follows either from Eq.(3.218) by putting $\omega_1 = \omega_2 = \omega$ or directly from Eq.(3.224) by observing that

$$\sum_n n\,e^{-nx} = -\frac{d}{dx}\left[\sum_n e^{-nx}\right]. \tag{3.226}$$

In the three-dimensional case we have

$$\varepsilon_n = \left(n + \tfrac{3}{2}\right)\hbar\omega, \qquad (\omega_1 = \omega_2 = \omega_3 = \omega), \tag{3.227}$$

$$Z_\mu^{\mathrm{qu}} = K' \sum_n \tfrac{1}{2}(n+1)(n+2)\,e^{-\beta\varepsilon_n}, \tag{3.228}$$

or

$$Z_\mu^{\mathrm{qu}} = K' \frac{e^{-\frac{3}{2}\beta\hbar\omega}}{\left(1 - e^{-\beta\hbar\omega}\right)^3}. \tag{3.229}$$

Fig.3.3. Partition of an integer n into three non-negative integers n_1, n_2, and n_3.

In Eq.(3.228) we used the fact that the number of different ways of writing n as the sum of three non-negative integers is $\tfrac{1}{2}(n+1)(n+2)$ (compare the derivation of Eq.(3.224)). Equation (3.229) then follows either from Eq.(3.222), or from Eq.(3.228) by using an equation similar to Eq.(3.226).

The easiest way to compute the number of ways of writing n as a sum of three non-negative integers is by thinking of it as a division of n dots by 2 strokes (Fig.3.3). Since there are $(n + 2)!$ different ways of arranging these n dots and 2 strokes, and since a rearrangement of the dots or of the strokes does not interest us, the total number of different divisions will be equal to $(n + 2)!/n!2! = \tfrac{1}{2}(n + 1)(n + 2)$.

As long as the oscillators are anisotropic, the situation is practically the same as in the case of the one-dimensional oscillator. For instance, in the three-dimensional case, μ-space has six dimensions and each energy level corresponds to one cell with a (six-dimensional) volume h^3. We could think of obtaining these cells (and their corresponding energy levels) by applying the Sommerfeld rules for each of the three independent oscillations.

However, as soon as we turn to the isotropic case, the situation becomes different. In the three-dimensional case an energy level ε_n does not correspond to one cell, but to $\frac{1}{2}(n + 1)(n + 2)$ cells. The Schrödinger equation in this case allows $\frac{1}{2}(n + 1)(n + 2)$ linearly independent solutions for the same energy level, and the energy level is said to be g-fold *degenerate* with $g = \frac{1}{2}(n + 1)(n + 2)$.

3.3.   Planck's Radiation Law

In this section we shall derive an expression for the energy density distribution in a radiation field. We shall assume with Planck and Rayleigh that the radiation field can be considered as consisting of one-dimensional oscillators with frequencies $\omega_1, \omega_2, \ldots$. If the first oscillator is in the state corresponding to a quantum number n_1, the second one in that corresponding to n_2, and so on, the total energy of the system will be given by the equation

$$E = \sum_i \varepsilon_i = \sum_i n_i \hbar \omega_i, \tag{3.301}$$

in which, it will be noticed, we have omitted the zero-point energies.[3] The reason why this leads to a correct result is that one really should discuss the radiation field in terms of light quanta (compare the discussion in § 4.1). From Eq.(3.212) we have for the average energy of the i-th oscillator

$$\overline{\varepsilon_i} = \frac{\hbar \omega_i}{e^{\beta \hbar \omega_i} - 1}. \tag{3.302}$$

We must draw attention to one important difference between Eqs.(3.212) and (3.302). Equation (3.212) gives us the average energy when the average is taken over a large number of identical oscillators, that is, we get a system average. In the present case, however, we have only one oscillator with a frequency ω_i and expression (3.302) must be considered to give an average over a large period rather than over a large number, that is, we have a time average. That we may use the time average and the system average interchangeably follows from the general equivalence of these two kinds of average. This equivalence will be discussed in §§ 5.7 and 5.14.

If the frequencies are lying very densely, the sum in Eq.(3.301) is virtually an integral. In order to evaluate this sum (or integral), we must know the

---

[3] For a discussion of the infinite zero-point self-energy of the electromagnetic field which would occur if we took the zero-point energies into account see, for instance, H.A.Kramers, *Quantum Mechanics*, North-Holland, Amsterdam, 1957, § 87.

number of frequencies $D(\omega)\,d\omega$ between ω and $\omega + d\omega$. This can be done in the following way.

In order to simplify the considerations, we assume that the volume V is a cube of edge length l. It can be proved that this simplification does not affect the final result for $D(\omega)$. Since we are dealing with standing waves, the wavelength λ of the waves in the cube must be such that the amplitude at the walls is zero. In the one-dimensional case (Fig.3.4) this leads to the equation

$$l = \tfrac{1}{2}n\lambda = \frac{\pi nc}{\omega}, \qquad (3.303)$$

where we have used the relation $2\pi c = \omega\lambda$ while c is the velocity of light.

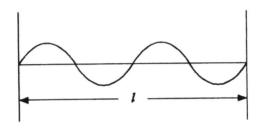

Fig.3.4. The fitting of a standing wave in a length l $(n = 4)$.

The three-dimensional case is much more complicated. Essentially, one must fit standing waves into the cube in such a way that they fit in the x-, the y-, and the z-direction. This leads to the following requirement for the frequency ω:

$$\frac{\omega^2 l^2}{\pi^2 c^2} = n_1^2 + n_2^2 + n_3^2, \qquad (3.304)$$

where n_1, n_2, and n_3 are positive integers.

To prove Eq.(3.304) we shall consider the solutions of the Maxwell equations for the case where the walls of the volume containing the radiation are metallic, so that the normal component of the electric field must vanish at the boundaries. As the electric field strength E must also satisfy the condition $(\nabla \cdot E) = 0$ in empty space, we find for the components of E the expressions

$$
\begin{aligned}
E_x &= A_x \cos\frac{n_1\pi x}{l}\sin\frac{n_2\pi y}{l}\sin\frac{n_3\pi z}{l}\sin(\omega t - \delta),\\[4pt]
E_y &= A_y \sin\frac{n_1\pi x}{l}\cos\frac{n_2\pi y}{l}\sin\frac{n_3\pi z}{l}\sin(\omega t - \delta),\\[4pt]
E_z &= A_z \sin\frac{n_1\pi x}{l}\sin\frac{n_2\pi y}{l}\cos\frac{n_3\pi z}{l}\sin(\omega t - \delta).
\end{aligned}
\qquad (3.305)
$$

Substituting expressions (3.305) into the wave equation,

$$\nabla^2 E = \frac{1}{c^2}\frac{\partial^2 E}{\partial t^2}, \qquad (3.306)$$

we find Eq.(3.304).

If we wish to know how many frequencies there are between ω and $\omega + d\omega$, we must calculate the number of lattice points with positive integer coordinates in a spherical shell of radius $\omega l/\pi c$ and thickness $l\,d\omega/\pi c$. In this way we get for $D(\omega)$ the equation

$$D(\omega)\,d\omega \;=\; 2 \cdot \frac{1}{8} \cdot \frac{4\pi \cdot \omega^2 l^2 \cdot l\,d\omega}{\pi^3 c^3} \;=\; \frac{\omega^2 V}{\pi^3 c^3}\,d\omega. \tag{3.307}$$

The factor 2 in the second member of Eq.(3.307) arises from the fact that the transverse light waves have two polarisation degrees of freedom. The factor $\frac{1}{8}$ arises, since we are dealing with that part of the spherical shell for which n_1, n_2, and n_3 are all positive.

The energy density per unit volume, $\varrho(\omega)$, due to oscillators with frequencies between ω and $\omega + d\omega$ can now be obtained from Eqs.(3.301), (3.302), and (3.307), and is given by the equation

$$\varrho(\omega)\,d\omega \;=\; \frac{\hbar}{\pi^2 c^3}\,\frac{\omega^3\,d\omega}{e^{\beta\hbar\omega}-1}. \tag{3.308}$$

Equation (3.308) was derived by Planck in 1900 and is called *Planck's radiation law*. For high frequencies $(\beta\hbar\omega \gg 1)$ it goes over into *Wien's law*,

$$\varrho(\omega) \;=\; \frac{\hbar}{\pi^2 c^3}\,\omega^3\,e^{-\beta\hbar\omega}, \tag{3.309}$$

while for low frequencies $(\beta\hbar\omega \ll 1)$ it becomes *Rayleigh's law*,

$$\varrho(\omega) \;=\; \frac{\hbar}{\pi^2 c^3}\,\omega^2, \tag{3.310}$$

the same result that Planck achieved by letting h tend to zero. Since in this way he could obviously not obtain Wien's law for high frequencies, Planck was obliged to let h remain finite.

For the total energy density per unit volume we get

$$\varrho \;=\; \int_0^\infty \varrho(\omega)\,d\omega \;=\; \frac{\hbar}{\pi^2 c^3} \int_0^\infty \frac{\omega^3\,d\omega}{e^{\beta\hbar\omega}-1},$$

or

$$\varrho \;=\; aT^4, \tag{3.311}$$

where a is given by the equation

$$a \;=\; \frac{\pi^2 k^4}{15\hbar^3 c^3} \;=\; 7.56 \times 10^{-16}\ \text{J K}^{-4}\ \text{m}^{-3}. \tag{3.312}$$

In the calculations leading to Eq.(3.311) we must evaluate the integral

$$\int_0^\infty \frac{x^3\,dx}{e^x-1} \;=\; \int_0^\infty e^{-x} x^3 \left(1 + e^{-x} + e^{-2x} + \cdots\right)\,dx$$

$$=\; \sum_{n=1}^\infty \int_0^\infty e^{-nx} x^3\,dx \;=\; 6\,\zeta(4) \;=\; \frac{\pi^4}{15},$$

where

$$\zeta(s) = \sum_{n=1}^{\infty} n^{-s} = \text{Riemann's } \zeta\text{-function.}$$

Equation (3.311) is called the *Stefan-Boltzmann law*, which states that the total energy density in a radiation field in temperature equilibrium is proportional to the fourth power of the temperature.

Let us now consider briefly the case where we have an equilibrium between atoms and a radiation field. From the known quantum-mechanical transition probabilities we can then again derive the radiation law (3.308). We assume that there is only one kind of atoms present that can be in stationary states with energies ε_k. According to Bohr's postulate (1913) the transition from a state k to a state l will be accompanied by the absorption or emission of a light quantum of frequency ω_{kl},

$$\omega_{kl} = \left| \frac{\varepsilon_k - \varepsilon_l}{\hbar} \right|, \tag{3.313}$$

where emission takes place when $\varepsilon_k > \varepsilon_l$, and absorption when $\varepsilon_k < \varepsilon_l$. If N_k denotes the number of atoms in the k-th state, the number of atoms $N_{k \to l}$ making a transition from k to l per unit time will be given by the equation

$$N_{k \to l} = N_k A_{kl} + N_k B_{kl} \varrho(\omega_{kl}), \tag{3.314}$$

where we have assumed that $\varepsilon_k > \varepsilon_l$. The first term on the right-hand side of Eq.(3.314) corresponds to spontaneous transitions and will be proportional to N_k, the proportionality factor being denoted by A_{kl}. The second term corresponds to induced — or stimulated — emissions (or negative absorptions) and will be proportional both to N_k and to the density of the radiation field at the frequency concerned, the factor here being denoted by B_{kl}.

The number of atoms $N_{l \to k}$ making the transition $l \to k$ per unit time is given by the equation

$$N_{l \to k} = N_l B_{lk} \varrho(\omega_{kl}), \tag{3.315}$$

since in this case no spontaneous transition can take place.

The coefficients A_{kl} and B_{kl} were introduced by Einstein in 1917 and are called the *Einstein transition probabilities*. In the quantum theory of radiation it is proved that[4]

$$B_{lk} = B_{kl}, \tag{3.316}$$

---

[4] See, for instance, H.A.Kramers, *Quantum Mechanics*, North-Holland, Amsterdam, 1957, §82; or A.S.Davydov, *Quantum Mechanics*, Pergamon, Oxford, 1990, §94.

and

$$A_{kl} = \frac{\hbar\omega^3}{\pi^2 c^3} B_{kl}. \tag{3.317}$$

We should perhaps mention here that the A_{kl} and B_{kl} are atomic quantities that do *not* depend on the temperature.

In equilibrium there will be as many transitions from k to l as from l to k (principle of detailed balancing; see § 6.5), so that we have the equation

$$N_k\big[A_{kl} + B_{kl}\varrho(\omega_{kl})\big] = N_l B_{lk}\varrho(\omega_{kl}). \tag{3.318}$$

As N_k and N_l in thermal equilibrium will satisfy Eq.(3.102), we have

$$N_k : N_l = e^{-\beta\varepsilon_k} : e^{-\beta\varepsilon_l}, \tag{3.319}$$

and, combining Eqs.(3.316) to (3.319), we can calculate the equilibrium energy density in the radiation field for which we again obtain Eq.(3.308).

3.4. The Transition to Classical Statistics

The Schrödinger equation of a particle will give us the energy levels ε_n of the stationary states and their degree of degeneracy g_n where n stands for all the quantum numbers that determine the energy level. We can introduce the partition function Z_μ^{qu} by the equation

$$Z_\mu^{\mathrm{qu}} = K' \sum_n g_n e^{-\beta\varepsilon_n}, \tag{3.401}$$

which comprises all expressions for Z_μ^{qu} that we have met in this chapter up to now: Eqs.(3.105), (3.211), (3.217), (3.221), (3.224), and (3.228).

Each level corresponds to a volume $g_n h^s$ in μ-space, where s is the number of degrees of freedom of the particle. If we accept, as we did in §§ 1.7 and 2.3 that the *a priori* probability of finding the representative point of the particle inside a specified volume of μ-space is proportional to that volume, we see that non-degenerate levels all have equal *a priori* probabilities of being realised, but that the *a priori* probability of a degenerate level is weighted with a factor g_n. For this reason the degree of degeneracy g_n is often called the *(statistical) weight* of the level ε_n.

Let us now return to the partition function and let us, as in § 3.1, introduce a function F by the equation

$$e^{-\beta F} = Z_\mu^{\mathrm{qu}}. \tag{3.402}$$

From Eqs.(3.104), (3.401), and (3.402) it follows that

$$\bar{\varepsilon} = \frac{\sum_n g_n e^{-\beta\varepsilon_n}\varepsilon_n}{\sum_n g_n e^{-\beta\varepsilon_n}} = -\frac{1}{Z_\mu^{\mathrm{qu}}}\frac{\partial Z_\mu^{\mathrm{qu}}}{\partial\beta} = \frac{\partial\beta F}{\partial\beta}, \tag{3.403}$$

and recognising this as the Gibbs-Helmholtz equation, we see again that F is the (Helmholtz) free energy per particle. In the same way as was done in §2.7 we can now derive all thermodynamic quantities from F or from Z_μ^{qu}.

In §2.4 we saw that often some degrees of freedom did not contribute to the average energy or the specific heat. In order to see how this can happen, let us consider the case of the anisotropic two-dimensional harmonic oscillator, and let us assume that ω_1 is much smaller than ω_2.

The partition function in this case is given by Eq.(3.218), and for the average energy we get from Eqs.(3.218) and (3.403)

$$\bar{\varepsilon} = \tfrac{1}{2}\hbar\omega_1 + \tfrac{1}{2}\hbar\omega_2 + \frac{\hbar\omega_1}{e^{\beta\hbar\omega_1} - 1} + \frac{\hbar\omega_2}{e^{\beta\hbar\omega_2} - 1}. \tag{3.404}$$

Let us now consider three temperature regions around T_1, T_2, and T_3 where

$$kT_1 \ll \hbar\omega_1 \ll kT_2 \ll \hbar\omega_2 \ll kT_3. \tag{3.405}$$

For $\bar{\varepsilon}$ we get, in these three regions,

$$T \sim T_1: \qquad \bar{\varepsilon} = \tfrac{1}{2}\hbar\omega_1 + \tfrac{1}{2}\hbar\omega_2 + \hbar\omega_1 e^{-\beta\omega_1} + \cdots, \tag{3.406}$$

$$T \sim T_2: \qquad \bar{\varepsilon} = \tfrac{1}{2}\hbar\omega_2 + kT + \hbar\omega_2 e^{-\beta\omega_2} + \frac{(\hbar\omega_1)^2}{12kT} + \cdots, \tag{3.407}$$

$$T \sim T_3: \qquad \bar{\varepsilon} = 2kT + \frac{(\hbar\omega_1)^2 + (\hbar\omega_2)^2}{12kT} + \cdots, \tag{3.408}$$

The specific heat follows from the equation

$$c_v = \frac{\partial\bar{\varepsilon}}{\partial T}, \tag{3.409}$$

or

$$T \sim T_1: \qquad c_v = k\left[(\beta\hbar\omega_1)^2 e^{-\beta\hbar\omega_1} + \cdots\right], \tag{3.410}$$

$$T \sim T_2: \qquad c_v = k\left[1 - \tfrac{1}{12}(\beta\hbar\omega_1)^2 + (\beta\hbar\omega_2)^2 e^{-\beta\hbar\omega_2} + \cdots\right], \tag{3.411}$$

$$T \sim T_3: \qquad c_v = k\left[2 - \tfrac{1}{12}(\beta\hbar\omega_2)^2 + \cdots\right], \tag{3.412}$$

Figure 3.5 shows the behaviour of c_v as a function of the temperature. From this figure we see the following:

(a) If the temperature is sufficiently high, the specific heat will reach its classical value $2k$ (compare Eq.(2.407) with $s = t = 2$).

(b) At temperatures large as compared to $\hbar\omega_1/k$, but small as compared to $\hbar\omega_2/k$, the specific heat is only k. It is as if one degree of freedom were frozen-in.

(c) Finally, at very low temperatures the specific heat is practically equal to zero.

That the classical behaviour follows at sufficiently high temperatures is again an instance of the correspondence principle. Indeed, we see from Eq.(3.404) that $\bar{\varepsilon}$ is the sum of the averages of the two terms on the right-hand side of Eq.(3.216). If we write

$$\bar{\varepsilon} \;=\; \overline{n_1}\,\hbar\omega_1 + \overline{n_2}\,\hbar\omega_2,$$

it follows from Eqs.(3.408) and (3.407) that if $kT \gg \hbar\omega_2$, both $\overline{n_1}$ and $\overline{n_2}$ will be large as compared to unity, whereas if $\hbar\omega_2 \gg kT \gg \hbar\omega_1$, although $\overline{n_1}$ is large as compared to unity, $\overline{n_2}$ is small, and only the first degree of freedom will give the classical contribution k.

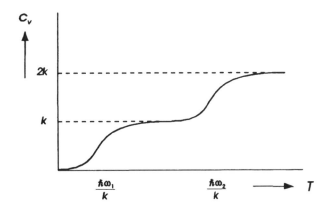

Fig.3.5. Specific heat of an anisotropic two-dimensional harmonic oscillator.

The problem with which we are thus left is to explain why at low temperatures the specific heat is so much smaller than its classical value. This can best be done by considering the transition from the quantum-mechanical formulæ to the classical ones.

We can start from the partition function

$$Z_\mu^{\mathrm{qu}} \;=\; K' \sum_n g_n e^{-\beta\varepsilon_n}. \tag{3.413}$$

The sum in Eq.(3.413) must be replaced by an integral over μ-space if we wish to go over to the classical case. Since each non-degenerate level corresponds to a volume h^s in μ-space, we should expect

$$\sum_n g_n \;\rightarrow\; \int \frac{d\omega}{h^s}. \tag{3.414}$$

We do not have to worry about degeneracy. A degenerate level corresponds to a volume $g \cdot h^s$ so that g disappears in the final result. We see that by making this transition in this way we get, indeed, the classical formula (2.718) for the partition function, and we have now a definite relation between K in Eq.(2.718) and K' in Eq.(3.413).

According to the old quantum theory one can justify the transition (3.414) rather easily as long as one is dealing with multiply periodic systems. One introduces the "phase integrals"

$$J_k = \oint p_k \, dq_k,$$

where the p_k and the q_k are the canonically conjugated generalised momenta and coordinates and where the integration is extended over one complete period.

It is possible to perform a canonical transformation from the p_k and q_k to the J_k and their conjugate coordinates w_k, the so-called *action and angle variables*.[5] The w_k measure the phase during the periodic motion and change by 1 over one period. Because the transformation is canonical, we have

$$\prod_k dp_k \, dq_k = \prod_k dJ_k \, dw_k.$$

The volume Ω of phase space corresponding to J_k values between $J_{k'}$ and $J_{k''}$ and w_k between w_k and $w_k + 1$ is given by the equation

$$\Omega = \prod_k \left(J_{k''} - J_{k'} \right).$$

If we are in the region of high quantum numbers, we have $J_{k''} = n_{k''}h$, $J_{k'} = n_{k'}h$ according to the Sommerfeld rules. The volume is thus equal to

$$\Omega = h^s \prod_k \left(n_{k''} - n_{k'} \right).$$

The number of stationary states in the volume under consideration is clearly

$$\prod_k \left(n_{k''} - n_{k'} \right).$$

We see thus that each stationary state corresponds to a volume h^s in μ-space. We do not wish to discuss here the importance of the theorem of the invariance of the statistical weights, but refer the reader to the literature.[6]

---

[5] See J.M.Burgers, *Het Atoommodel van Rutherford-Bohr*, Thesis, Leiden, 1918

[6] See, for instance, P.Ehrenfest, *Collected Scientific Papers*, North-Holland, Amsterdam, 1959.

The situation becomes more complicated when we consider modern quantum theory. In many textbooks on statistical mechanics one finds the statement that the transition (3.414) is connected with the Heisenberg relations, which are then written in the form

$$\Delta p \cdot \Delta q \geqslant h.$$

It can readily be seen that this statement can reveal only part of the true state of affairs, since the Heisenberg relations in their most rigorous form read

$$\Delta p \cdot \Delta q \geqslant \tfrac{1}{2}\hbar.$$

One is thus led to ask why the conversion factor should be h^s and not, for instance, $(\tfrac{1}{2}\hbar)^s$.

It seems to us, however, that one can justify the factor h^s in the following way. In classical mechanics the phase of the particle is given by the representative point in μ-space. One can consider this point in μ-space to be the combination of a point in coordinate- or q-space and a point in momentum- or p-space. In quantum mechanics, however, one must consider a probability density both in q- and in p-space. These two probability densities are not independent. The density in q-space is determined by the wavefunction ψ, and is, in fact, $|\psi|^2$. If one transforms the wavefunction from a coordinate representation to a momentum representation,[7] one obtains the probability amplitude in p-space, A. The probability density in p-space is then given by $|A|^2$.

We know that A and ψ are Fourier transforms of one another, if we consider q and p/h as the variables. (The occurrence of h is related to that of h in the Heisenberg relations.) Since ψ must be normalised to represent the actual state of a particle,

$$\int |\psi|^2 \, dq \; = \; 1,$$

it follows from the theory of Fourier transforms that[8]

$$\int |A|^2 \, dp \; = \; h^s.$$

The total volume in phase space corresponding to one stationary state is now obtained by multiplying the probability in q-space with the probability density in p-space and integrating over the whole of μ-space. The result is then h^s.[9]

---

[7] For a more thorough discussions of this transformation and proofs of the subsequent statements, we refer to W.Heisenberg, *The Physical Principles of Quantum Theory*, University of Chicago Press, 1932, or A.S.Davydov, *Quantum Mechanics*, Pergamon, Oxford, 1990, Chap.V.

[8] The functions ψ and A are functions of $q_1, \ldots, q_s$ and $p_1, \ldots, p_s$, respectively. We write abbreviatedly only one q or p. Similarly, dq and dp are abbreviated notations for volume elements in q- and p-space.

[9] Similar considerations are to be found in P.A.M.Dirac, *The Principles of Quantum Mechanics*, Clarendon Press, Oxford, 1935, § 37. See also, J.E.Moyal, *Proc. Cambridge Phil. Soc.* **45**, 99 (1949).

The next question is "Under what conditions is the approximation of the sum in Eq.(3.413) by an integral a good one?" The condition for this is that the successive terms in the sum do not differ appreciably or, in other words, that

$$\beta \left(\varepsilon_n - \varepsilon_{n-1} \right) \ll 1. \tag{3.415}$$

Condition (3.415) will certainly be realised in the limit as $h \to 0$, since ε_n and ε_{n-1} are the representative energies of neighbouring cells in μ-space and will become the energies of neighbouring representative points after we have let h tend to zero.

We also see that the higher the temperature the greater the chance that condition (3.415) will be satisfied. In general, then, we may expect that the higher the temperature the closer the approach to the classical formulæ.

If, however, condition (3.415) is not fulfilled, we are not allowed to substitute an integral for the sum. Let us assume for the sake of simplicity that $\varepsilon_0 = 0$ (no zero-point energy) and that all levels are non-degenerate. The partition function can then be written in the form

$$Z_\mu^{\mathrm{qu}} = K' \left[1 + e^{-\beta\varepsilon_1} + e^{-\beta\varepsilon_2} + \cdots \right]. \tag{3.416}$$

Since the terms in Eq.(3.416) are decreasing, we get to a first approximation and assuming that $\beta\varepsilon_1 \gg 1$ and that $\beta(\varepsilon_2 - \varepsilon_1) \gg 1$,

$$Z_\mu^{\mathrm{qu}} \approx K' \left[1 + e^{-\beta\varepsilon_1} \right], \tag{3.417}$$

and for the average energy and the specific heat we get from Eqs.(3.403), (3.409), and (3.417)

$$\bar{\varepsilon} \approx \varepsilon_1 \, e^{-\beta\varepsilon_1}, \tag{3.418}$$

and

$$c_v \approx k \left(\beta\varepsilon_1 \right)^2 e^{-\beta\varepsilon_1}. \tag{3.419}$$

Since we have assumed that $\beta\varepsilon_1 \gg 1$, $\bar{\varepsilon}$ and c_v will be very small as compared to ε_1 and k, respectively. We may therefore make the general statement that *a degree of freedom for which $\beta(\varepsilon_1 - \varepsilon_0) \gg 1$ will not make any appreciable contribution to the specfic heat.*[10] This explains why the specific heat of diatomic gases is $\frac{5}{2}k$ per molecule. The energies corresponding to vibrations along the axis of the molecules are of the order of electron volts, that is, they correspond to temperatures of the order of 10,000 K. The degree of freedom that such vibrations represent is thus completely frozen-in and will not contribute a term k to the specific heat.

---

[10]The statement in this form presupposes that the total energy of the particle is built up *additively* from the energies corresponding to the different degrees of freedom. That such a situation is often realised follows from a consideration of the Schrödinger equation in the case where the classical energy is constructed in this way.

3.5. The Rigid Rotator: the Hydrogen Molecule

Let us consider a dumbbell molecule. At room temperatures the electronic degrees of freedom and the degree of freedom corresponding to a vibration along the axis of the molecule are frozen-in and the molecule may be regarded as a *rigid rotator*, that is, a particle that has two rotational degrees of freedom around its centre of gravity. The three degrees of freedom relating to the motion of the centre of mass may be treated classically and will contribute $\frac{3}{2}kT$ to the average energy and $\frac{3}{2}k$ to the specific heat.

The energy levels of a point particle in a cubical box of edgelength a are given by the equation (compare Eq.(4.601))

$$\varepsilon = \varepsilon_0 \left(r^2 + s^2 + t^2\right), \quad \text{with} \quad \varepsilon_0 = \frac{h^2}{8ma^2}, \tag{3.501}$$

where r, s, and t are integers. Since $\varepsilon_0/k \approx 10^{-15}$ K, if $m = 10^{-26}$ kg and $a = 10^{-2}$ m, we see that we are always in the classical domain.

If we consider for the moment only the rotational part of the energy, we have (compare Eq.(2.414))

$$\varepsilon = \frac{p_\theta^2}{2A} + \frac{p_\varphi^2}{2A \sin^2 \theta}, \tag{3.502}$$

where A is the moment of inertia of the molecule with respect to an axis through the centre of mass, perpendicular to the axis of the molecule, and where θ and φ are two angles determining the direction of the axis of the molecule. In fact, we have taken for θ the angle between the axis of the molecule and the z axis and for φ the angle between the x axis and the projection of the molecular axis on the x, y plane. If m_1 and m_2 are the masses of the two atoms out of which the molecule is built up and if r_0 is their equilibrium distance apart, we have for A

$$A = \frac{m_1 m_2}{m_1 + m_2} r_0^2. \tag{3.503}$$

From the Schrödinger equation it follows that the energy levels of the rotator and their statistical weights are given by the equations

$$\varepsilon_j = \frac{\hbar^2}{2A} j(j+1), \quad g_j = 2j+1, \quad j = 0, 1, 2, \ldots, \tag{3.504}$$

where j is the rotational quantum number.

In order to determine the temperature above which classical behaviour may be expected, we introduce the "rotational" temperature $\Theta = \hbar^2/2Ak$. Expressing both m_1 and m_2 in units of the proton mass and r_0 in ångstrom, we find for this rotational temperature:

$$\Theta = 24.2 \frac{m_1 + m_2}{m_1 m_2 r_0^2} \text{ K.} \tag{3.505}$$

From Eqs.(3.504) and (3.415) it follows that classical behaviour may be expected when $T \gg \Theta$. From Eq.(3.505) it follows that above room temperatures the classical formulæ will probably be satisfactory, but at lower temperatures the quantum-mechanical formulæ may have to be used.

From Eqs.(3.504) it follows that the partition function of a rigid rotator is

$$Z_{\mu}^{\text{qu}} = K' \sum_{j=0}^{\infty} (2j+1) \, e^{-j(j+1)\beta k\Theta}. \tag{3.506}$$

At high temperatures $(T \gg \Theta)$ the sum can be replaced by an integral,

$$Z_{\mu}^{\text{qu}} \approx K' \int_{0}^{\infty} (2x+1) \, e^{-\beta k\Theta x(x+1)} \, dx = K' \frac{T}{\Theta}, \tag{3.507}$$

and at low temperatures $(T \ll \Theta)$ we get a power series,

$$Z_{\mu}^{\text{qu}} = K' \left[1 + 3\,e^{-2\beta k\Theta} + 5\,e^{-6\beta k\Theta} + \cdots \right]. \tag{3.508}$$

For the average rotational energy we get from Eqs.(3.507) and (3.508)

$$T \gg \Theta, \quad \bar{\varepsilon} \approx kT, \tag{3.509}$$

$$T \ll \Theta, \quad \bar{\varepsilon} \approx k\Theta \left[6\,e^{-2\beta k\Theta} - 18\,e^{-4\beta k\Theta} + \cdots \right]. \tag{3.510}$$

Finally, for the rotational specific heat we get

$$T \gg \Theta, \quad c_v \approx k, \tag{3.511}$$

$$T \ll \Theta, \quad c_v \approx k(\beta k\Theta)^2 \left[12\,e^{-2\beta k\Theta} - 72\,e^{-4\beta k\Theta} + \cdots \right]. \tag{3.512}$$

Equations (3.509) and (3.511) correspond to the classical expressions, since there is no potential energy and there are two degrees of freedom.

We shall now consider the case of the hydrogen molecule. At room temperatures and below, the vibrational degrees of freedom are frozen-in, and if we forget about the translational degrees of freedom, which contribute the usual $\frac{3}{2}k$ term to the specific heat, we can regard this molecule as a rigid rotator. In Fig.3.6 we have drawn (curve (b)) the specific heat of molecular hydrogen as determined by Eucken, and we see that there is a large discrepancy between this experimental curve and the theoretical curve (a) calculated from Eqs.(3.511) and (3.512). The difference between the calculated and observed curves cannot be due to the fact that we have neglected the vibrational and electronic energies. Firstly, these energies are far too large to have an effect at temperatures where the differences occur and, secondly, if they were important, the observed curve should lie higher, rather than lower, than the calculated one.

However, we have neglected two other factors. First of all, we have neglected the fact that both hydrogen nuclei have spin $\frac{1}{2}$.[11] Secondly, we

---

[11]The value of the spin is always expressed in units of $\hbar$.

have neglected the fact that a hydrogen molecule is built up out of two identical atoms. Since both the hydrogen nuclei have spin $\frac{1}{2}$, the resultant nuclear spin S of the molecule is either 0 (antiparallel spins) or 1 (parallel spins). The weight of a state with spin S is $2S + 1$ corresponding to the $2S + 1$ possibilities for the orientation of the total spin.[12] The state with $S = 0$, which is called the *para*-state, thus has a weight $g_S^P = 1$, while the *ortho*-state with $S = 1$ has a weight $g_S^o = 3$. To a very good approximation the energy of the molecule will be independent of the value of S and the spin weight factors therefore have no effect on the average energy, if the two atoms are not identical. If they are not identical, the partition function is multiplied by a constant factor $(2s_1 + 1)(2s_2 + 1)$ $(= 4 = g_S^P + g_S^o$ in our case), where s_1 and s_2 are the spins of the two atoms in the molecule. However, when the two atoms are identical, the Pauli principle intervenes.

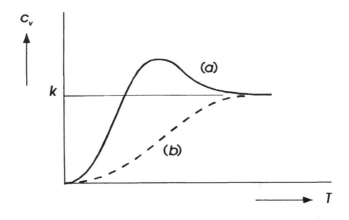

Fig.3.6. The specific heat (a) of a rigid rotator according to Eqs.(3.511) and (3.512), and (b) of hydrogen (experimentally).

This principle states that the final total wavefunction of the hydrogen molecule should be antisymmetrical in the two atoms. The vibrational and electronic parts of the wavefunction are symmetrical in the ground state, which is the only state of the electronic and vibrational parts that has to be considered. The spin function corresponding to para-states is antisymmetrical, while the spin functions corresponding to the ortho-states are symmetrical. The rotational wavefunctions are symmetrical or antisymmetrical according to whether j is even or odd. Since the total wavefunction has to

---

[12]More precisely, there are $2S+1$ linearly independent spin functions corresponding to the same total spin S.

be antisymmetrical, we see that we have the following combinations:

$$
\left.\begin{array}{llll}
j = 0, 2, 4, \cdots, & S = 0, & g_S^P = 1, \\
j = 1, 3, 5, \cdots, & S = 1, & g_S^o = 3,
\end{array}\right\} \tag{3.513}
$$

and the partition function is not given by Eq.(3.506) but by the equation

$$
Z_\mu^{(qu)} = Z_p + 3Z_o, \tag{3.514}
$$

where

$$
Z_p = K' \sum_{j=\text{even}} (2j + 1) e^{-\beta k \Theta j(j+1)} \tag{3.515}
$$

and

$$
Z_o = K' \sum_{j=\text{odd}} (2j + 1) e^{-\beta k \Theta j(j+1)}. \tag{3.516}
$$

Using expression (3.514) for the partition function, one can again calculate the specific heat; however, once more the calculated curve does not agree with the experimental curve.

The final solution to this problem was given by Dennison in 1927. In order to understand it, we must first calculate the ratio r, of the total number of ortho- to the total number of parahydrogen molecules at room temperature. To do this we use Eq.(3.102) for the number of particles in a given energy level and, remembering that that expression has to be multiplied by the degree of degeneracy, we get

$$
r = \frac{N_{\text{ortho}}}{N_{\text{para}}} = \frac{\displaystyle\sum_{j=\text{odd}} 3(2j + 1) e^{-\beta k \Theta j(j+1)}}{\displaystyle\sum_{j=\text{even}} (2j + 1) e^{-\beta k \Theta j(j+1)}}. \tag{3.517}
$$

From Eq.(3.517) we see, first of all, that as $T \to 0$, r tends to zero and, secondly, that as soon as T is well above Θ, so that the sums can be replaced by integrals, r becomes equal to 3, since the two sums lead to the same integrals, apart from the spin degeneracy factor. At room temperature, then, as this is well above Θ (compare Eq.(3.505)), there will be three times as many orthohydrogen as parahydrogen molecules present in the gas. If now the temperature is lowered, one would expect the ratio to decrease in the appropriate manner according to Eq.(3.517). However, one runs into the following difficulty. Since the transition from ortho- to parahydrogen involves the turning-over of the spin of one of the atoms, the transition probability will be small and one cannot expect equilibrium with regard to the ortho-para ratio to be attained quickly. Indeed, the periods concerned are of the order of years, even at relatively high temperatures such as room temperature. If one therefore performs a measurement of the specific heat

of hydrogen at low temperatures, one is dealing with a mixture of ortho-
and parahydrogen in the ratio of 3 to 1, and one should expect for the
specific heat

$$c_v = \tfrac{3}{4}c_{vo} + \tfrac{1}{4}c_{vp}, \tag{3.518}$$

where c_{vo} and c_{vp} are the specific heats calculated from the ortho- and para-
partition functions (3.516) and (3.515). The specific heat curve calculated
in this way agrees well with the observed specific heat curve.

Further evidence that Denison's explanation is the correct one can be
obtained by performing experiments with ortho-para mixtures where r has
a value different from 3. One can speed up the ortho-para conversion by
passing the hydrogen over activated charcoal. By doing this at various
temperatures and afterwards removing the catalyst, one can fix r at any
desired value between 0 and 3. The specific heat will then follow a curve
obtained by mixing c_{vo} and c_{vp} with appropriate weights. If one measures
c_v in such a way that r has at every temperature the value corresponding
to Eq.(3.517), it will follow the curve calculated from the expression for the
partition function given by Eq.(3.514).

Analogous considerations apply to the case of heavy hydrogen D_2. In the
case of HD there are no complications, since the two constituent atoms are
different and the specific heat can be calculated from expression (3.506) for
the partition function.

Problems

1. Derive Eq.(3.229) directly from Eq.(3.228).

2. Prove Eqs.(3.305).

3. Use Eqs.(3.319) and (3.308) to prove the relations (3.316) and (3.317)
   between the Einstein transition probabilities.

4. Prove the following relation between the entropy S and the energy E of
   the radiation field:

   $$S = \frac{4E}{3T}.$$

5. Consider radiation in a dispersive medium, in which the refractive index
   $n(\omega)$ depends on the frequency ω. Derive an expression for the radiation
   energy density in such a medium.

6. Sometimes the energy levels of a quantum-mechanical system will be such
   that the first excited level lies close to the lowest level, while all other
   levels are far away. In that case, we can at most temperatures neglect all
   levels except the lowest and the first excited one. Find an expression for
   the specific heat of a system with two levels only. Find the value of the
   temperature for which the specific heat has a maximum, and discuss the
   physical reason for the shape of the specific heat curve, which represents
   the so-called *Schottky anomaly*.

7. Consider a system with three energy levels, 0, ε, and 2ε, of which the middle one is doubly degenerate. Derive an expression for the specific heat of this system.

8. Consider a Debye solid. The Debye solid is a model in which the eigenvibrations of the N atoms in a solid are replaced by the elastic vibrations of an isotropic continuum. However, in order to have the same number of degrees of freedom as for the real solid, one assumes that one needs consider only the modes with the $3N$ lowest eigenfrequencies. Of these $3N$ eigenvibrations, $2N$ will be transverse and N longitudinal vibrations. If, as is the case in an isotropic elastic continuum, the velocity of the waves is independent of the frequency, prove that the number of vibrational modes $g(\omega)\,d\omega$ with frequencies between ω and $\omega + d\omega$ is given by (compare Eq.(3.307))

$$g(\omega) \;=\; \frac{3V}{2\pi^2}\frac{\omega^2}{s^3} \;=\; \frac{9N}{\omega_m^3}\,\omega^2, \qquad \omega < \omega_m; \tag{A}$$

$$=\; 0 \qquad\qquad , \qquad \omega > \omega_m; \tag{B}$$

here s is the velocity of the waves and ω_m is the cut-off frequency, defined by the relation

$$\int_0^{\omega_m} g(\omega)\,d\omega \;=\; 3N. \tag{C}$$

Discuss the change to be made in Eq.(A) when the velocities s_{trans} and s_{long} of the transverse and the longitudinal waves are different, but Eqs.(B) and (C) are retained.

Use the above expression for $g(\omega)$ to evaluate the (Debye) specific heat of a solid, assuming that the average energy of an eigenvibration with frequency ω is again given by Eq.(3.212).

Discuss the temperature dependence of the specific heat at low temperatures $(T \ll \Theta)$ and at high temperatures $(T \gg \Theta)$, where the *Debye temperature* Θ is defined by the relation

$$k\Theta \;=\; \hbar\omega_m. \tag{D}$$

9. Discuss the specific heats of two- and one-dimensional Debye solids.

10. Figure 3.7 shows the specific heat curve of a real solid. The horizontal line represents the classical limit of $6N$ (see problem 8); the classical behaviour is called the Dulong-Petit law.

By expressing the specific heat of the solid as the sum of the specific heats of the $3N$ normal modes, each of which has an energy given by Eq.(3.212), prove that the area between the specific heat curve and the horizontal line equals the zero-point energy of the solid. Note that no special assumptions need be made about the shape of $g(\omega)$.

11. Assume that instead of a dispersion relation (relation between frequency ω and wavevector $\boldsymbol{q}$) $\omega = sq$ $(q = |\boldsymbol{q}|)$, which is valid for the Debye solid, we have a dispersion relation

$$\omega = Cq^n, \qquad C = \text{const.}$$

Find the low-temperature behaviour of the specific heat of a solid with such a dispersion relation.

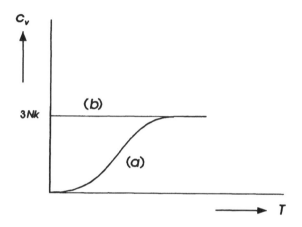

Fig.3.7. The specific heat of a solid: (a) actual curve; (b) curve corresponding to the classical Dulong-Petit law.

12. Consider a system of independent spin-$\frac{1}{2}$ point particles that have a magnetic moment μ. Calculate the entropy and the magnetic susceptibility of this system at a temperature T in a magnetic field $\boldsymbol{B}$. What is the final temperature of the system, if the magnetic field is now brought to zero adiabatically?

13. An atom with angular momentum J (in units $\hbar$) has in a magnetic field $\boldsymbol{B}$ an energy equal to $g\mu_\mathrm{B}MB$, where M is the magnetic quantum number which can be equal to $-J, -J+1, \cdots, J-1, J$ and μ_B is the Bohr magneton. Prove that the magnetic moment $\mathcal{M}$ of a system of N such atoms, assumed to be non-interacting with one another, is equal to $\mathcal{M} = Ng\mu_\mathrm{B}JB_J(x)$, where $B_J(x)$ is the so-called *Brillouin function* and $x = \beta g\mu_\mathrm{B}JB$. The Brillouin function is given by the equation

$$B_J(x) = \frac{2J+1}{2J}\coth\left(\frac{2J+1}{2J}x\right) - \frac{1}{2J}\coth\frac{x}{2J}.$$

Figure 3.8 shows $B_J(x)$ as function of x for $J = \frac{1}{2}, 1, \frac{5}{2}$, and ∞. In the last case, one puts in the expression for $\mathcal{M}$: $\mu_\mathrm{B} \to 0$ and $g\mu_\mathrm{B}J \to \mu_0$.

Consider the high-temperature limit ($\beta g \mu_{\mathrm{B}} J B \ll 1$) of the magnetic susceptibility.

14. A crystal contains ions in a state $J = 1$. An internal electric field partly lifts the degeneracy of this state, leaving a doublet lowest and a singlet higher by an energy D. When a small magnetic field $\boldsymbol{B}$ is applied at right angles to the crystal axis, the energies of these states become: $-(g \mu_{\mathrm{B}} B)^2/D$, 0, and $D+(g \mu_{\mathrm{B}} B)^2/D$, where we assume that $g \mu_{\mathrm{B}} B \ll D$. Find the partition function for such an ion. Find the magnetic moment $\mathcal{M}$ of a system of N such ions, assuming them to be non-interacting. Find both the high-temperature ($\beta D \ll 1$) and the low-temperature ($\beta D \gg 1$) behaviour of the susceptibility.

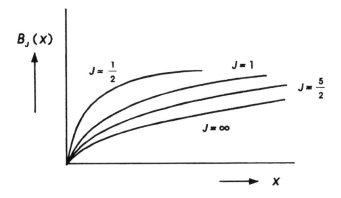

Fig.3.8. The Brillouin function $B_J(x)$ for $J = \frac{1}{2}, 1, \frac{5}{2}$, and ∞.

15. Discuss the specific heat of a gas of non-interacting heteronuclear diatomic molecules, stating clearly what approximations are made.

Consider the case of nitric oxide NO. In this case the rotational temperature given by Eq.(3.505) is equal to 2.5 K and the vibrational temperature, defined by $\hbar\omega/k$, where ω is the frequency of the vibrations in the direction of the molecular axis that are assumed to be harmonic, is equal to 2700 K. At the same time, the electronic energy spectrum is such that apart from high-lying levels, which we can neglect, there is one excited level at a distance δ from the ground state level with $\delta/k = 45$ K. Assuming that the rotational and the vibrational properties of the molecules are unaffected by the electronic state give a rough sketch of the specific heat curve in its dependence on temperature.

16. Find an expansion in powers of Θ/T, where Θ is the rotational temperature, of the rotational partition function (3.507) and hence estimate how accurately Eq.(3.511) gives the specific heat at $T/\Theta = 10$. To find the

expansion, use the Euler-MacLaurin summation formula:[13]

$$\sum_{n=0}^{\infty} f(n) = \int_0^{\infty} f(x)\,dx + \frac{1}{2}f(0)$$
$$- \frac{1}{12}f'(0) + \frac{1}{720}f'''(0) - \frac{1}{30240}f'''''(0) + \cdots .$$

17. Discuss the specific heat of a gas of non-interacting HD molecules and of a gas of non-interacting D_2 molecules.

18. To a very high degree of approximation the energy levels of the Morse potential (1.301), which is a reasonable approximation for the vibrational potential energy, are given by the equation

$$\varepsilon_n = \hbar\omega(n + \tfrac{1}{2})[1 - \gamma(n + \tfrac{1}{2})] - D,$$

where $\gamma = a^2\hbar/2\mu\omega$ with μ the reduced mass.

For most diatomic molecules γ is a small quantity. Calculate the effect of the term with γ in the expression for the energy levels, that is, the effect of the anharmonicity of the vibrations upon the vibrational specific heat, retaining only the first-order term in γ.

Bibliographical Notes

General references for this chapter are

1. P.Jordan, *Statistische Mechanik auf Quantentheoretischer Grundlage*, Vieweg, Braunschweig, 1933.

2. M.Planck, *Theorie der Wärme*, Teubner, Leipzig, 1930; *Theory of Heat*, London, 1932.

Section 3.1. The story of the introduction of h is beautifully told by Planck himself in

3. M.Planck, *Naturwiss.* **31**, 153 (1943)

For an account of the old quantum theory we can refer to.

4. A.Sommerfeld, *Atombau und Spektrallinien*, Vol.I, Vieweg, Braunschweig, 1919; *Atomic Structure and Spectral Lines*, London, 1924.

5. D.ter Haar, *The Old Quantum Theory*, Pergamon, Oxford, 1966.

In Reference 5 some of the papers (References 16 and 22) mentioned in these Bibliographical Notes are reprinted.

The Sommerfeld quantisation rules were independently proposed by Sommerfeld and by Wilson:

6. A.Sommerfeld, *Münchener Ber.*, **1915**, 425, 459; **1916**, 131.

7. W.Wilson, *Phil. Mag.* **29**, 795 (1915).

The "halbzahlige" quantisation was introduced in 1920:

8. A.Sommerfeld, *Ann. Physik* **63**, 221 (1920).

The correspondence principle was first clearly stated by Bohr in 1918:

---

[13]See, for example, M.Abramowitz and I.A.Stegun, *Handbook of Mathematical Functions*, Dover, New York, 1965, p.16.

9. N.Bohr, *Trans. Roy. Dan. Acad. Sci.* (8)4, Nr 1 (1918).

A discussion of the reasons why and when the Sommerfeld rules are exactly correct is usually given by the W-K-B or semiclassical method:

10. G.Wentzel, *Z. Physik* **38**, 518 (1926).

11. H.A.Kramers, *Z. Physik* **39**, 828 (1926).

12. L.Brillouin, *Compt. Rend.* **183**, 24 (1926).

See also

13. A.Zwaan, *Intensitäten im Ca-Funkenspektrum*, Thesis, Utrecht (1929).

For a general account of this method see, for instance,

14. A.S.Davydov, *Quantum Mechanics*, Pergamon, Oxford, 1990, Ch.III.

The mathematical method had previously been introduced by Jeffreys, and the method is therefore sometimes called the J-W-K-B method:

15. H.Jeffreys, *Proc. London Math. Soc.* **23**, 428 (1923).

Sections 3.2 and 3.4. See References 1 and 2.

Section 3.3. See References 3 and 5. Planck's law and Rayleigh's law were both derived in 1900:

16. M.Planck, *Verh. Deuts. Physik. Ges.* **2**, 202, 237 (1900).

17. Lord Rayleigh, *Phil. Mag.* **49**, 539 (1900).

A numerical error in Rayleigh's expression was corrected by Jeans, and Rayleigh's law is therefore often referred to as the Rayleigh-Jeans law:

18. J.H.Jeans, *Phil. Mag.* **10**, 91 (1905).

Wien's law is older and was stated in 1896:

19. W.Wien, *Ann. Physik* **58**, 662 (1896).

Stefan-Boltzmann's law was first deduced in 1879 by Stefan and deduced theoretically by Boltzmann in 1884:

20. J.Stefan, *Wien. Ber.* **79**, 391 (1879).

21. L.Boltzmann, *Ann. Physik* **29**, 31, 291, 616 (1884).

The derivation of Planck's radiation law from atomic considerations can be found in the paper in which Einstein introduced his transition probabilities:

22. A.Einstein, *Physik. Z.* **18**, 121 (1917).

Section 3.5. An approximate formula that can be used to evaluate the partition function of the three-dimensional rigid rotator was given by Mulholland:

23. H.P.Mulholland, *Proc. Cambridge Phil. Soc.* **24**, 280 (1928).

For an account of the history of the hydrogen rotational specific heat, see:

24. A.Farkas, *Orthohydrogen, Parahydrogen and Heavy Hydrogen*, Cambridge Univ. Press, 1935.

The specific heat of normal hydrogen was measured by Eucken:

25. A.Eucken, *Sitz. Ber. Preuss. Akad. Wiss.* **1912**, 41.

The specific heat curves of mixtures of ortho- and parahydrogen were measured by Eucken and Hiller and by Clusius and Hiller:

26. A.Eucken and K. Hiller, *Z. Physik. Chem.* **B4**, 142 (1929).

27. K.Clusius and K. Hiller, *Z. Physik. Chem.* **B4**, 158 (1929).

The theoretical explanation of the specific heat curve of normal hydrogen was given by Dennison in

28. D.M.Dennison, *Proc. Roy. Soc.* (*London*) **A115**, 483 (1927).

CHAPTER 4

BOSE-EINSTEIN AND
FERMI-DIRAC STATISTICS

4.1. Deviations from Boltzmann Statistics

In § 3.3 we discussed the radiation field and derived Planck's radiation law (3.308) for the energy density $\varrho(\omega)$ in the radiation field. It was mentioned at that time that Planck had been obliged to introduce his quantum of action h, different from zero, since Eq.(3.308) only with finite h agreed with the observational evidence. Let us examine for a moment the basic assumptions used in deriving Eq.(3.308). It was assumed that the radiation field was built up out of one-dimensional harmonic oscillators all of which possessed different frequencies. A given situation was completely described by giving the quantum numbers n_i of the stationary states of the various oscillators. When we wished to obtain the expression for the total energy density of the system in temperature equilibrium, we remarked that it can be shown that the time average of the energy of one oscillator is equal to the system average of the energy of a system of many identical oscillators. In calculating this last average one uses the assumption of equal *a priori* probabilities for different (non-degenerate) stationary states. In other words, in order to derive Planck's radiation law — which agrees well with experimental data — we must assume that any sequence $n_1, n_2, \ldots, n_i, \ldots$ of the quantum numbers of the oscillators has the same probability. However, we know that a modern picture of the radiation field must take into account the fact that light is quantised, that is, that it occurs as light quanta. This means that the situation which we just described by the sequence $n_1, n_2, \ldots, n_i, \ldots$ in reality corresponds to a radiation field in which there are n_1 quanta with energy $\hbar\omega_1$, n_2 quanta with energy $\hbar\omega_2$, and so on. Instead of having one oscillator corresponding to a given frequency ω_i we have n_i quanta with energy $\hbar\omega_i$. If we wished to calculate the energy density of the field without having recourse to our previous considerations, our first task would be to find the *a priori* probability for finding n_1 quanta with energy $\hbar\omega_1$, n_2 quanta with energy $\hbar\omega_2$, and so on. This probability[1] $W(n_1, n_2, \ldots, n_i, \ldots)$ of distributing $n_1 + n_2 + \ldots$ particles over the possible

---

[1] We shall mostly use the abbreviated notation $W(n_i)$ for $W(n_1, n_2, \ldots, n_i, \ldots)$.

energy levels $\hbar\omega_1$, $\hbar\omega_2$, ... was calculated in § 3.1 (Eq.(3.101)). We found there the relation

$$W(n_i) = A \prod_i \frac{1}{n_i!}, \tag{4.101}$$

where A is a normalising constant which we shall leave undetermined for the time being. In § 2 of the present chapter we shall make some particular choice for A and then we shall also give the reasons for our choice. If we are dealing with a system such that we can apply Eq.(4.101) we say that the system obeys *Boltzmann statistics*, which are sometimes called *classical statistics*.

A moment ago, however, we saw that in order to obtain the radiation law which agrees with experiments we must assign the value of 1 to $W(n_i)$, irrespective of the values of the n_i, or

$$W(n_i) = 1. \tag{4.102}$$

We see here a departure from the formulæ that were used in the preceding chapters. If we try to use Eq.(4.101) for the *a priori* probability, we end up with a formula for the radiation density which is in disagreement with the observational evidence. This is often expressed in the following way. Light quanta do not obey Boltzmann statistics, but they obey *Bose-Einstein statistics*. Formula (4.102) for the *a priori* probability was introduced by Bose in 1924 for the case of light quanta and applied by Einstein in the same year to the case of an ideal gas.

> It must be remarked here that the term "Bose-Einstein *statistics*" (or Boltzmann *statistics* or Fermi-Dirac *statistics*) could imply the use of different statistical methods. This is not the case. It would thus be less confusing to talk about Boltzmann *formulæ*, Bose-Einstein *formulæ*, or Fermi-Dirac *formulæ*. However, the other nomenclature has been and is used so extensively that we shall also continue to use it, the more readily since there seems to be no danger of serious misunderstandings.

There exists a second kind of statistics that differs from the classical or Boltzmann statistics which we discussed in the preceding chapters. These statistics were introduced by Fermi and extensively discussed by Dirac in 1926. They are called the *Fermi-Dirac statistics*[2] and are characterised by the following values of $W(n_i)$:

$$\left.\begin{array}{ll} W(n_i) = 1, & \text{if all the } n_i \text{ are} \leqslant 1; \\ W(n_i) = 0, & \text{if there is at least one } n_i \geqslant 1. \end{array}\right\} \tag{4.103}$$

We shall discuss presently the connection between this choice of $W(n_i)$ and Pauli's exclusion principle.

---

[2] Fermi-Dirac statistics are often called Fermi statistics and similarly Bose-Einstein statistics are often called Bose statistics; we shall use the longer terms.

Apart from discussing to what particles one must apply Bose-Einstein or Fermi-Dirac statistics and whether those are the only kind of statistics apart from Boltzmann statistics, we shall also consider briefly in the last section of the present chapter a few cases where Bose-Einstein or Fermi-Dirac statistics are applied and why one can no longer apply classical statistics in those cases. We may mention here that Bose-Einstein statistics have been applied to a discussion of the behaviour of helium at low temperatures, to the case of light quanta, and to the case of various elementary excitations in solid state physics. Fermi-Dirac statistics are used in the description of the behaviour of conduction electrons in metals, nucleons in nuclei, and also of some kinds of elementary excitations.

In the present chapter we shall treat the two "quantum" statistics, as the Bose-Einstein and Fermi-Dirac statistics are often called, in a way analogous to that used for the Boltzmann statistics in the preceding chapters. The reason for the name "quantum" statistics will become clear in later sections of this chapter. The name is rather misleading, since one can quite often get a good approximation to the behaviour of a system by applying Boltzmann formulæ to systems where the energy levels have been calculated by quantum-mechanical methods, as we did, for instance, in Chapter 3. The point is that Bose-Einstein and Fermi-Dirac statistics appear as a consequence of the symmetry effects of quantum mechanics, whereas the diffraction effects of quantum mechanics by themselves merely slightly modify the Boltzmann formulæ, usually by replacing integrals by sums, as we saw in the preceding chapter. In the limit as $h \to 0$ — which is commonly called the classical limit — "quantum" statistics go over into Boltzmann statistics, that is, all Bose-Einstein and Fermi-Dirac formulæ go over into the corresponding Boltzmann formulæ; and as in the limit as $h \to 0$ the quantum-mechanical partition function goes over into the classical one — as we saw in § 3.4 — we are, indeed, led back to the classical case, where both diffraction and symmetry effects can be neglected.

4.2.   The Probability Aspect of Statistics

We shall use here a formulation which is slightly different from the one used in Chapter 1, since we wish to apply quantum mechanics from the start.

Let us assume that we are dealing with a system consisting of N identical independent particles. The Hamiltonian of the system will be the sum of the Hamiltonians of the individual particles,

$$\widehat{\mathcal{H}}_{\text{system}} = \sum_i \widehat{\mathcal{H}}_i(p_k^{(i)}, q_k^{(i)}). \qquad (4.201)$$

In Eq.(4.201) k runs from 1 to s, where s is the number of degrees of freedom of one particle. The $\widehat{\mathcal{H}}_i$ differ only in the indices enumerating the particles, but are otherwise identical functions of their arguments.

The time-independent Schrödinger equation for the system will be

$$\widehat{\mathcal{H}}_{\text{system}} \Psi = E\Psi, \tag{4.202}$$

where E and Ψ are, respectively, the total energy of the system and its wavefunction. Since the Hamiltonian is a sum, Eq.(4.202) can be solved by the method of the separation of coordinates and we can look for solutions of Eq.(4.202) in the form

$$\Psi = \prod_i \psi(i), \tag{4.203}$$

$$E = \sum_i \varepsilon(i), \tag{4.204}$$

where $\psi(i)$ and $\varepsilon(i)$ are the eigenfunctions and eigenvalues of the single-particle Schrödinger equation

$$\widehat{\mathcal{H}}_i \psi_m(i) = \varepsilon_m \psi_m(i). \tag{4.205}$$

We have written here for the eigenfunction and eigenvalue of the m-th eigenstate of the i-th particle $\psi_m(i)$ and ε_m rather than $\psi_m(q_k^{(i)})$ and $\varepsilon_m(i)$.

A stationary state of the system can be described by giving the number of particles in each stationary state corresponding to the single-particle Schrödinger equation. If there are n_1 particles in the first stationary state, n_2 particles in the second stationary state, and so on, we have from Eq.(4.203)

$$\Psi = \psi_1(1)\psi(2)\cdots\psi_1(n_1)\psi_2(n_1+1)\psi_2(n_1+2)\cdots\psi_2(n_1+n_2)\psi_3(\cdots)\cdots, \tag{4.206}$$

where we have numbered the particles starting from those occupying the first stationary state, and so on.

Since the particles in the system are supposed to be identical, the state of the system corresponding to the wavefunction (4.206) looks highy degenerate — even if we assume that there are no accidental degeneracies due to linear relations with integral coefficients between the ε_m — since all wavefunctions $\widehat{P}\Psi$ given by the equation

$$\widehat{P}\Psi = \psi_1(k_1)\psi(k_2)\cdots\psi_1(n_1)\psi_2(k_{n_1+1})\cdots\psi_2(k_{n_1+n_2})\psi_3(k_{...})\cdots, \tag{4.207}$$

where $k_1, k_2, \cdots, k_i, \cdots, k_N$ is a permutation P of the numbers $1, 2, \cdots, i,$ $\cdots, N$, will also be eigenfunctions of the Schrödinger equation (4.202) with the same eigenvalue E given by Eq.(4.204).

In classical mechanics it was regarded as permissible to imagine labels to be attached to each of the N particles, even though they are identical. In that way each of them could be followed along its orbit and localised at every moment. This possibility of distinguishing between the various (identical) particles would entail that Ψ and $\widehat{P}\Psi$ describe two different stationary states of the system. We shall call such a stationary state a *micro-situation*. Both Ψ and $\widehat{P}\Psi$, though they correspond to different micro-situations, describe the same *macro-situation* which is characterised by the numbers n_1, n_2, $\cdots$. The *a priori* probability or *weight* $W(n_i)$ of this macro-situation will, as long as we are only considering the classical case, be given by the number of permutations of 1, 2, $\cdots$, N which give n_1 numbers in group 1, n_2 numbers in group 2, and so on. We calculated this weight in § 3.1 and found expression (3.101). If, indeed, we have some way of distinguishing the different micro-situations, we would have for the degeneracy (weight) of the macro-situation the expression

$$W(n_i) = N! \prod_i \frac{1}{n_i!}. \qquad (4.208)$$

Such a situation will arise when we are dealing with a crystalline system where, although the particles are indistinguishable, the lattice points are not, and that means that we can use the distinguishable lattice points to distinguish the particles by stating which lattice point they occupy. In that case we can use Eq.(4.208) for the degeneracy or weight of the situation characterised by the set of numbers n_1, n_2, n_3, $\cdots$, which we shall call an n_i-distribution.

If we are dealing with a situation where we have no means of keeping track of each separate particle, we are overestimating the weight of a micro-situation by a factor $N!$ insofar as we could think of $N!$ different ways of attaching labels to the N particles. We should therefore divide expression (4.208) by $N!$ to obtain the correct weight of an n_i-distribution. In this way we avoid counting each distinguishable macro-situation $N!$ times instead of once. The weights we obtain in this way will be called, for reasons that will become obvious later in the discussion, the Boltzmann weights, and they are given by the equation

$$W_{\text{Bo}}(n_i) = \prod_i \frac{1}{n_i!}. \qquad (4.209)$$

The derivation of Eq.(4.209) is not very convincing and the main justification lies in the fact that it is the correct "classical" limit of the Bose-Einstein and Fermi-Dirac formulæ which we shall encounter presently.

> We have followed mainly Rushbrooke's arguments[3] in going from Eq.(4.208) to (Eq.(4.209). He points out that labellling in the case of a crystal occurs because we can label the lattice sites. Even though the atoms in the cystal can and do change places, we can describe the situation by giving the quantum state of the atom on a well-defined position in space.

---

[3] G.S.Rushbrooke, *Introduction to Statistical Mechanics*, Clarendon Press, Oxford, Chap.3, § 1.

Rushbrooke states that the transition from Eq.(4.208) to Eq.(4.209) in the case of a gas is *not* a quantum-mechanical effect. I would beg to differ here. So long as we are dealing with classical systems, the $N!$ micro-situations are in fact different and can be distinguished from one another. In quantum-mechanical systems, however, we can no longer distinguish them. This is certainly due to the fact that we cannot follow the atoms so closely along their orbits that we can, for instance, know which atom is which after a collision. This again is a consequence of Heisenberg's famous relations. That this impossibility of following the individual particles along their orbits has a physical meaning follows, for instance, from a discussion of experiments where α-particles are scattered by α-particles.[4] Another case in which the indistinguishability plays an important rôle is in the helium atom where the two orbital electrons are identical and indistinguishable, leading to the *exchange integral* which in fact gives to some extent the "period of exchange".[5] I agree, of course, with Rushbrooke that in deriving Eq.(4.209) only part of the complete quantum-mechanical argument has been used. The exclusion principle has not been used and should have been used in a complete quantum-mechanical treatment. (Compare also Kramers' and Davydov's remarks on the interpretation of the exchange integral.)

We shall have many opportunities to discuss the consequences and importance of the omission of the $N!$ factor.

So far we have not taken into account any symmetry effects. For quite a large fraction of the systems with which we are dealing in physics and for the majority of the systems we shall be considering in this book, the degeneracy discussed a moment ago is only an apparent one, inasmuch as for them the wavefunction describing a system of identical particles must either be completely symmetrical or completely antisymmetrical in all the arguments corresponding to those identical particles. Particles for which we need symmetrical wavefunctions are called *bosons* and those for which we need antisymmetrical wavefunctions *fermions*. In the last section of the present chapter we shall be discussing to what systems this statement applies.

If the wavefunction is symmetrical in all its arguments, the only possible wavefunction (or *accessible* state, as Fowler calls it) Ψ_{BE} will be given by the equation

$$\Psi_{BE} = C \sum_{P} \hat{P}\Psi. \tag{4.210}$$

where C is a normalisation constant and where the summation is extended over all $N!$ possible permutations P of the arguments.

---

[4] Compare J.R.Oppenheimer, *Phys. Rev.* **32**, 361 (1928); N.F.Mott, *Proc. Roy. Soc. (London)* **A125**, 222 (1929); **A126**, 259 (1930).

[5] Compare H.A.Kramers, *Quantum Mechanics*, North-Holland, Amsterdam, 1957; see also A.S.Davydov, *Quantum Mechanics*, Pergamon, Oxford, 1991, § 74.

We see that in this case there corresponds only *one* (symmetrical) wavefunction to each n_i-distribution. The weight or probability of the distribution will thus be independent of the distribution[6] and we have

$$W_{\mathrm{BE}} = 1. \tag{4.211}$$

Since Eq.(4.211) is the same as Eq.(4.102), we see the reason for the index "BE".

On the other hand, in the case of an antisymmetrical wavefunction the only possibility is given by

$$\Psi_{\mathrm{FD}} = C \sum_{\mathrm{P}} \varepsilon_{\mathrm{P}} \hat{P} \Psi, \tag{4.212}$$

where ε_{P} is $+1$ or -1 according to whether the permutation is even or odd. Expression (4.212) is often called a *Slater determinant*, since it was used extensively by Slater[7] in the form of a determinant,[8]

$$\Psi_{\mathrm{FD}} = C \begin{vmatrix} \psi_{l_1}(1) & \psi_{l_1}(2) & \cdots & \psi_{l_1}(N) \\ \psi_{l_2}(1) & \psi_{l_2}(2) & \cdots & \psi_{l_2}(N) \\ \vdots & \vdots & \ddots & \vdots \\ \psi_{l_N}(1) & \psi_{l_N}(2) & \cdots & \psi_{l_N}(N) \end{vmatrix}. \tag{4.213}$$

From Eq.(4.213) it is seen immediately that as soon as one of the n_i is larger than 1, we have $\Psi_{\mathrm{FD}} = 0$. In the Fermi-Dirac case no two particles can be in the same state. This corresponds to the Pauli principle which was originally stated for electrons.

For the weight of an n_i-distribution we now have

$$\left. \begin{aligned} W_{\mathrm{FD}}(n_i) &= 1, & \sum_i n_i^2 &= N; \\ W_{\mathrm{FD}}(n_i) &= 0, & \sum_i n_i^2 &> N, \end{aligned} \right\} \tag{4.214}$$

which is the same as Eqs.(4.103).

We note that the $W(n_i)$ given by Eqs.(4.209), (4.211), and (4.214) all have the property that they are the product of factors pertaining to the single levels.

In the remainder of this chapter we shall mainly consider cases where the energy levels are lying so densely that we can practically speak of a continuous energy spectrum. In that case we can bundle the energy levels

---

[6] We equate here the weight of a distribution to the number of accessible states.

[7] J.C.Slater, *Phys. Rev.* **34**, 1293 (1929).

[8] Instead of numbering the rows n_1 times 1, n_2 times 2, and so on, as in Eq.(4.206), we have numbered them $l_1, l_2, \cdots, l_N$.

into groups. The number of levels in the j-th group will be denoted by Z_j and the number of particles with energies in that interval will be denoted by N_j. Since the energy levels are supposed to lie very densely, it will be possible to ascribe to each group a rather well-defined energy value E_j. We shall assume that we are dealing with cases where, without losing the accuracy with which the E_j are defined, we can choose the Z_j so large that the N_j and, if necessary, also the differences $Z_j - N_j$ are large numbers as well.

For the total energy E and the total number of particles N we have the equations

$$\sum_j N_j E_j = E \tag{4.215}$$

and

$$\sum_j N_j = N. \tag{4.216}$$

We now wish to calculate the probability $W(N_1, N_2, \cdots, N_j, \cdots)$ or, more succinctly, $W(N_j)$ that there are N_1 particles in group 1, N_2 particles in group 2, and so on. Figure 4.1 depicts one possible distribution.

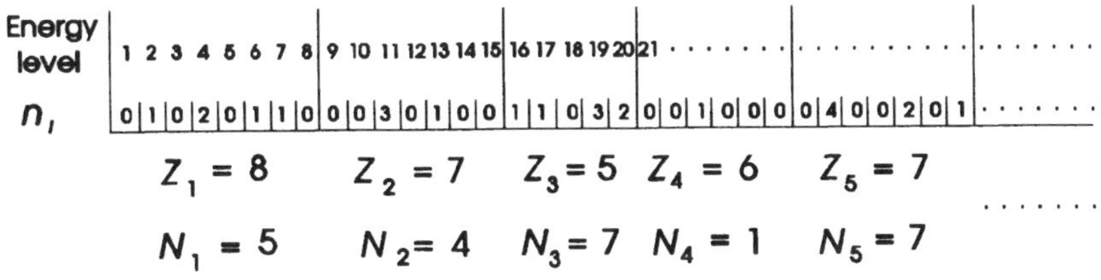

Fig.4.1. A possible distribution of particles over energy levels that are taken together in groups.

The $W(N_j)$ are related to the $W(n_i)$ through the equation

$$W(N_j) = \sum_{N_i = \text{const}} W(n_i),$$

where the summation extends over all possible n_i-distributions corresponding to a given N_i-distribution. Because the $W(n_i)$ are products of factors pertaining to single levels, the $W(N_j)$ will also be products of factors, this time pertaining to different groups of levels. We can thus write generally, independently of what kind of statistics we are dealing with,

$$W(N_j) = \prod_j W_j, \tag{4.217}$$

where each of the W_j is given by the relation

$$W_j = \sum W_j(n_i), \qquad (4.218)$$

with the $W(n_i)$ containing only those n_i referring to the ε_i in the j-th group and where the summation is over all combinations of n_i that add up to N_j.

We have three different cases corresponding to formulæ (4.209), (4.211), and (4.214) for $W(n_i)$. First of all, we shall consider the Boltzmann case. Let us consider the group (or cell) Z_j and let the number of particles in the first level be n_1, in the second level be n_2, $\cdots$, in the last level be n_{Z_j}. If we keep the other n_i constant, the contribution from this group will be

$$\sum_{n_1+n_2+\cdots+n_{Z_j}=N_j} \frac{1}{n_1!n_2!\cdots n_{Z_j}!} = \frac{1}{N_j!} \sum \frac{N_j!}{n_1!n_2!\cdots n_{Z_j}!}$$

$$= \frac{Z_j^{N_j}}{N_j!},$$

where we have used Eq.(1.704) for the polynomial expansion. For the $W_{j\mathrm{Bo}}$ we thus get

$$W_{j\mathrm{Bo}} = \frac{Z_j^{N_j}}{N_j!}. \qquad (4.219)$$

Apart from a constant factor, expression (4.217) with W_j given by Eq. (4.219) is identical with the expression from probability calculus for the probability that N points are distributed among cells Z_j in such a way that there are N_j points in cell Z_j. We met the same expression in §1.7 when we discussed the distribution of representative points over volumes in velocity space.

Since $W_{\mathrm{BE}}(n_i)$ does not depend on the values of the n_i, we obtain $W_{j\mathrm{BE}}$ if we evaluate the number of different ways by which N_j particles can be distributed over Z_j levels. This number is equal to $(N_j + Z_j - 1)!/N_j!$ $(Z_j - 1)!$ and we thus get

$$W_{j\mathrm{BE}} = \frac{(N_j + Z_j - 1)!}{N_j!(Z_j - 1)!}. \qquad (4.220)$$

The expression for the number of ways by which N_j particles can be distributed among Z_j levels can be obtained by a generalisation of the argument given in small type in §3.2 for the derivation of Eq.(3.228). We now have to write N_j as a sum of Z_j non-negative integers. That means that we have to divide N_j dots by $Z_j - 1$ strokes. We then get immediately the final result.

If we wish to calculate $W_{j\mathrm{FD}}$ we have to evaluate the number of ways by which N_j particles can be distributed over Z_j levels without having more than one particle in any one level. This number is simply the binomial coefficient

$$\binom{Z_j}{N_j}.$$

We thus get

$$W_{jFD} = \frac{Z_j!}{(Z_j - N_j)!N_j!}. \tag{4.221}$$

If $N_j \ll Z_j$, we can write Eqs.(4.220) and (4.221) in the following form

$$W_{jBE} = \frac{(N_j + Z_j - 1)(N_j - Z_j - 2)\cdots(Z_j + 1)Z_j}{N_j!} \approx \frac{Z_j^{N_j}}{N_j} = W_{jBo}$$

and

$$W_{jFD} = \frac{Z_j(Z_j - 1)\cdots(Z_j - N_j + 2)(Z_j - N_j + 1)}{N_j!} \approx \frac{Z_j^{N_j}}{N_j} = W_{jBo}.$$

We see that in the limit where $N_j \ll Z_j$ the Bose-Einstein and the Fermi-Dirac formulæ go over into the Boltzmann formulæ. This can be understood as follows. If $N_j \ll Z_j$, most levels will be empty and practically all n_i will be equal to 0 or 1, which means that $W_{Bo} \cong W_{FD} \cong W_{BE} = 1$.

Let us for a moment consider two completely separated systems composed of the same kind of particles. Let the total number of particles in the two systems be $N^{(1)}$ and $N^{(2)}$ with distributions $N_j^{(1)}$ and $N_j^{(2)}$. Let $W^{(1)}$ be the probability of the $N_j^{(1)}$-distribution and $W^{(2)}$ that of the $N_j^{(2)}$-distribution. The total probability that we have the $N_j^{(1)}$-distribution in the first and the $N_j^{(2)}$-distribution in the second system will be given by the expression $W = W^{(1)}W^{(2)}$. This is a consequence of the product property of the $W(N_i)$, and that in turn is a consequence of the product property of the $W(n_i)$. If we had used Eq.(4.208) instead of Eq.(4.209), we would not have been able to state that $W = W^{(1)}W^{(2)}$. Since the multiplicative property of the $W(N_j)$ ensures that the entropy which, as we shall see presently, is given by the equation $S = k \ln W$, will have the necessary *additive* property, the multiplicative property is most desirable.

Both the fact that the choice of Eq.(4.209) for $W_{Bo}(n_i)$ has as a consequence that $W_{Bo}(N_i)$ is the limiting value of both $W_{BE}(N_i)$ and $W_{FD}(N_i)$, and the multiplicative property of our $W_{Bo}(N_i)$ are strong arguments in favour of the omission of the $N!$.

4.3. The Elementary Method of Statistics

The elementary method of statistics consists of determining the values of N_j for which $W(N_j)$ is a maximum for a given total number of particles and a given total energy. It is then assumed that this N_j-distribution, for which $W(N_j)$ is maximum, will be realised in an actual physical system in equilibrium.

We shall now proceed to find the N_j-distributions for which $W(N_j)$, or to be precise $\ln W(N_j)$, is a maximum for given values of the total energy E and the total number of particles N; these two quantities are related to the N_j through the equations

$$\sum_j N_j E_j = E, \qquad (4.301)$$

$$\sum_j N_j = N. \qquad (4.302)$$

The argument is completely analogous to that in § 1.7. We shall assume that all Z_j, N_j, and (in the Fermi-Dirac case) $Z_j - N_j$ are large as compared to unity, so that we can use Stirling's formula for the factorial in its simplest form (1.708). We then have

$$\ln W_{\text{Bo}} = \sum_j N_j \left[\ln \frac{Z_j}{N_j} + 1 \right], \qquad (4.303)$$

$$\ln W_{\text{BE}} = \sum_j \left[(N_j + Z_j) \ln \left(\frac{Z_j}{N_j} + 1 \right) - Z_j \ln \frac{Z_j}{N_j} \right], \qquad (4.304)$$

$$\ln W_{\text{FD}} = \sum_j \left[(N_j - Z_j) \ln \left(\frac{Z_j}{N_j} - 1 \right) + Z_j \ln \frac{Z_j}{N_j} \right]. \qquad (4.305)$$

In order not to repeat every equation three times, we shall combine Eqs. (4.303) to (4.305) as follows

$$\ln W(N_j) = \sum_j \left[N_j \left\{ \ln \left(\frac{Z_j}{N_j} + \gamma \right) + 1 - \gamma^2 \right\} + \gamma Z_j \ln \left(1 + \frac{\gamma N_j}{Z_j} \right) \right], \qquad (4.306)$$

where we have introduced a parameter γ which is defined by the equations

$$\gamma_{\text{Bo}} = 0, \qquad \gamma_{\text{BE}} = 1, \qquad \gamma_{\text{FD}} = -1. \qquad (4.307)$$

Since the total energy E and the total number of particles are fixed, the conditions for a maximum of $\ln W$ are that, for any variation δN_j of the N_j, we have

$$\delta \ln W = 0, \qquad \delta N = 0, \qquad \text{and} \qquad \delta E = 0. \qquad (4.308)$$

From Eq.(4.306) we now have

$$\delta \ln W = \sum_j \ln \left(\frac{Z_j}{N_j} + \gamma \right) \delta N_j, \qquad (4.309)$$

while Eqs.(4.301) and (4.302) lead to

$$\delta N = \sum_j \delta N_j = 0, \tag{4.310}$$

$$\delta E = \sum_j E_j \, \delta N_j = 0. \tag{4.311}$$

Using Lagrange's method of undetermined multipliers, we have that for any arbitrary choice of δN_j

$$\delta \ln W + \alpha \, \delta N - \beta \, \delta E = 0. \tag{4.312}$$

Hence it follows that

$$\ln \left(\frac{Z_j}{N_j} + \gamma \right) = -\alpha + \beta E_j. \tag{4.313}$$

Equation (4.313) determines the N_j as functions of the Lagrangian multipliers α and β and of the external parameters a_k (through the E_j; compare the discussion in § 2.6). If we wish to do so, we can use Eqs.(4.301) and (4.302) to express α and β in terms of N and E.

From Eqs.(4.313) and (4.307) we get for the N_j-distributions which make $\ln W(N_j)$ a maximum the following expressions:

$$\text{Boltzmann}: \qquad N_j = Z_j e^{\alpha} e^{-\beta E_j}, \tag{4.314}$$

$$\text{Bose-Einstein}: \qquad N_j = \frac{Z_j}{e^{-\alpha + \beta E_j} - 1}, \tag{4.315}$$

$$\text{Fermi-Dirac}: \qquad N_j = \frac{Z_j}{e^{-\alpha + \beta E_j} + 1}. \tag{4.316}$$

Equation (4.314) gives us once more the Maxwell-Boltzmann distribution. Equation (4.315) reminds us of Eq.(3.302), but with the difference that we have to put $\alpha = 0$ in order to get Eq.(3.302). This corresponds to the fact that for light quanta we have to use Bose-Einstein statistics, but without any restrictions as to the total number of particles, so that the restriction $\delta N = 0$ leading to Eq.(4.310) is not operative. This means that there is only one Lagrangian multiplier, β.

From Eqs.(4.314) to (4.316) we see that the Bose-Einstein and Fermi-Dirac formulæ will go over into the Boltzmann formula if

$$e^{-\alpha + \beta E_j} \gg 1. \tag{4.317}$$

This is equivalent, however, to the condition $N_j/Z_j \ll 1$ as can immediately be seen from Eq.(4.314). One way of satisfying condition (4.317) is by choosing α to be large and negative.

From Eqs.(4.314) and (4.316) we see that any value of α will correspond to an N_i-distribution in the case of Boltzmann or Fermi-Dirac statistics. In the case of Bose-Einstein statistics, however, we see that α is restricted to such values that

$$e^\alpha < e^{\beta \varepsilon_0},\tag{4.318}$$

where ε_0 is the lowest energy level of a particle in the system.

4.4.   Connection with Thermodynamics

In this section we shall discuss the connection between the probability arguments given in the previous section and thermodynamics, following much the same lines of argument as in §§ 1.6 and 2.7.

We first of all remind ourselves that the energy levels ε_i, and hence the representative energy values E_j, may depend on external parameters,

$$\varepsilon_i = \varepsilon_i(a_1, a_2, \cdots),\tag{4.401}$$

and thus

$$E_j = E_j(a_1, a_2, \cdots).\tag{4.402}$$

We now introduce the force A_{ik} exerted by one particle in the i-th single-particle eigenstate on the external parameter a_k (compare the discussion in § 2.6),

$$A_{ik} = - \frac{\partial \varepsilon_i}{\partial a_k}.\tag{4.403}$$

The total force[9] a_k exerted by the system on the parameter A_k will then be given by the equation

$$A_k = \sum_i n_i A_{ik} = - \sum_i n_i \frac{\partial \varepsilon_i}{\partial a_k} = - \sum_j N_j \frac{\partial E_j}{\partial a_k}.\tag{4.404}$$

We now compare two equilibrium states of the system for which α, β, and the external parameters a_k have slightly different values, the differences being denoted by $\delta\alpha$, $\delta\beta$, and δa_k.

From Eqs.(4.309) and (4.313) we have

$$\delta \ln W_{\max} = -\alpha \sum_j \delta N_j + \beta \sum_j E_j \, \delta N_j.\tag{4.405}$$

We must emphasise here the difference between the $\delta \ln W$ and the δN_j considered at this moment and the quantities denoted by the same symbols in the preceding section. In that section we were comparing for given values of N and E (or, what is equivalent, of α and β) and given values of the a_k,

---

[9] It must be noted that, while the A_i in § 2.6 were the forces per molecule, the A_k of Eq.(4.404) are the total forces exerted by the system.

that is, of the ε_i or E_j, different N_j-distributions and hence different W-values, where one value corresponded to a maximum of W and the varied one to a value of W which was slightly off the maximum. Now, however, we are comparing different equilibrium distributions, but for varying values of α, β, and the a_k. In the remainder of this section we shall understand by N_j the values corresponding to the equilibrium distribution for given values of α, β, and the a_k, and by W (or $W_{\max}$) the maximum (equilibrium) value for those values of α, β, and the a_k.

Denoting by δA the work done by the system on the external parameters when these parameters change from a_k to $a_k + \delta a_k$, and using Eq.(4.404), we have

$$\delta A = \sum_k A_k \, \delta a_k = - \sum_j \sum_k N_j \frac{\partial E_j}{\partial a_k} \delta a_k = - \sum_j N_j \, \delta E_j. \quad (4.406)$$

From Eqs.(4.302) and (4.301) we have also

$$\delta N = \sum_j \delta N_j, \quad (4.407)$$

$$\delta E = \sum_j E_j \, \delta N_j + \sum_j N_j \, \delta E_j = \sum_j E_j \, \delta N_j - \delta A. \quad (4.408)$$

We now define a quantity δQ by the equation

$$\delta Q = \sum_j E_j \, \delta N_j. \quad (4.409)$$

From Eqs.(4.408) and (4.409) it follows that

$$\delta Q = \delta E + \delta A, \quad (4.410)$$

or: δQ is the heat added to the system, that is, the increase of energy not accounted for by macroscopically measurable work (compare the definition of δA).

Combining Eqs.(4.405) and (4.407) to (4.410), we have

$$\delta \ln W_{\max} = -\alpha \delta N + \beta \delta Q = -\alpha \delta N + \beta(\delta E + \delta A). \quad (4.411)$$

In connection with Eq.(4.313) we mentioned that the N_j were functions of α, β, and the a_k, or if we preferred it, of N, E, and the a_k. In Eq.(4.411) we have, indeed, used N, E, and the a_k as independent variables, since these seem to be the natural variables to use in this case. From thermodynamics we know[10] that the thermodynamic potential for which N, E, and the a_k

---

[10]See, for instance, D.ter Haar and H.Wergeland, *Elements of Thermodynamics*, Addison-Wesley, Reading, Mass., 1966.

are the natural variables is the entropy, and we are thus led to identify $\ln W_{\max}$, apart from possible multiplicative and additive constants, with the entropy S. We can also see this by considering two equilibrium states of the system for which $\delta N = 0$. In that case we see from Eq.(4.411) that β is an integrating factor of δQ, which means that it must be proportional to the reciprocal of the absolute temperature T and that $\beta \delta Q$ must be proportional to δS. Thus we have

$$\beta = \frac{1}{kT}, \tag{4.412}$$

$$S = k \ln W_{\max}. \tag{4.413}$$

The constant k that occurs in Eqs.(4.412) and (4.413) is again Boltzmann's constant. This can be proved, for instance, by comparing the equation of state following from Eq.(4.411) with the perfect gas law (see Eq.(4.613)).

> The complete proof that k is really Boltzmann's constant is slightly more complicated. It involves the consideration of a system consisting of the system under consideration combined with a "gas thermometer". One then first proves that β is the same for the two systems and, secondly, one assumes that the gas in the "thermometer" is so rarefied that it can be treated as a perfect gas. We refer to the literature for a more extensive discussion.[11]

From Eqs.(4.411) and (4.413) we see that the entropy is a function not only of the a_k and β (or T), but also of α. We know from thermodynamics that the entropy is an extensive or additive quantity, that is that for two systems (1) and (2) which are independent the total entropy will be equal to $S^{(1)} + S^{(2)}$. This property follows from Eq.(4.413), since the $W_{\max}$ are multiplicative (compare the discussion at the end of § 4.2).

Let us now consider the physical meaning of α. For the variation of the (Helmholtz) free energy F $(= E - ST)$ of the system we have from Eqs.(4.411) to (4.413)

$$\delta F = \delta(E - ST) = -S\,\delta T + \frac{\alpha}{\beta}\,\delta N - \delta A. \tag{4.414}$$

We thus get

$$\frac{\alpha}{\beta} = \left(\frac{\partial F}{\partial N}\right)_{T,a_k} = \mu, \tag{4.415}$$

where μ is the free energy per particle, or the thermal potential per particle. For α we thus have

$$\alpha = \beta\mu = \frac{\mu}{kT}. \tag{4.416}$$

That μ can also be interpreted as the partial thermal potential follows easily when we consider a system where the volume V is one of the external

---

[11]See, for example, R.C.Tolman, *The Principles of Statistical Mechanics*, Oxford University Press, 1938, p.85.

parameters.   Without loss of generality we can assume for the sake of simplicity that it is actually the only one, so that $\delta A = P\,\delta V$.

Introducing the thermal potential[12] or Gibbs free energy G by the equation

$$G = F + PV = E - ST + PV, \tag{4.417}$$

we have for its variation

$$\delta G = \delta F + P\,\delta V + V\,\delta P = -S\,\delta T + \frac{\alpha}{\beta}\,\delta N + V\,\delta P, \tag{4.418}$$

whence

$$\mu = \left(\frac{\partial G}{\partial N}\right)_{T,P}. \tag{4.419}$$

Before concluding this section we wish to introduce a new thermodynamical quantity, called by Kramers the q-potential. This q-potential is dimensionless and defined by the equation

$$q = \ln W_{\max} + \alpha N - \beta E = \frac{ST + \mu N - E}{kT}. \tag{4.420}$$

Using Eq.(4.411) we get for its variation

$$\delta q = N\,\delta\alpha - E\,\delta\beta + \beta\delta A. \tag{4.421}$$

We see that in this case α, β, and the a_k are the natural choice of independent variables. From Eqs.(4.421) and (4.406) we get the equations

$$N = \frac{\partial q}{\partial \alpha}, \tag{4.422}$$

$$E = -\frac{\partial q}{\partial \beta}, \tag{4.423}$$

and

$$A_k = \frac{1}{\beta}\frac{\partial q}{\partial a_k}. \tag{4.424}$$

If we are dealing with a homogeneous sytem of which the volume is one of the external parameters, we have

$$\mu N = G = E - ST + PV, \tag{4.425}$$

and hence from Eq.(4.420)

$$q = \frac{PV}{kT}. \tag{4.426}$$

---

[12]The thermal potential is also sometimes called the chemical potential or simply the thermodynamic potential.

4.5.   The Darwin-Fowler Method

Before we apply the formulæ derived in the two preceding sections, we shall give a different derivation of Eqs.(4.314) to (4.316), using the so-called Darwin-Fowler method. The reason is partly that this is a method which has been widely used in the statistical mechanics literature and partly that it uses mathematical techniques which can be applied to other problems. We shall see in Chapter 5 that there are strong physical arguments to use ensemble theory to derive Eqs.(4.314) to (4.316) for the case where the Z_j are not large as compared to unity, and inasmuch as we accept these arguments there is no reason to use the Darwin-Fowler method, and good reasons to consider it merely a mathematical trick. Boltzmann[13] emphasised that there are two ways of defining the equilibrium situation: either by identifying it with the most probable distribution — as we did in § 4.3, for instance — or by identifying it with the average distribution. As long as we are dealing with systems consisting of a large number of particles, the most probable situation will be so much more probable that it is also the average situation.[14] Furthermore, by taking the energy levels in groups, which was necessary in order that we could use Stirling's formula for the factorial, we are in fact calculating the average and not the most probable situation.[15]

We shall now look for the average situation rather than the most probable one, restricting ourselves to the case of a system in which there is only one kind of independent particles. Defining $W(n_i)$ as in § 4.2, we first of all

---

[13]L.Boltzmann, *Wien. Ber.* **76**, 373 (1877).

[14]This is a sketchy argument which we do not wish to make more rigorous at this point.

[15]See E.Schrödinger, *Berliner Ber.* **1925**, 434. The most probable situation is found by writing for the entropy

$$S = k \ln W_{\text{max}}, \tag{A}$$

where W_{max} should be the maximum value of $W(n_i)$. The average situation, on the other hand, corresponds to the equation

$$S = k \ln \sum W, \tag{B}$$

as follows after a moment's reflection. In § 4.3 we lumped the energy levels into groups where all levels in one group had about the same energy E_j, and again using Eq.(A), but now with W being the probability where N_j particles are in the j-th group, we calculated the entropy. We now must take into account the following two facts. (a) The maximum of $W(N_j)$ is so steep that, apart from a negligible constant, we may in Eq.(B) substitute $W_{\text{max}}(N_j)$ for the sum. (b) $W_{\text{max}}(N_j)$ itself is, however, a sum of several $W(n_i)$ (see Eqs.(4.217) and (4.218))). We see thus that by using the $W(N_j)$ instead of the $W(n_i)$ we can reduce Eq.(B) to Eq.(A).

remind ourselves that $W(n_i)$ can always be written in the form

$$W(n_i) = \prod_i \eta(n_i),$$ (4.501)

where the actual form of the $\eta(n_i)$ depends on the nature of the particles and is different for the three cases of Boltzmann, Bose-Einstein, and Fermi-Dirac statistics.

We now wish to calculate the average values of the $\bar{n}_i$ under the supplementary conditions of a given total number of particles and a given total energy,

$$N = \sum n_i,$$ (4.502)

$$E = \sum n_i \varepsilon_i.$$ (4.503)

The average values $\bar{n}_i$ clearly satisfy the equation

$$\bar{n}_i = \frac{1}{G} \sum{}'' n_i W(n_k),$$ (4.504)

where

$$G = \sum{}'' W(n_k).$$ (4.505)

The summations extend over all possible combinations of the n_k compatible with the conditions (4.502) and (4.503); this is indicated by the double prime on the summation signs.

In order to be able, if necessary, to consider fluctuations and correlations, we are also interested in expressions of the type

$$\overline{n_i^p n_j^q} = \frac{1}{G} \sum{}'' n_i^p n_j^q W(n_k).$$ (4.506)

The first probem is the evaluation of G. This is done by introducing a generating function[16] $F(x, y; \varepsilon_i)$ by the equation

$$F(x, y; \varepsilon_i) = \sum W(n_k) \, x^{n_1 + n_2 + \cdots} \, y^{n_1 \varepsilon_1 + n_2 \varepsilon_2 + \cdots},$$ (4.507)

where the summation is over all values of the n_k without any restrictions. In order to obtain G from $F(x, y; \varepsilon_i)$ we must pick from the sum those terms of which the exponent of x is equal to N and the exponent of y is equal to E. This is done by using the theory of complex functions; one has, in fact,

$$G = \left(\frac{1}{2\pi i}\right)^2 \oint dx \oint dy \, x^{-N-1} y^{-E-1} F(x, y; \varepsilon_i),$$ (4.508)

---

[16]This method goes back to Laplace, who introduced it in his *Théorie Analytique des Probabilités*.

where both integrations are counterclockwise along a closed contour around the origin in the complex x- and y-planes, respectively.[17]

In a similar way one can show that

$$\overline{n_i^p n_j^q} = \frac{1}{G} \left(\frac{1}{2\pi i}\right)^2 \oint dx \oint dy\, x^{-N-1} y^{-E-1}$$
$$\times \left(\frac{1}{\ln y} \frac{\partial}{\partial \varepsilon_i}\right)^p \left(\frac{1}{\ln y} \frac{\partial}{\partial \varepsilon_j}\right)^q F(x, y; \varepsilon_i), \quad (4.509)$$

We may remark here that if we use Eq.(4.501) for $W(n_k)$ we can write $F(x, y; \varepsilon_i)$ as

$$F(x, y; \varepsilon_i) = \prod_i f(xy^{\varepsilon_i}), \quad (4.510)$$

where

$$f(z) = \sum_{n=0}^{\infty} \eta(n)\, z^n. \quad (4.511)$$

Equations (4.508) and (4.509) are exact. In some cases one can evaluate the integrals and thus obtain exact expressions for the $\overline{n}_i$. However, in most cases these integrals cannot be evaluated and we must have recourse to approximation methods.

Method of Steepest Descents.[18] One such approximate method is the method of steepest descents, which is used to obtain an approximate evaluation of integrals of the kind

$$I = \int_A^B \chi(z)\, e^{tf(z)}\, dz, \quad (4.512)$$

where t is large, real, and positive, and $f(z)$ is an analytic function of the complex variable z ($= x + iy$; note that now x and y are the real and imaginary parts of a single complex variable). The real and imaginary parts of $f(z)$ will be called φ and ψ,

$$f = \varphi + i\psi, \quad (4.513)$$

and since $f(z)$ is an analytic function, φ and ψ both have to satisfy Laplace's equation

$$\frac{\partial^2 \varphi}{\partial x^2} + \frac{\partial^2 \varphi}{\partial y^2} = \frac{\partial^2 \psi}{\partial x^2} + \frac{\partial^2 \psi}{\partial y^2} = 0. \quad (4.514)$$

---

[17]The contours are supposed to lie within the circles of convergence of F. It must be noted that here x and y are both complex variables and not the real and imaginary parts of one complex variable.

[18]The method of steepest descents is due to P.Debye (*Math. Ann.* **67**, 535 (1909)). For an account, see, for instance, H. and B.S.Jeffreys, *Methods of Mathematical Physics*, Cambridge University Press, 1946, §§ 17-03 and 17-04; or E.Schrödinger, *Statistical Thermodynamics*, Cambridge University Press, 1948, p.29.

In the integral on the right-hand side of Eq.(4.512) the largest contribution will come from regions where φ is large. From the theory of complex functions it follows that φ has nowhere an absolute maximum, but that there are points where

$$\frac{\partial \varphi}{\partial x} = \frac{\partial \varphi}{\partial y} = 0. \qquad (4.515)$$

Since

$$\frac{\partial \varphi}{\partial x} = \frac{\partial \psi}{\partial y} \quad \text{and} \quad \frac{\partial \varphi}{\partial y} = -\frac{\partial \psi}{\partial x}, \qquad (4.516)$$

it then follows that

$$\frac{\partial \psi}{\partial x} = \frac{\partial \psi}{\partial y} = 0 \quad \text{and} \quad f'(z) = 0, \qquad (4.517)$$

where $f'(z)$ stands for df/dz. These points are called *saddle points*. It is possible to draw through such a saddle point at least two curves along which φ is constant. Between those curves, φ will alternately be larger and smaller than at the saddle point; in other words, there are at least two hills and two valleys.

Let us now consider a path in the complex plane. The line element of such a path will be denoted by ds and the angle between ds and the positive x-axis by θ. We then have

$$\frac{\partial \varphi}{\partial s} = \cos \theta \, \frac{\partial \varphi}{\partial x} + \sin \theta \, \frac{\partial \varphi}{\partial y}. \qquad (4.518)$$

If we are looking for paths along which $|\partial \varphi / \partial s|$ is maximum, θ has to be solved from the equation

$$0 = -\sin \theta \, \frac{\partial \varphi}{\partial x} + \cos \theta \, \frac{\partial \varphi}{\partial y}. \qquad (4.519)$$

From Eq.(4.516) it then follows that on these paths $\partial \psi / \partial s = 0$, or that ψ is constant. Such a path will be called a *line of steepest descent*. In each valley there will be such a line. The path over which we shall now integrate from A to B will be the following one. From A along a line of constant φ to the line of steepest descent, then via this line over the saddle point to the valley in which B is situated, and finally again along a line of constant φ to B. (We do not consider here the case where A and B are in the same valley.)

Let z_0 be the coordinate of the saddle point. In the neighbourhood of z_0 we can expand $f(z)$ in the form (compare Eq.(4.517)

$$f(z) = f(z_0) = \tfrac{1}{2}(z - z_0)^2 f''(z_0) + \cdots. \qquad (4.520)$$

We take the path in the direction of the valley such that $(z-z_0)^2 f''(z_0)$ is real and negative, because we *descend* along a line of constant ψ. If we introduce a new variable ζ by putting

$$f(z) - f(z_0) = -\tfrac{1}{2}\zeta^2,$$

Eq.(4.512) takes the form

$$I = e^{tf(z_0)} \int_A^B \chi(z) e^{\frac{1}{2}t\zeta^2} \frac{dz}{d\zeta} d\zeta. \tag{4.522}$$

Introducing

$$z - z_0 = r e^{-i\alpha}, \tag{4.523}$$

we have

$$\frac{dz}{d\zeta} = e^{i\alpha} |f''(z_0)|^{-1/2}, \tag{4.524}$$

where we have made use of the fact that $(z - z_0)^2 f''(z_0)$ is real and negative.

Before substituting (4.524) into Eq.(4.522), we will consider the integral

$$J = \int_{-\infty}^{+\infty} g(z) e^{-\frac{1}{2}a^2 z^2} dz, \tag{4.525}$$

where $g(z)$ is analytic and bounded on the real axis, while a is real and large.

The function $g(z)$ can be expanded as follows:

$$g(z) = b_0 + b_1 z + b_2 z^2 + \cdots = \sum_{n=0}^{\infty} b_n z^n. \tag{4.526}$$

Substituting this expansion into J we get

$$J = \sum_{n=0}^{\infty} \int_{-\infty}^{+\infty} b_n z^n e^{-\frac{1}{2}a^2 z^2} dz, \tag{4.527}$$

or

$$J = \sqrt{2\pi} \sum_{n=0}^{\infty} \frac{(2n-1)! \, b_{2n}}{(n-1)! \, 2^n a^{2n+1}}. \tag{4.528}$$

Since a is supposed to be large, we break off after only a few terms. Taking the first term only and remembering the meaning of b_0, we have, asymptotically,

$$J \approx \sqrt{2\pi} \frac{g(0)}{a}. \tag{4.529}$$

Now substituting expression (4.524) into Eq.(4.522) and using Eq. (4.525), we get

$$I \approx e^{tf(z_0)} \chi(z_0) e^{i\alpha} \left| \frac{2\pi}{tf''(z_0)} \right|^{1/2}. \tag{4.530}$$

The other terms in the expansion are difficult to obtain and are very often not needed.

We now return to Eqs.(4.508) and (4.509). It can be shown that the integrands of the integrals under consideration possess a steep minimum on the real axis which lies inside the circle of convergence. The path of integration can then be taken through this minimum. In order to have our integrals in a suitable form, we write

$$F(x, y; \varepsilon_i) \, x^{-N-1} = e^{Ng(x)}. \tag{4.531}$$

Concentrating on the integration over x, we are concerned with the integral

$$I(x) = \frac{1}{2\pi i} \oint dx \, e^{Ng(x)}, \tag{4.532}$$

which is approximately given by the equation (the path is along a circle around the origin so that $\alpha = \frac{1}{2}\pi$)

$$I(x) = e^{Ng(x_0)} \sqrt{\frac{1}{2\pi N g''(x_0)}}, \qquad g''(x) = \frac{\partial^2 g}{\partial x^2}, \tag{4.533}$$

where x_0 satisfies the equation

$$\frac{\partial g}{\partial x} = 0, \qquad \text{or} \qquad x \frac{\partial F}{\partial x} - (N+1)F = 0. \tag{4.534}$$

The last factor in the expression for $I(x)$ is only of the order of $N^{1/2}$, while the first factor contains a factor N in the exponent. We may therefore to a first approximation neglect the second factor[19] and write

$$I(x) = e^{Ng(x_0)} = F(x_0, y; \varepsilon_i) \, x_0^{-N-1}. \tag{4.535}$$

The integration over y can be performed in a similar way, and we finally find

$$G \approx x_0^{-N-1} \, y_0^{-E-1} \, F(x_0, y_0; \varepsilon_i), \tag{4.536}$$

where x_0 and y_0 are the roots of Eq.(4.534) and the equation

$$y \frac{\partial F}{\partial y} - (E+1)F = 0. \tag{4.537}$$

Since the method of steepest descent provides only a reasonable approximation, if N and E are large, we may neglect the 1 in Eqs.(4.534) and (4.537) and have thus

$$N = \frac{x}{F} \frac{\partial F}{\partial x}, \tag{4.538}$$

$$E = \frac{y}{F} \frac{\partial F}{\partial y}, \tag{4.539}$$

---

[19]Since in averages like the ones in Eq.(4.509) we are concerned with the ratio of two integrals, the neglect of this factor is completely negligible.

By the same method we find for $\overline{n_j^p n_k^q}$

$$
\begin{aligned}
\overline{n_j^p n_k^q} &= \frac{1}{G} x_0^{-N-1} y_0^{-E-1} \left(\frac{1}{\ln y_0} \frac{\partial}{\partial \varepsilon_j}\right)^p \left(\frac{1}{\ln y_0} \frac{\partial}{\partial \varepsilon_k}\right)^q F(x_0, y_0; \varepsilon_i) \\
&= \frac{1}{F} (\ln y_0)^{-p-q} \left(\frac{\partial}{\partial \varepsilon_j}\right)^p \left(\frac{\partial}{\partial \varepsilon_k}\right)^q F,
\end{aligned} \tag{4.540}
$$

where we used Eq.(4.536) for G.

If we use Eq.(4.510) for F, we see immediately that $\overline{n_j^p n_k^q}$ vanishes unless either p or q is equal to zero, and if this condition is satisfied, we have

$$
\overline{n_j^p} = (\ln y_0)^{-p} \frac{1}{f_j} \frac{\partial^p f_j}{\partial \varepsilon_j^p}, \qquad f_j = f(x_0 y_0^{\varepsilon_j}), \tag{4.541}
$$

where f is given by Eq.(4.511).

Equations (4.536), (4.541), (4.538), and (4.539) together give us all the necessary information which we shall now apply to the case of a system of independent particles, obeying either Boltzmann, Bose-Einstein, or Fermi-Dirac statistics. The difference between the statistics lies in the expressions for the $\eta(n_i)$. We have from Eqs.(4.209), (4.211), and (4.214)

$$
\text{Boltzmann}: \qquad \eta(n_i) = \frac{1}{n_i!}; \tag{4.542}
$$

$$
\text{Bose-Einstein}: \qquad \eta(n_i) = 1; \tag{4.543}
$$

$$
\left.\begin{aligned}
\text{Fermi-Dirac}: \qquad \eta(n_i) &= 1, \quad n_i = 0 \quad \text{or} \quad 1, \\
&= 0, \quad n_i > 1;
\end{aligned}\right\} \tag{4.544}
$$

and for the functions f we get from Eq.(4.511)

$$
f_{\text{Bo}} = \sum_0^\infty \frac{z^n}{n!} = e^z, \tag{4.545}
$$

$$
f_{\text{BE}} = \sum_0^\infty z^n = \frac{1}{1-z}, \tag{4.546}
$$

$$
f_{\text{FD}} = \sum_0^1 z^n = 1 + z. \tag{4.547}
$$

We then get from Eq.(4.510) for F:

$$
F_{\text{Bo}} = \exp\left[x \sum_i y^{\varepsilon_i}\right], \tag{4.548}
$$

$$
F_{\text{BE}} = \prod_i \frac{1}{1 - xy^{\varepsilon_i}}, \tag{4.549}
$$

$$F_{\mathrm{FD}} = \prod_i \left(1 + xy^{\varepsilon_i}\right). \tag{4.550}$$

The values of x_0 and y_0 follow from Eqs.(4.538) and (4.539), or

$$N = \sum_i \frac{\partial \ln f_i}{\partial \ln x_0}, \tag{4.551}$$

$$E = \sum_i \frac{\partial \ln f_i}{\partial \ln y_0}, \tag{4.552}$$

Using Eqs.(4.541) with $p = 1$, we see that Eqs.(4.551) and (4.552) can be written in the following translucent form:

$$N = \sum_i \bar{n}_i, \tag{4.553}$$

$$E = \sum_i \bar{n}_i \varepsilon_i. \tag{4.554}$$

From Eq.(4.541) we have

$$\bar{n}_i = \frac{x_0}{f_i} \frac{\partial f_j}{\partial x_0}, \tag{4.555}$$

or, using Eqs.(4.545) to (4.547)

$$(\bar{n}_i)_{\mathrm{Bo}} = x_0 y_0^{\varepsilon_i}, \tag{4.556}$$

$$(\bar{n}_i)_{\mathrm{BE}} = \frac{x_0 y_0^{\varepsilon_i}}{1 - x_0 y_0^{\varepsilon_i}}, \tag{4.557}$$

$$(\bar{n}_i)_{\mathrm{FD}} = \frac{x_0 y_0^{\varepsilon_i}}{1 + x_0 y_0^{\varepsilon_i}}. \tag{4.558}$$

If we compare expressions (4.556) to (4.558) with Eqs.(4.314) to (4.316) with $Z_j \equiv 1$, $N_j \equiv n_i$, we see that we have the same equations, provided

$$y_0 = e^{-\beta}, \tag{4.559}$$

$$x_0 = e^{\alpha}. \tag{4.560}$$

These equations give us the physical meaning of x_0 and y_0. They can be derived directly, as was done by Darwin and Fowler in their original papers.

4.6. The Perfect Boltzmann Gas

We shall consider in the present and the subsequent sections the case of a perfect gas. In order to simplify the expressions for the energy levels, we assume that the gas is enclosed in a cube of edgelength L and of volume V $(= L^3)$.

If m is the mass of one particle, the energy levels are given by the equation

$$\varepsilon = \frac{\varepsilon_0}{3}\left(k_1^2 + k_2^2 + k_3^2\right), \qquad \varepsilon_0 = \frac{3h^2}{8mL^2}, \qquad (4.601)$$

where k_1, k_2, and k_3, which are positive integers, are the quantum numbers corresponding to the three translational degrees of freedom. The quantity ε_0 is the lowest energy level, corresponding to $k_1 = k_2 = k_3 = 1$.

Each stationary state corresponds to a point in the three-dimensional k_1, k_2, k_3-space which is situated in the positive octant. Introducing k by the equation

$$k^2 = k_1^2 + k_2^2 + k_3^2, \qquad (4.602)$$

we find for the number of energy levels dZ with k between k and $k + dk$ the expression

$$dZ = \tfrac{1}{8} \cdot 4\pi k^2\, dk, \qquad (4.603)$$

where we have assumed that the particle has no spin and that the energy levels are lying so closely together that we can treat the set of points corresponding to the stationary states as being a continuum. If the particle had spin S, expression (4.603) should be multiplied by $2S+1$. Using Eqs.(4.601) and (4.602) we can write this expression for dZ in the form[20]

$$dZ = 2\pi\left(\frac{2m}{h^2}\right)^{3/2} V\sqrt{\varepsilon}\, d\varepsilon. \qquad (4.604)$$

If we consider a Boltzmann gas, we have from Eqs.(4.303) and (4.313)

$$\ln W_{\mathrm{Bo}} = \sum_j N_j\left(-\alpha + \beta E_j + 1\right) = -(\alpha - 1)N + \beta E, \qquad (4.605)$$

where we have dropped the index "max" of W and where the N_j are once more the equilibrium values.

For the q-potential we find

$$q_{\mathrm{Bo}} = \ln W + \alpha N - \beta E = N = \sum_j N_j, \qquad (4.606)$$

or

$$q_{\mathrm{Bo}} = e^\alpha \sum_j Z_j\, e^{-\beta E_j} = e^\alpha\, Z(\beta), \qquad (4.607)$$

where $Z(\beta)$ which, apart from the factor K', is the same as $Z_\mu^{(\mathrm{qu})}$ of Eq.(3.413) is the *partition function* which is a function of β (and of the external parameters),

$$Z(\beta) = \sum_j Z_j\, e^{-\beta E_j}. \qquad (4.608)$$

---

[20]It has to be remarked that Eq.(4.604) remains valid even if the volume V is not a cube, provided its shape is not too pathological (see, for instance, R.Courant and D.Hilbert, *Methods of Mathematical Physics*, Vol.I, New York, 1957).

For the group of levels Z_j we can in the case of a gas take energy intervals dZ, and then all sums will go over into integrals.

Using expression (4.604) for dZ we get for the partition function

$$Z = 2\pi \left(\frac{2m}{h^2}\right)^{3/2} V \int_0^\infty \varepsilon^{1/2} e^{-\beta\varepsilon} d\varepsilon = \left(\frac{2\pi m}{\beta h^2}\right)^{3/2} V, \qquad (4.609)$$

and for the q-potential

$$q_{Bo} = \left(\frac{2\pi m}{\beta h^2}\right)^{3/2} V e^\alpha. \qquad (4.610)$$

From Eq.(4.422) we get

$$N = \frac{\partial q_{Bo}}{\partial \alpha} = q_{Bo}, \qquad (4.611)$$

which is the same as Eq.(4.606).

If we asume that V is the only external parameter so that $\delta A = P \delta V$, we have from Eq.(4.424) (see also Eq.(4.426))

$$\beta P = \frac{\partial q_{Bo}}{\partial V} = \frac{q_{Bo}}{V} = \frac{N}{V}, \qquad (4.612)$$

or

$$PV = \frac{N}{\beta} = NkT, \qquad (4.613)$$

which is once more the perfect gas law.

The energy of the system follows from Eq.(4.423):

$$E = -\frac{\partial q_{Bo}}{\partial \beta} = \frac{3}{2}\frac{q_{Bo}}{\beta} = \frac{3}{2}NkT, \qquad (4.614)$$

corresponding to the equipartition of the kinetic energy, giving a contribution $\frac{1}{2}kT$ for each degree of freedom.

The partial thermal potential divided by kT, α, is given by

$$\alpha = \ln \frac{q_{Bo}}{Z} = \ln P - \frac{5}{2} \ln T - \mathcal{C} - \frac{3}{2} \ln M, \qquad (4.615)$$

where M is the molecular weight of the gas and $\mathcal{C}$ a constant which is connected with the so-called chemical constants[21] and which is given by the equation

$$\mathcal{C} = \ln \left[k \left(\frac{2\pi m_H k}{h^2}\right)^{3/2} \right], \qquad (4.616)$$

---

[21] See, for instance, D.ter Haar, *Elements of Thermostatistics*, Holt, Rinehart and Winston, 1966, § 9.3.

where m_H is the molecular weight unit.

Finally, using Eqs.(4.611) and (4.614) we get from Eq.(4.420) for the entropy

$$S = kq_{Bo} - k\alpha N + \frac{E}{T} = Nk\left(\tfrac{5}{2} - \alpha\right), \qquad (4.617)$$

or, using Eq.(4.615),

$$S = kN \left[\tfrac{5}{2}\ln T - \ln P + \mathcal{C} + \tfrac{3}{2}\ln M + \tfrac{5}{2}\right], \qquad (4.618)$$

the so-called Sackur-Tetrode equation.[22]

4.7.   The Perfect Bose-Einstein Gas

We now turn to quantum gases. First of all, we shall investigate under what circumstances we can safely apply Boltzmann statistics.

Since the energy eigenvalues given by Eq.(4.601) are all positve and since $\beta\varepsilon_0$ is very small as compared to unity for all attainable temperatures, as we saw in §3.5, the condition for the applicability of Boltzmann statistics (Eq.(4.317)) reduces to the condition

$$e^{-\alpha} \gg 1. \qquad (4.701)$$

If inequality (4.701) is satisfied, the total number of particles will be given to a fair approiximation by the same equation for the three statistics,

$$N = \sum_j N_j \approx \sum_j Z_j e^{\alpha - \beta E_j},$$

or

$$N \approx e^{\alpha} \left(\frac{2\pi mkT}{h^2}\right)^{3/2} V, \qquad (4.702)$$

where we have used Eqs.(4.314) to (4.316), (4.701), (4.608), and (4.609).

We thus get for e^{α} approximately

$$e^{-\alpha} \approx \left(\frac{2\pi mk}{h^2}\right)^{3/2} T^{3/2} \left(\frac{N}{V}\right)^{-1}. \qquad (4.703)$$

From Eqs.(4.703) and (4.701) we see, first of all, that in the limit as $h \to 0$ Boltzmann statistics may be applied. We can thus understand why Boltzmann statistics are often called the classical statistics. Secondly, we see that in the limit of high temperatures or low densities condition (4.701) is satisfied and again Boltzmann statistics can be applied.

---

[22]O.Sackur, *Ann. Physik* **36**, 958 (1911); **40**, 67 (1912); H.Tetrode, *Ann. Physik* **38**, 434 (1912) (corrections in *Ann. Physik* **39**, 255 (1912)); *Proc. Kon. Nederl. Akad. Wet.* **17**, 1167 (1915).

We should like to write Eq.(4.703) in a slightly different form. The average energy per particle—which in the present case is the same as the average kinetic energy—is in the classical limit equal to $\frac{3}{2}kT$ (compare Eq.(4.614)). It follows from the usual quantum-mechanical arguments that the thermal de Broglie wavelength λ_{dBr} of the particle will be given by the equation

$$\lambda_{\mathrm{dBr}} = \frac{h}{p} = \frac{h}{\sqrt{2m\varepsilon}} = \left(\frac{h^2}{3mkT}\right)^{1/2}, \qquad (4.704)$$

where p is the average momentum of the particle.

Combining Eqs.(4.703) and (4.704) we have

$$e^\alpha \approx \left(\frac{3}{2\pi}\right)^{3/2} \lambda_{\mathrm{dBR}}^3 \frac{N}{V} \equiv \zeta. \qquad (4.705)$$

We see from Eq.(4.705) that condition (4.701) is equivalent to the condition that the thermal de Broglie wavelength of the particles be small compared to the average distance apart in the gas.

The quantity ζ introduced by Eq.(4.705) is essentially what Sommerfeld[23] calls the *degeneracy parameter*. We shall see presently that it is the natural parameter in which to expand for the case when the perfect gas is nearly classical, while in the case of extreme quantum behaviour, when $\zeta \gg 1$, the natural expansion parameter is an inverse power of ζ. Of course, Eqs.(4.703) and (4.705) are valid only in the case when $e^\alpha \ll 1$, and they actually represent the first terms in a series expansion in ζ.

We now turn to a perfect Bose-Einstein gas, and to simplify our formulæ we assume that the particles have zero spin. From Eqs.(4.304) and (4.315) we have

$$\ln W_{\mathrm{BE}} = \sum_j N_j\,(-\alpha + \beta E_j) - \sum_j Z_j\,\ln\left(1 - e^{\alpha - \beta E_j}\right). \qquad (4.706)$$

The q-potential follows from Eq.(4.420) and is given by the equation

$$q_{\mathrm{BE}} = -\sum_j Z_j\,\ln\left(1 - e^{\alpha - \beta E_j}\right). \qquad (4.707)$$

In most cases we can again use an integral instead of the sum, and then we can apply Eq.(4.604) and get

$$q_{\mathrm{BE}} = -2\pi\left(\frac{2m}{h^2}\right)^{3/2} V \int_0^\infty \sqrt{\varepsilon}\,\ln\left(1 - e^{\alpha - \beta\varepsilon}\right)d\varepsilon, \qquad (4.708)$$

or

$$q_{\mathrm{BE}} = \left(\frac{2\pi m}{\beta h^2}\right)^{3/2} V\,g(\alpha), \qquad (4.709)$$

---

[23]A.Sommerfeld, *Z. Physik* **47**, 1 (1928).

where

$$g(\alpha) = \sum_{n=1}^{\infty} \frac{e^{n\alpha}}{n^{5/2}}. \tag{4.710}$$

We have for $g(\alpha)$, as long as $e^{\alpha} \leqslant 1$,

$$g(\alpha) = \frac{-2}{\sqrt{\pi}} \int_0^{\infty} \sqrt{x} \, \ln\left(1 - e^{\alpha - x}\right) dx$$

$$= \frac{2}{\sqrt{\pi}} \sum_{n=1}^{\infty} \frac{1}{n} \int_0^{\infty} \sqrt{x} \, e^{n\alpha - nx} \, dx,$$

whence follows Eq.(4.710).

For the total number of particles we have from Eq.(4.422)

$$N = \frac{\partial q_{\text{BE}}}{\partial \alpha} = \left(\frac{2\pi m}{\beta h^2}\right)^{3/2} V \, g'(\alpha) = q_{\text{BE}} \frac{g'(\alpha)}{g(\alpha)}, \tag{4.711}$$

where

$$g'(\alpha) = \frac{dg(\alpha)}{d\alpha} = \sum_{n=1}^{\infty} \frac{e^{n\alpha}}{n^{3/2}}. \tag{4.712}$$

The pressure and the total energy are given by the equations

$$P = \frac{q_{\text{BE}}}{\beta V}, \tag{4.713}$$

$$E = -\frac{\partial q_{\text{BE}}}{\partial \beta} = \frac{3}{2} \frac{q_{\text{BE}}}{\beta}. \tag{4.714}$$

We see that it follows from Eqs.(4.713) and (4.714) that

$$P = \frac{2E}{3V}. \tag{4.715}$$

This equation could have been derived much more easily. From Eq.(4.601) we see that the energy of a particle is proportional to $V^{-2/3}$. As $-\partial \varepsilon / \partial V$ is the pressure exerted by one particle, it follows that this partial pressure is equal to $2\varepsilon/3V$ and Eq.(4.715) follows immediately. It also follows from this derivation that this equation must hold irrespective of the kind of statistics. Indeed, from Eqs.(4.613) and (4.614) it follows in the Boltzmann case, and it will also be found to hold in the Fermi-Dirac case.

From Eqs.(4.711) and (4.713) we get

$$PV = NkT \frac{g(\alpha)}{g'(\alpha)}. \tag{4.716}$$

Since in the Bose-Einstein case Eq.(4.318) must always be satisfied, we can always expand in powers of e^α. If we expand $g(\alpha)/g'(\alpha)$ in powers of e^α we get from Eq.(4.716)

$$PV \; = \; NkT \; \left[1 - \frac{e^\alpha}{2^{5/2}} + \cdots\right].$$ (4.717)

We observe here deviations from the perfect gas law. These deviations tend to zero as α goes to $-\infty$.

Introducing now the degeneracy parameter ζ through Eq.(4.705) we get from Eq.(4.711) the following series expansion for e^α:

$$e^\alpha \; = \; \zeta \; \left(1 - \frac{\zeta}{2^{3/2}} + \cdots\right),$$ (4.718)

and we see that, indeed, Eq.(4.703) follows in the limit as $\zeta \to 0$.

From Eqs.(4.717) and (4.718) we then get for the equation of state:

$$PV \; = \; NkT \; \left(1 - \frac{\zeta}{2^{5/2}} + \cdots\right).$$ (4.719)

From Eqs.(4.714), (4.711), and (4.718) we get the following expansion for the energy:

$$E \; = \; \frac{3}{2} NkT \; \left(1 - \frac{\zeta}{2^{5/2}} + \cdots\right).$$ (4.720)

Equation (4.720) could, of course, also have been obtained from Eqs.(4.715) and (4.719).

The entropy follows from Eq.(4.420):

$$S \; = \; k \left(q - \alpha N + \beta E\right).$$ (4.721)

Using Eqs.(4.713), (4.718), (4.719), and (4.720) we get for S the following expansion:

$$S \; = \; Nk \; \left[\frac{5}{2} - \ln \zeta - \frac{\zeta}{2^{7/2}} + \cdots\right].$$ (4.722)

We now wish to calculate the specific heats C_v and C_P and their ratio $\kappa = C_P/C_v$. For C_P we have from Eqs.(4.705) and (4.720):

$$C_v \; = \; \left(\frac{\partial E}{\partial T}\right)_V \; = \; \frac{3}{2} Nk \; \left[1 + \frac{\zeta}{2^{7/2}} + \cdots\right].$$ (4.723)

In order to obtain C_P we must expres δQ in terms of δT under the condition that the pressure is kept constant. This can be done in the following way. First of all, we use Eq.(4.715) to get

$$\delta Q \; = \; \delta E + P \, \delta V \; = \; \delta E + \frac{2E}{3V} \, \delta V \; = \; \frac{5}{3} \, \delta E,$$ (4.724)

since it follows from Eq.(4.715) that in the case of constant pressure we
have $\delta \ln E = \delta \ln V$.

From Eqs.(4.720) and (4.705) we have

$$\delta E = \frac{3}{2} NkT \left[1 - \frac{\zeta}{2^{5/2}} + \cdots \right] \delta T - \frac{3}{2} NkT \frac{\delta \zeta}{2^{5/2}} + \cdots, \qquad (4.725)$$

and

$$\delta \ln \zeta = -\tfrac{3}{2} \delta \ln T - \delta \ln V. \qquad (4.726)$$

Combining Eqs.(4.724) to (4.726) we finally get for C_P the expansion

$$C_P = \left(T \frac{\partial S}{\partial T} \right)_P = \frac{5}{2} Nk \left[1 + \frac{3\zeta}{2^{7/2}} + \cdots \right], \qquad (4.727)$$

and for the ratio of the specific heats

$$\kappa = \frac{C_P}{C_v} = \frac{5}{3} \left[1 + \frac{\zeta}{2^{5/2}} + \cdots \right]. \qquad (4.728)$$

All series converge at sufficiently high temperatures, and everywhere the
first term gives the expression for the perfect Boltzmann gas.

We mentioned in §4.3 that in the Bose-Einstein case α should be restricted to values smaller than $\beta \varepsilon_0$. Let us consider for a moment the case
where α approaches this limiting value. We shall assume that α is so close
to $\beta \varepsilon_0$ that we have

$$-\alpha + \beta \varepsilon_0 \ll 1 \qquad \text{and} \qquad -\alpha + \beta \varepsilon_0 \ll \beta(\varepsilon_1 - \varepsilon_0), \qquad (4.729)$$

where ε_1 is the lowest level but one (it follows from Eq.(4.601) that $\varepsilon_1 = 2\varepsilon_0$).

If we use Eqs.(4.216) and (4.315) to calculate the total number of particles, we have

$$N = \sum_j \frac{Z_j}{e^{-\alpha + \beta E_j} - 1}. \qquad (4.730)$$

We shall ignore for the moment the fact that in our derivations we have
used the assumption $Z_j \gg 1$ and we shall assume all Z_j to be equal to 1,
so that the E_j are the energy levels of the particles. Using the inequalities
(4.729) we find that Eq.(4.730) reduces to

$$N = \frac{1}{e^{-\alpha + \beta \varepsilon_0} - 1} + \frac{1}{e^{-\alpha + \beta \varepsilon_1} - 1} + \cdots, \qquad (4.731)$$

or

$$N \approx \frac{1}{-\alpha + \beta \varepsilon_0}. \qquad (4.732)$$

Since the different terms on the right-hand side of Eq.(4.730) represent
the number of particles occupying the various energy levels, Eq.(4.732)

expresses the fact that as $\alpha \to \beta\varepsilon_0$ all particles tend to occupy the lowest level. This phenomenon is the so-called *Einstein condensation*. It is a typical quantum effect, since we find that $\alpha \to -\infty$ as $h \to 0$.

Although we shall discuss some aspects of condensation phenomena in more detail in Chapter 9 we shall briefly describe here the way the isotherms of a perfect Bose-Einstein gas behave and show that they have a horizontal part.

It is convenient to rewrite the equations we have just derived. If we introduce the activity z by the equation

$$z = e^{\alpha}, \tag{4.733}$$

we can write Eqs.(4.713) and (4.709) in the form

$$q = \beta PV = \frac{V}{v_0} f(z), \tag{4.734}$$

where

$$f(z) = \sum_{n=1}^{\infty} \frac{z^n}{n^{5/2}}, \tag{4.735}$$

while v_0 is, apart from a numerical factor of order of magnitude unity, the cube of the thermal de Broglie wavelength (compare Eq.(4.704)),

$$v_0 = \left(\frac{\beta h^2}{2\pi m}\right)^{3/2}. \tag{4.736}$$

Introducing the specific volume $v = V/N$ we find from Eq.(4.711) the following relation between v and z:

$$N = z\frac{\partial q}{\partial z} = \frac{V}{v_0} zf'(z), \qquad \left(f'(z) = \sum_{n=1}^{\infty} \frac{z^n}{n^{3/2}}\right), \tag{4.737}$$

or

$$\frac{1}{v} = \frac{1}{v_0} zf'(z). \tag{4.738}$$

The power series on the right-hand sides of Eqs.(4.734) and (4.738) converge for $z \leqslant 1$, but are no longer convergent for $z > 1$.[24] As long as the specific volume v is larger than a critical value v_c given by the equation

$$\frac{1}{v_c} = \frac{1}{v_0} f'(1) = \frac{2.61}{v_0}, \tag{4.739}$$

---

[24]Opechowski (*Physica* 4, 722 (1937)) has shown that the following expansions are valid when z lies in the neighbourhood of 1:

$$f(z) = 2.36\,(-\ln z)^{3/2} + 1.34 + 2.61\ln z - 0.73\,(\ln z)^2 + \cdots,$$

$$zf'(z) = -3.54\,(-\ln z)^{1/2} + 2.61 - 1.46\ln z - 0.10\,(\ln z)^2 + \cdots.$$

we can use Eqs.(4.734) and (4.738) to obtain the equation of state. However, one has to treat the case where $v < v_c$ with special care.

Let us now consider the isotherms. They will consist of two parts. The first part, where $v \geqslant v_c$ is given by Eq.(4.734):

$$P = \frac{kT}{v_0} f(z), \qquad v \geqslant v_c, \tag{4.740}$$

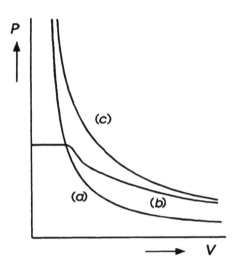

Fig.4.2. The isotherms of a perfect boson gas: (a) is the curve on which all condensation points lie; (b) the isotherm of a boson gas; and (c) the isotherm of a Boltzmann gas at the same temperature.

where z as a function of v follows from Eq.(4.738). The second part of the isotherm is given by the equation

$$P = \frac{kT}{v_0} f(1), \qquad v < v_c. \tag{4.741}$$

Curve (b) of Fig.4.2 depicts one of these isotherms. We have drawn in the same figure for comparison the isotherm of a perfect Boltzmann gas at the same temperature (curve (c)). Finally, curve (a) is the curve on which all condensation points lie. The equation of this curve is obtained by eliminating T from Eqs.(4.734) and (4.738) and putting $z = 1$; the result is

$$Pv^{5/3} = \frac{h^2}{2\pi m} \frac{f(1)}{[f'(1)]^{5/3}}. \tag{4.742}$$

Equations (4.740) and (4.741) are exactly valid only in the limiting case as $N \to \infty$. They can be proved most easily by first considering the case of finite N and afterwards letting N go to infinity. We shall briefly sketch the argument. We must make use of the fact that although Eqs.(4.734) and (4.738) have been derived by replacing sums over energy levels by integrals, the energy spectrum will in reality be discrete as long as the volume remains finite, and instead of Eqs.(4.734) and (4.738) we should, in fact, use Eqs.(4.707) and (4.422) with all Z_j put equal to 1:

$$q = - \sum_s \ln \left(1 - e^{\alpha - \beta \varepsilon_s} \right), \tag{4.743}$$

$$N = \sum_s \frac{1}{e^{-\alpha + \beta \varepsilon_s} - 1}. \tag{4.744}$$

If v approaches v_c, α will approach 0, and it is dangerous to replace the sums in Eqs.(4.743) and (4.744) by integrals. We shall, for the sake of simplicity, shift the zero of the energy levels in such a way that the lowest energy level, ε_0, is zero. Since α will be in the vicinity of zero, the first term in the sum on the right-hand side of Eq.(4.744) will become much more important than the other terms and we shall therefore write Eq.(4.744) as

$$N = \frac{1}{e^{-\alpha} - 1} + \sum_{s \neq 0} \frac{1}{e^{-\alpha + \beta \varepsilon_s} - 1}. \tag{4.745}$$

Replacing the sum over all energy levels but the lowest by an integral and using the fact that the first term on the right-hand side of Eq.(4.745) represents the average number, n_0, of particles in the lowest energy level (compare Eq.(4.315)), we can write Eq.(4.745) in the form

$$N = n_0 + \frac{V}{v_0} z f'(z). \tag{4.746}$$

Introducing the specific volume v ($= V/N$) and its critical value v_c, given by Eq.(4.739), we have

$$1 = \frac{n_0}{N} + \frac{v f'(z)}{v_c f'(1)}. \tag{4.747}$$

As long as $v > v_c$ one can easily satisfy Eq.(4.747) and, indeed, only a small fraction of the particles will occupy the lowest level so that Eqs.(4.747) and (4.738) are identical. However, if $v < v_c$, we must replace Eq.(4.738) by Eq.(4.747). In that case we should expect z to be about equal to one — which can be verified by a careful analysis — and using Eq.(4.315) for n_0 and the fact that $z \cong 1$ we can write Eq.(4.747) in the form

$$1 = \frac{1}{N(1 - z)} + \frac{v}{v_c}, \qquad v < v_c, \tag{4.748}$$

or

$$z = 1 - \frac{1}{N} \frac{v_c}{v_c - v}, \tag{4.749}$$

and we see that, indeed, in the limit as $N \to \infty$, z will be equal to unity for all values of v smaller than v_c, and Eq.(4.741) follows.

We shall not give here a discussion of the isotherm either for $v < v_c$ or for v in the neighbourhood of v_c,[25] or a discussion of the possible relation between the Einstein condensation and the λ-transition of liquid helium.[26]

Before leaving the subject of the Einstein condensation we wish to see what happens when, instead of decreasing the specific volume while keeping the temperature constant, we lower the temperature while keeping the density (or v) constant. Condensation will then set in as soon as the temperature becomes equal to a critical temperature T_0 given by the equation (compare Eqs.(4.739) and (4.736))

$$T_0 = \frac{h^2}{2\pi mk} (2.61\, v)^{-2/3} = \frac{h^2}{2\pi mk} \left(\frac{\varrho}{2.61\, m}\right)^{2/3}, \qquad (4.750)$$

or

$$T_0 = 1.15\, \varrho^{2/3} M^{-5/3} \text{ K}, \qquad (4.751)$$

where ϱ is the density of the system in kg m^{-3} and where M is the molecular weight of the gas.

Let us now briefly summarise the behaviour of the pressure, energy, and specific heat as functions of the temperature for given values of N and V.

First consider the pressure, from which the energy follows immediately if we use Eq.(4.715). As long as the temperature is above T_0 we have Eq.(4.740), or

$$P = CT^{5/2} f(z), \qquad T > T_0, \qquad (4.752)$$

where

$$C = \left(\frac{2\pi mk}{h^2}\right)^{3/2} k, \qquad (4.753)$$

and where z as function of T is given by Eq.(4.738).

At temperatures below T_0, when z will be equal to unity, we have

$$P = CT^{5/2} f(1), \qquad T \leqslant T_0. \qquad (4.754)$$

The specific heat per particle, c_v, follows from Eqs.(4.752), (4.754), (4.740), and the relation

$$c_v = \frac{1}{N} \frac{\partial E}{\partial T}, \qquad (4.755)$$

and we have

$$c_v = \tfrac{3}{2} v \frac{\partial P}{\partial T}. \qquad (4.756)$$

[25] See, for instance, D.ter Haar, *Proc. Roy. Soc. (London)* **A212**, 552 (1952).

[26] See, for example, D.ter Haar, *Elements of Statistical Mechanics*, Rinehart, New York, 1954, Chapter 9.

At temperatures below T_0 Eq.(4.756) leads to the equation

$$c_v = \tfrac{15}{4} vCT^{3/2} f(1), \qquad T \leqslant T_0. \tag{4.757}$$

The value of c_v on the condensation curve, where $z = 1$ and $T = T_0$, follows from Eqs.(4.757) and (4.750), and we have

$$c_v(T_0) = \frac{15\, f(1)}{4\, f'(1)}\, k = 1.926\, k, \tag{4.758}$$

which is actually larger than the classical value of $1.5k$. From Eq.(4.756) it follows that c_v is a continuous function of T, though $\partial c_v/\partial T$ is discontinuous. Figure 4.3 shows P and c_v as functions of T.

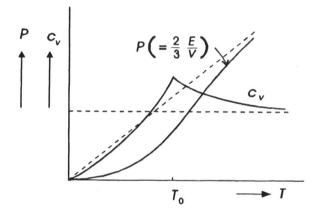

Fig.4.3. Pressure and specific heat of a perfect boson gas.

To what extent can the Einstein condensation be compared to the condensation of an imperfect gas? In the latter case, part of the system will form the liquid and thus form a different phase. This is not the case when we are considering a perfect Bose-Einstein gas. The particles that are too many, that is, those particles in excess of Nv/v_c, will all be found in the lowest energy level (compare Eq.(4.747)), as was first pointed out by Einstein himself. London[27] makes the distinction between a condensation in momentum space in the case of the perfect boson gas and a condensation in coordinate space in the case of an imperfect gas. This does not seem to be justified, since the condensed phase of an imperfect gas also does not contribute to the pressure and thus also corresponds to a condensation in momentum space. Moreover, as has been pointed out by Becker[28] one can also consider Einstein condensation to be a condensation in coordinate space.

[27]F.London, *Phys. Rev.* **54**, 947 (1938).

[28]R.Becker, *Z. Physik* **128**, 120 (1950).

4.8.   The Perfect Fermi-Dirac Gas

The discussion in this section develops parallel to that given in the preceding section as long as $e^\alpha \ll 1$. For that case we shall give the results without detailed comments. Instead of Eqs.(4.706) to (4.714) we now have[29]

$$\ln W_{FD} = \sum_j N_j (-\alpha + \beta E_j) + \sum_j Z_j \ln(1 + e^{\alpha - \beta E_j}), \quad (4.801)$$

$$q_{FD} = \sum_j Z_j \ln(1 + e^{\alpha - \beta E_j}), \quad (4.802)$$

$$q_{FD} = 4\pi \left(\frac{2m}{h^2}\right)^{3/2} V \int_0^\infty \sqrt{\varepsilon} \ln(1 + e^{\alpha - \beta \varepsilon}) \, d\varepsilon, \quad (4.803)$$

$$q_{FD} = 2 \left(\frac{2\pi m}{\beta h^2}\right)^{3/2} V h(\alpha), \quad (4.804)$$

$$h(\alpha) = -\sum_{n=1}^\infty \frac{(-e^\alpha)^n}{n^{5/2}}, \quad (4.805)$$

$$h'(\alpha) = -\sum_{n=1}^\infty \frac{(-e^\alpha)^n}{n^{3/2}}, \quad (4.806)$$

$$N = q_{FD} \frac{h'(\alpha)}{h(\alpha)}, \quad (4.807)$$

$$P = \frac{q_{FD}}{\beta V}, \quad (4.808)$$

$$E = \frac{3}{2} \frac{q_{FD}}{\beta}. \quad (4.809)$$

From Eqs.(4.808) and (4.809) it follows again that Eq.(4.715) is satisfied.

Introducing once more a quantity ζ by Eq.(4.705), but with $(3/8\pi)^{3/2}$ instead of $(3/2\pi)^{3/2}$ in the second member of that equation to take into account the spin of the fermions, we now get instead of Eqs.(4.718) to (4.720), (4.722), (4.723), (4.727), and (4.728) the expansions

$$e^\alpha = \zeta \left(1 + \frac{\zeta}{2^{3/2}} + \cdots\right), \quad (4.810)$$

$$PV = NkT \left(1 + \frac{\zeta}{2^{5/2}} + \cdots\right), \quad (4.811)$$

$$E = \frac{3}{2} NkT \left(1 + \frac{\zeta}{2^{5/2}} + \cdots\right), \quad (4.812)$$

$$S = Nk \left[\frac{5}{2} - \ln\zeta + \frac{\zeta}{2^{7/2}} + \cdots\right], \quad (4.813)$$

---

[29]We have assumed for the sake of simplicity that we are dealing with spin-$\frac{1}{2}$ particles. This means that expression (4.603) for dZ must be multiplied by 2, and that there occur extra factors 2 in equations such as (4.803) or (4.804).

$$C_v = \frac{3}{2} Nk \left[1 - \frac{\zeta}{2^{7/2}} + \cdots\right], \qquad (4.814)$$

$$C_P = \frac{5}{2} Nk \left[1 - \frac{3\zeta}{2^{7/2}} + \cdots\right], \qquad (4.815)$$

$$\kappa = \frac{5}{3} \left[1 - \frac{\zeta}{2^{5/2}} + \cdots\right]. \qquad (4.816)$$

All series converge as long as $e^\alpha \leqslant 1$. We see that to the lowest approximation Bose-Einstein and Fermi-Dirac statistics differ only in the *sign* of the deviation from the Boltzmann statistics.

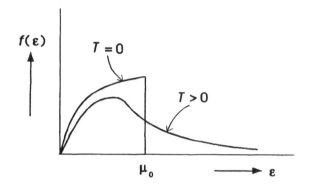

Fig.4.4. The distribution function $f(\varepsilon)$ of Eq.(4.818) at $T = 0$ and at a finite temperature.

We shall now discuss the case where $e^\alpha \gg 1$. This is called the case of *strong degeneracy* and the gas is called a *degenerate quantum gas*. We shall find as $T \to 0$ that $\alpha \to \infty$, and we shall say that at absolute zero the degeneracy is complete. Let us first discuss the behaviour at absolute zero.

From Eqs.(4.803) and (4.422) we get for the total number of particles

$$N = \frac{\partial q}{\partial \alpha} = 4\pi \left(\frac{2m}{h^2}\right)^{3/2} V \int_0^\infty \frac{\sqrt{\varepsilon}\, d\varepsilon}{1 + e^{-\alpha + \beta\varepsilon}}. \qquad (4.817)$$

From this equation we see that the number of particles dN with energies between ε and $\varepsilon + d\varepsilon$ is given by the equation

$$dN = f(\varepsilon)\, d\varepsilon = 4\pi \left(\frac{2m}{h^2}\right)^{3/2} V \frac{\sqrt{\varepsilon}\, d\varepsilon}{1 + e^{-\alpha + \beta\varepsilon}}. \qquad (4.818)$$

This equation is actually the same as Eq.(4.316) if we use for Z_j the expression (4.604) — with an extra spin-factor 2 — for the number of levels

dZ with energies between ε and $\varepsilon + d\varepsilon$. We note that, whereas $f(\varepsilon)$ varies as $\sqrt{\varepsilon}$ for small values of ε, for high energies it behaves as $\sqrt{\varepsilon}\,e^{-\beta\varepsilon}$: the Fermi distribution (4.818) has a "Maxwellian tail" (compare Eq.(1.104) and take into account that for a spherically symmetric distribution d^3c can be written as $4\pi c^2\,dc = 4\pi\sqrt{(2\varepsilon/m^3)}\,d\varepsilon$). We show in Fig.4.4 the function $f(\varepsilon)$ both for a finite temperature and for $T = 0$, when there is no Maxwellian tail. It is often useful to consider instead of $f(\varepsilon)$ a function $F(\varepsilon)$, sometimes called the *Fermi function*, which differs from $f(\varepsilon)$ by a factor $dZ/d\varepsilon$:

$$F(\varepsilon) \;=\; \frac{1}{e^{-\alpha+\beta\varepsilon}+1}. \tag{4.819}$$

This function and the negative of its derivative with respect to ε are shown in Fig.4.5 for the case of a finite temperature. The dashed line in Fig.4.5 shows $F(\varepsilon)$ for $T = 0$; in that case $-\partial F/\partial\varepsilon$ becomes a delta-function.

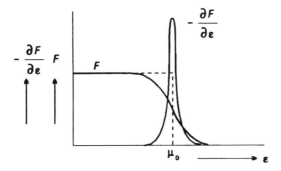

Fig.4.5. The Fermi function $F(\varepsilon)$ of Eq.(4.819) and its derivative at a finite temperature. The dashed line corresponds to $T = 0$.

We now use Eq.(4.415) to introduce the partial chemical potential $\mu = \alpha/\beta$. We denote its value for $T = 0$ by μ_0. In order to determine the value of μ_0 we split the integral in Eq.(4.817) into two parts and write

$$N \;=\; 4\pi\left(\frac{2m}{h^2}\right)^{3/2} V\left\{\int_0^\mu \frac{\sqrt{\varepsilon}\,d\varepsilon}{1+e^{-\beta(\mu-\varepsilon)}} + \int_\mu^\infty \frac{\sqrt{\varepsilon}\,d\varepsilon}{1+e^{-\beta(\mu-\varepsilon)}}\right\}. \tag{4.820}$$

This equation is valid for any value of T (or β). In the limit as $T \to 0$, or $\beta \to \infty$, the second integral vanishes and the first one is equal to $\frac{2}{3}\mu_0^{3/2}$. We have thus in that limit

$$N \;=\; \frac{8\pi}{3}\left(\frac{2m}{h^2}\right)^{3/2} V\,\mu_0^{3/2}, \tag{4.821}$$

or

$$\mu_0 = \frac{h^2}{2m}\left(\frac{3N}{8\pi V}\right)^{2/3}. \tag{4.822}$$

We note, by the way, that in the limit as $T = 0$ the functions $f(\varepsilon)$ $(= dN/d\varepsilon)$ and $F(\varepsilon)$ have a discontinuity at $\varepsilon = \mu_0$.

The energy of the system follows from Eqs.(4.803) and (4.423) and is given by the equation

$$E = \frac{\partial q}{\partial \beta} = 4\pi\left(\frac{2m}{h^2}\right)^{3/2} V \int_0^\infty \frac{\varepsilon\sqrt{\varepsilon}\,d\varepsilon}{1 + e^{-\alpha + \beta\varepsilon}}. \tag{4.823}$$

At $T = 0$ we have

$$E_0 = \frac{8\pi}{5}\left(\frac{2m}{h^2}\right)^{3/2} V \mu_0^{5/2} = \frac{3}{5}\frac{h^2}{2m} N \left(\frac{3N}{8\pi V}\right)^{2/3}, \tag{4.824}$$

or

$$E_0 = \tfrac{3}{5} N \mu_0. \tag{4.825}$$

Equations (4.824) or (4.825) give us the so-called zero-point energy of a Fermi-Dirac gas. We could also have calculated E_0 by using the Pauli principle and filling up the lowest energy levels until all particles had been assigned.

Combining Eqs.(4.715) and (4.824) we find for the *zero-point pressure*

$$P_0 = \frac{2E}{3V} = \left(\frac{3}{8\pi}\right)^{2/3} \frac{h^2}{5m}\left(\frac{N}{V}\right)^{5/3}, \tag{4.826}$$

and we see that P_0 varies as the $\frac{5}{3}$ power of the density.

That this zero-point pressure is not always negligible can be seen when we consider the "gas" of the conduction electrons in a metal. The density N/V will be of the order of 10^{30} m^{-3} and with $m \cong 10^{-30}$ kg we have

$$P_0 \sim 10^6 \text{ atm.} \tag{4.827}$$

With the same values of N/V and m we find that

$$\mu_0 \approx 35 \text{ eV}, \tag{4.828}$$

corresponding to a temperature μ_0/k of about 4×10^5 K.

We note that we could also have obtained Eq.(4.826) by combining Eqs. (4.803) and (4.426), performing an integration by parts, and taking the limit as $T \to 0$. It follows from the occurrence of h in the expression for P_0 that the zero-point pressure, like the Einstein condensation, is a typical quantum effect.

We should at this point mention that μ is often called the *Fermi level*,[30] especially in metal and semiconductor theory.[31] This Fermi level is the *electrochemical potential* and is equal to the partial thermal potential μ as can be seen as follows. Addition of δN electrons will result in a change δF of the free energy equal to $\mu \delta N$ (compare Eq.(4.414)). Consider now two systems A and B at electrical potentials V_A and V_B. An adiabatic transfer of δN electrons each with charge e from A to B will result in a change δF of the free energy of the combined system which, on the one hand, is equal to $e(V_B - V_A)$ and, on the other hand, is equal to $\mu_B - \mu_A$. This shows that μ is the electrochemical potential.

We note in passing that, as it follows from thermodynamics[32] that in the case of equilibrium between two systems which can exchange particles the partial thermal potentials in the two systems must be equal, we find that in equilibrium two metals or semiconductors in contact should have the same value for their Fermi levels. When we use the term Fermi level, μ is often denoted by E_F. The name "Fermi level" derives from the fact that at absolute zero all levels up to E_F will be occupied (compare Eq.(4.820)). The particles occupying these levels up to E_F are sometimes called the *Fermi sea*.

From the result of Eq.(4.828) it is clear that for the case of the conduction electrons in a metal for any temperature the inequality $\beta\mu \gg 1$ will hold, which means that we are then always dealing with a strongly degenerate gas. We can therefore again expect that we might be able to give series expansions for the various quantities of interest in terms of inverse powers of $\beta\mu$ or of $\beta\mu_0$. We shall use as our expansion parameter a quantity η defined by the equation

$$\eta = \frac{1}{\beta\mu_0} = \frac{2mkT}{h^2}\left(\frac{8\pi V}{3N}\right)^{2/3}\left[=\left(\frac{8}{9\pi}\right)^{1/3}\zeta^{-2/3}\right], \qquad (4.829)$$

where ζ is defined by Eq.(4.705) including the spin factor 2.

To find the expansions we use the general theorem that

$$\int_0^\infty G(\varepsilon)\,\frac{\partial F}{\partial\varepsilon}\,d\varepsilon = -G(\mu) - \frac{\pi^2}{6\beta^2}\left.\frac{\partial^2 G}{\partial\varepsilon^2}\right|_{\varepsilon=\mu} + \cdots. \qquad (4.830)$$

Equation (4.830) can be proved as follows.[33] First of all, we introduce a new variable x by the equation

$$x = \beta(\varepsilon - \mu), \qquad (4.831)$$

---

[30]See, for example, J.C.Slater, *Quantum Theory of Matter*, McGraw-Hill, New York, 1951, Chapter 12; W.Shockley, *Electrons and Holes in Semiconductors*, New York, 1951, p.231 and Chapter 16.

[31]Sometimes the Fermi level is called the Fermi energy; this can be confusing because this term is also used for the average energy of the electrons in the system (see Slater, *op.cit*, p.420).

[32]See, for instance, D.ter Haar and H.Wergeland, *Elements of Thermodynamics*, Addi-son-Wesley, Reading, Mass., § VI.1.

[33]See, for instance, A.Sommerfeld and H.Bethe, *Handb. Phys.* 24_2, 346 (1933).

so that we have

$$F(x) \rightarrow F(x) = \frac{1}{e^x + 1}. \tag{4.832}$$

The integral now becomes

$$\int_{-\beta\mu}^{\infty} G(x) \frac{\partial F}{\partial x} dx \equiv I. \tag{4.833}$$

We now expand $G(x)$ in a Taylor series:

$$G(x) = G(0) + xG'(0) + \tfrac{1}{2}x^2 G''(0) + \cdots . \tag{4.834}$$

Defining a set of quantities I_i through the equations

$$I_i = \frac{1}{i!} \int_{-\beta\mu}^{\infty} \frac{\partial F}{\partial x} x^i dx, \tag{4.835}$$

we find from Eqs.(4.833) to (4.835)

$$I = \sum_{n=0}^{\infty} G^{(n)}(0) I_n. \tag{4.836}$$

The problem is thus reduced to determining the I_i. We now first of all note that as $\beta\mu \gg 1$, we can replace the lower limit in all integrals I_i by $-\infty$; the errors introduced in this way are of the order of magnitude of $e^{-\beta\mu}$ and thus totally negligible in the case of interest to us. We further note that $\partial F/\partial x$ is an even function of x so that all I_i with i odd vanish, while for the I_i with i even we have[34]

$$I_i = -\frac{2}{i!} \int_0^{\infty} \frac{\varepsilon^i e^{-\varepsilon} d\varepsilon}{(1 + e^{-\varepsilon})^2} = -\frac{2}{i!} \int_0^{\infty} \varepsilon^i \sum_{n=1}^{\infty} (-1)^n n e^{-n\varepsilon} d\varepsilon$$

$$= 2 \sum_{n=1}^{\infty} \frac{(-1)^{n+1}}{n^i} = \frac{2\pi^i (2^{i-1} - 1)}{i!} B_i, \tag{4.837}$$

where the B_i are the Bernoulli numbers. Changing back from x to ε as variable now gives us Eq.(4.830).[35]

Applying Eq.(4.830) we get for N from Eq.(4.817)

$$N = \frac{8\pi}{3} \left(\frac{2m}{h^2}\right)^{3/2} V\mu^{2/3} \left[1 + \frac{\pi^2}{8}(\beta\mu)^{-2} + \cdots\right], \tag{4.838}$$

---

[34]See, for instance, I.N.Bronshtein and K.A.Semendyayev, *A Guidebook to Mathematics*, Pergamon, Oxford, 1963, p.354. (Note that there is a misprint in their formula 20: 2^{2k} should be 2^{2k-1}.)

[35]We note that in the limit as $\beta \rightarrow \infty$, when only the first term on the right-hand side of (4.830) has to be taken into account, $-\partial F/\partial\varepsilon$ acts, indeed, as a δ-function.

and hence for μ:

$$\mu = \mu_0 \left[1 - \frac{\pi^2}{12} \eta^2 + \cdots \right]. \tag{4.839}$$

Using Eqs.(4.830) and (4.839) we then get for E, P, C_v, and S:

$$E = E_0 \left[1 + \frac{5\pi^2}{12} \eta^2 + \cdots \right], \tag{4.840}$$

$$P = P_0 \left[1 + \frac{5\pi^2}{12} \eta^2 + \cdots \right], \tag{4.841}$$

$$C_v = \frac{\partial E}{\partial T} = \frac{\pi^2}{2} Nk\eta + \cdots, \tag{4.842}$$

$$S = \int_0^T \frac{C_v \, dT}{T} = \frac{\pi^2}{2} Nk\eta + \cdots. \tag{4.843}$$

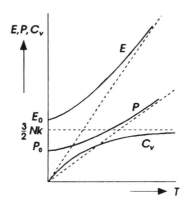

Fig.4.6. The energy, pressure, and specific heat of a perfect fermion gas.

In Fig.4.6 we have drawn E, P, and C_v as functions of the temperature. The dashed lines indicate the behaviour of a Boltzmann gas.

It is easy to see the physical meaning of Eqs.(4.840) and (4.842). If we study Figs.4.4 and 4.5 we see that only a fraction of the particles is excited when we go from absolute zero to a finite temperature. Calculating the slope of $F(\varepsilon)$ at the point of inflection and drawing a straight line through the inflection point at a slope half that of the tangent to $F(\varepsilon)$ in that point, we find that the line will intersect the ε-axis at a distance $4kT$ from μ_0. Replacing the shaded areas in Fig.4.7 by the triangles ABC and CDE we find that the total number of particles involved in the thermal properties of the system will be of the order of $(kT/\mu)N$. As the energy of each of these particles is increased roughly by an amount kT, we find for the total increase in energy an amount of the order of $(kT/\mu)^2 N\mu$, in agreement with Eq.(4.840). If instead of the shaded areas we take the triangles, we find for the numerical factor $\frac{8}{3}$ which is, indeed, extremely close to the factor $\pi^2/4$ of Eq.(4.840). As only a fraction kT/μ is involved in the thermal properties, we can understand that the specific heat is less than the classical value Nk by a factor of that order of magnitude.

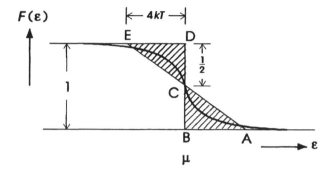

Fig.4.7. The Fermi function at a finite temperature. The two shaded areas correspond, respectively, to the electrons missing from the totaly degenerate Fermi sea and to those which lie above the Fermi level.

We note also that it follows from Eq.(4.841) that the quantity $\partial P/\partial T$, which is constant in the classical case, is now proportional to T.

4.9.   Are All Particles Bosons or Fermions?

In the present chapter we have so far considered systems of non-interacting identical point particles and assumed that the wavefunction describing such a system is either completely symmetric or completely antisymmetric. We have not considered at all either whether all systems in nature are of this kind or the question under what circumstances we must consider deviations from classical behaviour. The present section will be devoted to these two questions. The second question is the more straightforward one, but we shall start with a discussion of the first one.

The question of what particles obey what statistics is a very profound and difficult one. There is a popular misconception that just as, in Gilbert's words[36] "... every boy and every gal that's born into the world alive is either a little Liberal or else a little Conservative!" (Iolanthe, Act II), so every particle in nature is either a boson or a fermion.[37]   A moment's

---

[36] *Plays and Poems of W.S.Gilbert*, Random House, New York, 1932, p.266.

[37] For instance, slightly quoting out of context, I.J.R.Aitchison, *SERC Bulletin*

reflection shows that this is certainly not true. A hydrogen atom or a nitrogen molecule, for instance, are certainly neither bosons nor fermions. One finds many extremely glib and simple statements that any particle composed of an even number of electrons or nucleons is a boson and any particle composed of an odd number of electrons or nucleons a fermion. First of all, it is certainly not true that all particles which occur in nature are either fermions or bosons. As a first, rough approximation one hopes that all truly elementary particles[38] are bosons, if they have integral spin, and fermions, if they have half-odd-integral spin. This would mean that photons, gravitons, pions, or kaons are bosons, and electrons, neutrinos, muons, protons, and neutrons fermions. However, some of these particles are themselves compound particles, being built up out of quarks and gluons. This raises, on the one hand, the question of the statistics of compound particles and to a very rough approximation it is probably safe to say that as long as the energy with which the more elementary particles are bound into the compound particle is high as compared to other energies in the problem, a compound particle containing an even number of fermions will obey Bose-Einstein statistics (examples are the α-particle and the ^{4}He atom), while those containing an odd number of fermions will obey Fermi-Dirac statistics (for example ^{3}He atoms).[39] On the other hand, it raises the question of whether one can prove that there is a one-to-one relation between statistics and spin.

The question of what statistics are obeyed by particles — or by quasi-particles — becomes simpler if one describes the particles (or quasi-particles) in terms of annihilation and creation operators $\widehat{a}_i$ and $\widehat{a}_i^\dagger$ (see Chapter 8) in the so-called "second quantisation" representation,[40] the particles are bosons, if these operators obey the so-called symmetric commutation rules, that is, if they satisfy the equations

$$\widehat{a}_i\widehat{a}_j^\dagger - \widehat{a}_j^\dagger\widehat{a}_i = \delta_{ij}, \qquad \widehat{a}_i\widehat{a}_j - \widehat{a}_j\widehat{a}_i = 0, \qquad \widehat{a}_i^\dagger\widehat{a}_j^\dagger - \widehat{a}_j^\dagger\widehat{a}_i^\dagger = 0, \qquad (4.91)$$

where δ_{ij} is the Kronecker δ-function which is 0, if $i \neq j$ and equals 1, if $i = j$.

4, No 8, 21 (1991):"Is there some deep reason why there are only two types of particle?"

[38]This leaves us with the question of what is a truly elementary particle. After all, many so-called elementary particles contain "more elementary" particles such as quarks and gluons as building blocks.

[39]A discussion of the statistics of compound particles was given in P.Ehrenfest and J.R.Oppenheimer, *Phys. Rev.* **37**, 333 (1931).

[40]P.Jordan and O.Klein, *Z. Physik* **45**, 751 (1927); P.Jordan and E.Wigner, *Z. Physik* **47**, 631 (1928); compare also G.Wentzel, *Quantum Theory of Fields*, New York, 1949, § 22; H.A.Kramers, *Quantum Mechanics*, North-Holland, Amsterdam, 1957, § 72; A.S.Davydov, *Quantum Mechanics*, Pergamon, Oxford, 1991, Chapters 10 and 11.

If, on the other hand, the $\hat{a}_i$ and $\hat{a}_i^\dagger$ satisfy the anticommutation relations

$$\hat{a}_i\hat{a}_j^\dagger + \hat{a}_j^\dagger\hat{a}_i = \delta_{ij}, \qquad \hat{a}_i\hat{a}_j + \hat{a}_j\hat{a}_i = 0, \qquad \hat{a}_i^\dagger\hat{a}_j^\dagger + \hat{a}_j^\dagger\hat{a}_i^\dagger = 0, \qquad (4.92)$$

the particles are fermions. This, of course, begs the question, since one then asks why Eqs.(4.91), or Eqs.(4.92), hold for a particular particle. It is interesting to note, for instance, that the commutation relation for the Cooper pairs in the BCS theory of superconductivity[41] which are often quoted as being bosons are only approximately Eqs.(4.91).

The advantage of approaching the statistics of particles, or quasi-particles, through the commutation relations they satisfy is that the question whether or not they are point particles — and we know that even elementary particles such as the electron or the proton are strictly speaking not point particles — elementary particles, or quasi-particles does not have to be addressed.

The relation between spin and statistics is a very complex one. In 1940 Pauli and Belinfante[42] gave the following general principles (we quote here their summary and have inserted further references): "Es wird untersucht, in wie weit Spekulationen über die Statistik willkürlicher, hypothetischer Teilchen in einer relativistisch invarianten Theorie möglich sind, wenn alle oder ein Teil der folgenden drie Postulate vorausgesetzt werden: (I) die Energie ist positiv; (II) Observable an verschiedenen Raum-Zeitpunkten mit raumartiger Verbindungslinie sind kommutativ; (III) es gibt zwei äquivalente Beschreibungen der Natur, in welchen die Elementarladungen entgegengesetzte Vorzeichen haben und in welchen einander entsprechende Feldgrössen sich bei Lorentz-Transformationen in gleicher Weise transformieren[43] .

"Der eine von uns[44] hat bereits gezeigt, dass allgemein bei ganzzahligem Spin aus (II) allein Einstein-Bose-Statistik, bei halbzahligem Spin aus (I) allein Fermi-Dirac-Statistik folgt. Ferner hat der andere von uns[45] gezeigt, dass für eine gewisse Klasse von Teilchen, (die alle bisher in der Natur beobachteten umfasst, und die dadurch charakterisiert ist, dass sie höchtens durch *einen* Undor[46] von gegebener Stufe beschrieben wird), aus dem durch ein spezielles Transformationsgesetz spezialisierten Postulat (III) für ganzen Spin E.-B.-Statistik, für halbzahligen Spin F.-D.-Statistik gefolgert werden kann. In der vorliegenden Note wird in den typischen Fällen von Spin 0 und Spin $\frac{1}{2}$ durch Beispiele gezeigt, dass im allgemeinen Fall mehrerer Undoren von gleicher Stufe aus (III) nicht mehr eindeutig auf die Statistik der Teilchen geschlossen werden kann, während (II), bzw. (I) für ganzen bzw. halben Spin hiezu stets hinreichend bleiben. In den speziellen Fällen des Skalarfeldes, des Vektorfeldes und des Dirac-Elektrons, wo nur ein

[41] J.Bardeen, L.N.Cooper, and J.R.Schrieffer, *Phys. Rev.* **108**, 1175 (1957).

[42] W.Pauli and F.J.Belinfante, *Physica* **7**, 177 (1940).

[43] Cf. H.A.Kramers, *Proc. Kon. Ned. Akad. Wet. (Amsterdam)* **40**, 814 (1937).

[44] W.Pauli, *Phys. Rev.* **58**, 716 (1940).

[45] F.J.Belinfante, *Physica* **6**, 870 (1939).

[46] Cf. F.J.Belinfante, *Physica* **6**, 849 (1939).

einziger Undor gegebener Stufe in die Theorie eingeht, folgt dagegen das Transformationsgezetz der Ladungskonjugierung eindeutig, so dass hier (III) zur Festlegung der Statistik ausreicht."[47]

As long as we are dealing with particles for which we can forget their structure and if we consider the wavefunction of two such particles, interchanging the two identical particles must leave all expectation values unchanged so that under the interchange the wavefunction must be multiplied by a phase factor, $e^{i\varphi}$ (φ real). If we interchange the two particles for a second time we are back where we started so that we expect that the only possible choices for φ are

$$\varphi = \pi \quad \text{or} \quad \varphi = 2\pi. \tag{4.93}$$

For the wavefunctions of a system of N identical particles this would lead to the totally antisymmetric or the totally symmetric wavefunctions which we considered in § 4.2. This argument which seems absolutely watertight is, however, not true for one- or two-dimensional systems. It was pointed out by Laidlaw and de Witt[48] and by Leinaas and Myrheim[49] that for two-dimensional systems with particles which have impenetrable cores φ could have any value. Such particles were called *anyons* by Wilczek.[50] The reason for this is that in two-dimensional space the fact that the two identical particles cannot occupy the same point in space leads to the fact that the two-dimensional space for two identical particles is no longer simply connected. Anyons do not

---

[47]We investigate in how far it is possible to speculate about the statistics of any hypothetical particles in a relativistically invariant theory when one or more of the following assumptions are made: (I) the energy is positive; (II) observables at different space-time points with a space-like connection commute; (III) there exist two equivalent descriptions of nature in which the elementary charges have opposite signs and in which corresponding field quantities transform in the same way under a Lorentz transformation.

One of us (W.P.) has shown already that in general for integral spin Bose-Einstein statistics follows from (II) and for half-odd-integral spin Fermi-Dirac statistics follows from (I). Moreover, the other of us (F.J.B.) has shown that one can conclude from assumption (III) specialised by a special transformation rule for a certain class of particles (including all particles observed in nature and characterised by the fact that they are described by at most a single undor of given rank) that E.B. statistics follows for integral and F.D. statistics for half-odd-integral spin. In the present note we show through examples that in the special case where one needs several undors of the same rank one can no longer uniquely determine the statistics of the particles from (III) whereas (II) or (I), respectively, remain always sufficient for integral or half-odd-integral spin. In the special cases of a scalar field, a vector field, or the Dirac electron where only a single undor of given rank occurs in the theory the space conjugation transformation follows uniquely and therefore (III) is sufficient to determine the statistics.

[48]M.G.G.Laidlaw and C.M.de Witt, *Phys. Rev.* **D3**, 1375 (1971).

[49]J.M.Leinaas and J.Myrheim, *Nuovo Cim.* **37B**, 1 (1977).

[50]F.Wilczek, *Phys. Rev. Lett.* **49**, 957 (1982).

obey either Bose-Einstein or Fermi-Dirac statistics but *intermediate* or *fractional* statistics. Recently there has been increased interest in their properties since both the fractional quantum Hall effect[51] and some cases of high-temperature superconductivity[52] occur in essentially two-dimensional systems. It should be stressed that anyons are quasi-particles rather than actual particles.

We shall not consider in any more detail the question of whether at least all elementary particles are either bosons or fermions.

We now turn to the question: "When can we expect deviations from Boltzmann statistics?" Let us confine ourselves to gaseous systems, or systems which to a first approximation may be treated as gases. We saw (Eq.(4.701)) that the condition for the applicability of Boltzmann statistics was that e^α be small compared to unity. We can calculate e^α for various gases at their boiling point, using Eq.(4.703), and the result is given in Table 4.1.

TABLE 4.1

Gas	Boiling Point	e^α
Argon	87.4 K	2×10^{-6}
Neon	27.2 K	10^{-4}
Hydrogen	20.3 K	7×10^{-3}
^{4}He	4.2 K	0.13
^{3}He	3.2 K	0.4

We see that the only gas for which deviations from classical statistics could conceivably be detected should be helium. In the case of ^{4}He there seem to be some indications that the equation of state follows more closely that of a Bose-Einstein gas than that of a Fermi-Dirac gas. We should expect Bose-Einstein behaviour from ^{4}He since the nucleus is built up out of 4 nucleons and there are 2 orbital electrons. Since e^α in the case of ^{3}He is even larger due to the smaller mass of the ^{3}He nucleus and its lower boiling point, an accurate measurement of its equation of state should give a definite confirmation that ^{3}He obeys Fermi-Dirac statistics.

We must remark here that the statistics of ^{4}He and ^{3}He have been much more often been connected with the behaviour of their liquid phases. Liquid ^{3}He and ^{4}He are called *quantum liquids*. We have already mentioned that the λ-transition from helium I to helium II resembles in some respects the Einstein condensation. Similarly, liquid ^{3}He is a typical *Fermi liquid*.

The radiation field is, as we discussed at the beginning of the present chapter, a system to which Bose-Einstein statistics must be applied. This has, for instance, important consequences for fluctuations in this field.

We see from Eq.(4.703) that a small mass or a high density will make it necessary to abandon Boltzmann statistics. For instance, in the case

[51]D.C.Tsui, H.L.Störmer, and A.C.Gossard, *Phys. Rev. Lett.* **48**, 1559 (1982).

[52]See R.B.Laughlin, *Science* 242, 525 (1988).

of the "gas" of the conduction electrons in a metal, we find that for an average spacing of 1 Å for the atoms in a metal and one conduction electron per atom, we find at room temperatures a value of about 10^5 for e^α. In fact, many poperties of metals can be understood, if we apply Fermi-Dirac statistics to this case (compare the problems on metals and semiconductors at the end of the present chapter).

A second case where Fermi-Dirac statistics must be applied is the nuclear "gas." To a first rough approximation the nucleons inside a nucleus may be treated as a perfect Fermi-Dirac gas, or rather a mixture of two perfect Fermi-Dirac gases, that of the protons and that of the neutrons. We have here a high value of e^α due to the very high nuclear density $(\varrho > 10^{15}$ kg m$^{-3})$. In the interior of some stars densities may occur that are so high that, notwithstanding the very high stellar temperatures, the gas is degenerate. Both Fermi-Dirac and Bose-Einstein degeneracy may occur.

We may mention also here that the various quasi-particles that enter so prolifically into modern theoretical physics, such as the magnons and phonons in solid state physics, as well as many of the (more or less) elementary particles from high-energy physics usually obey well-defined statistics.

Problems

1. Prove the equivalence of Eqs.(4.213) and (4.212).

2. Gentile[53] introduced in 1940 the so-called intermediate statistics in which the maximum-allowed number of particles in any one energy level is d. Derive the equilibrium value of the number of particles N_j in the j-th group of levels, containing Z_j levels.

3. Use the method of §§ 4.2 and 4.3 to express the concentration ratio $c_A c_B / c_{AB}$ in terms of Z_A, Z_B, and Z_{AB} for the case of the dissociative equilibrium

$$A + B \leftrightarrows AB.$$

Assume that there are no interactions between the particles and that there is a binding energy for the compound AB equal to χ, that is, that the ground state of AB lies lower than the ground state of A + B by an amount χ.

4. Use the Darwin-Fowler method to derive the result obtained in the preceding problem.[54]

---

[53]G.Gentile, *Nuovo Cimento* 17, 493 (1942); see also A.Sommerfeld, *Ber. Dtsch. Chem. Ges.* 75, 1988 (1942), H.Wergeland, *Proc. Roy. Norwegian Acad. Sc.* 17, 51 (1944), G.Schubert, *Z.Naturf.* 1, 113 (1946), H.Müller, *Ann. Physik* 7, 420 (1950), and D.ter Haar, *Physica* 18, 199 (1952).

[54]See, for example, R.H.Fowler, *Statistical Mechanics*, Cambridge University Press, 1929, pp. 104ff.

5. Write down the integral representation for the factorial,

$$z! = \int_0^\infty e^{-t} t^z \, dt,$$

and introduce a new variable v through the equations

$$(1+u)\,e^{-u} = e^{-\frac{1}{2}v^2}, \qquad t = z(1+u).$$

Then use the method of steepest descents to find Stirling's asymptotic formula

$$z! \approx \sqrt{2\pi z} \left(\frac{z}{e}\right)^z.$$

The factor $\sqrt{2\pi z}$ can in most cases be dropped when we are using the expression for $\ln z!$.

6. Consider a perfect gas in a volume V without an external field of force. Dividing the volume into m equal cells and denoting the number of particles in the k-th cell by n_k, use the Darwin-Fowler method to find an expression for the average value $\overline{n_k}$ of n_k.

7. Find the equation for the curve a in Fig.4.2.

8. Find the way the number of bosons in the ground state of a perfect Bose-Einstein gas depends on the temperature for temperatures below the condensation temperature T_0.

9. Prove that a two-dimensional or a one-dimensional perfect boson gas does not undergo an Einstein condensation.

10. Consider a perfect gas of bosons, each of which can be either in the ground state or in an excited state Δ above the ground state. Assume that $\Delta/kT_0^{(0)} \gg 1$ ($T_0^{(0)}$ is the condensation temperature for the case where $\Delta = \infty$). Find an approximate expression for the condensation temperature of this gas, and explain qualitatively the result obtained.

11. Find an equation for the isentropics in the $P-V$ diagram for a perfect monatomic quantum gas.

12. Prove that the specific heat C_v of a perfect two-dimensional gas of fermions is identically the same as that of a perfect two-dimensional boson gas.[55]

13. Find the next terms in the series expansions in ζ for the various quantities given in §§ 4.7 and 4.8.

14. Find the first few terms in a power series expansion in the temperature for the energy of a two-dimensional perfect gas of fermions.

15. Find the next terms in the series expansions in η for the various quantities given in § 4.8.

---

[55] R.M.May, *Phys. Rev.* **135**, A1515 (1964).

16. Prove that the tangent at the point of inflection in Fig.4.5 cuts the ε-axis at $\mu + 2kT$.

17. Prove that, if we replace the shaded areas in Fig.4.7 by the areas ABC and DEF, the total energy of the system becomes $E_0 + \frac{8}{3}N\mu(\beta\mu)^{-2}$.

18. Evaluate the compressibility of a strongly degenerate, perfect Fermi-Dirac gas.

19. Show that the specific heat of a strongly degenerate, perfect fermion gas is given by

$$C_v \approx \frac{1}{3}\pi^2 k^2 T \varrho(\mu_0),$$

where $\varrho(\varepsilon)$ is the single-particle density of states.

The next few problems deal with metals and semiconductors; first we shall briefly review the basic ideas upon which the elementary theory of metals and semiconductors is based. We can only give an outline of these ideas and must refer to the literature[56] for details and for complications such as overlapping bands. We also refer to the literature for a discussion of such concepts as the effective mass and the rôle of interactions between the electrons.

First we must enter into a slightly detailed discussion of the band structure of the energy levels in crystals. If we are concerned with the movement of electrons in a crystal and if we treat the electrons as non-interacting with one another, the first problem is to find the energy eigenvalues of a particle moving in a one-dimensional potential energy field with periodic structure. One can show — at any rate, if one is dealing with an infinite crystal — that the energy spectrum consists of bands.[57] This means that there exist energy values E_0, E_1, E_2, $\cdots$ such that all energies satisfying the inequalities

$$E_0 \leqslant E \leqslant E_1, \qquad E_2 \leqslant E \leqslant E_3, \qquad E_4 \leqslant E \leqslant E_5, \qquad \cdots \qquad (A)$$

are eigenvalues, whereas none of the energies satisfying the inequalities

$$E < E_0, \qquad E_1 < E < E_2, \qquad E_3 < E < E_4, \qquad \cdots \qquad (B)$$

are eigenvalues. The energy values satisfying Eq.(A) are called *allowed* energies and the energy values satisfying Eq.(B) are called *forbidden* energies; one also speaks of *allowed* and *forbidden* bands.

Let us consider the situation at absolute zero. There are two possibilities. The first possibility is that the electrons fill up the lowest bands completely, but only partly fill the last band that is at all occupied: we are dealing with a metal. In this case any amount of energy can remove electrons from the top level occupied at absolute zero. The

---

[56]See, for example, A.H.Wilson, *The Theory of Metals*, Cambridge University Press, 1953; F.Seitz, *The Modern Theory of Solids*, McGraw-Hill, New York, 1940; A.J.Dekker, *Solid State Physics*, Englewood Cliffs, 1957.

[57]We refer to a paper by Kramers (*Physica* 2, 483 (1935)) for a general discussion of this behaviour.

second possibility is that the electrons exactly fill up the lowest bands, but that the next allowed band is completely empty. This is the case of an *intrinsic semiconductor* or an *insulator*. The difference between these two types of materials is only quantitative and depends on the energy gap $\Delta\varepsilon$ between the last filled band and the first empty band. In insulators this gap is of the order of 5 to 10 eV, while in intrinsic semiconductors it is of the order of 1 eV. For instance, the gap in tellurium is 0.4 eV and in germanium 0.76 eV.

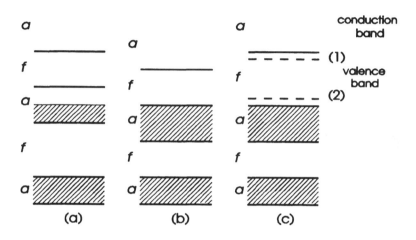

Fig.4.8. The band structures of (a) a metal, (b) an intrinsic semiconductor, and (c) an impurity semiconductor.

Let us now consider the case of a semiconductor, the structure of which is not completely periodic due, for instance, to impurities in the crystal. In that case it is possible that there are a few energy levels that lie in the forbidden band between the last occupied and the first empty bands (see Fig.4.8). These levels can either lie close to the conduction band (Fig.4.8, case (1)), that is, the first band in which there are empty levels, or close to the valence band (Fig.4.8, case (2)). In the first case they are called *donor levels*. Only a relatively small amount of energy will lift electrons from a donor level into the conduction band. Such a case will occur when germanium, for instance, is doped with an element from the fifth column in the periodic table, such as phosphor. In the second case they are called *acceptor levels*. Only a relatively small amount of energy is needed to lift electrons from the valence band into an acceptor level, leading to hole conduction. We must note that usually the wavefunctions corresponding to the donor and the acceptor levels are localised. Indeed, in Fig.4.8, in which we have indicated the case of a metal (a), of an intrinsic semiconductor (b), and of an impurity semiconductor (c), and where a and f indicate, respectively, allowed and forbidden bands, we have given two-fold information. First of all, one can read from the figure the values of the eigenenergies. However, this could also be done from a figure such as given by Fig.4.9 without leading to confusion.[58] Secondly, one gives information about the wavefunctions corresponding to the various values of the energy: if the shaded areas, indicating the levels occupied at absolute zero, or the empty areas of the allowed energy bands extend all the way, it is implied that the corresponding wavefunction has an appreciable amplitude throughout that region of space, while the extent of the lines indicating the impurity levels is limited, corresponding to the localised nature of the corresponding wavefunctions. This kind of information is particularly useful when one considers the band structure near the surface of a solid or near the interface of two solids in contact.[59]

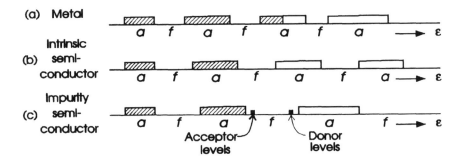

Fig.4.9. The energy spectrum of (a) a metal, (b) an intrinsic semiconductor, and (c) an impurity semiconductor.

For the discussions in the problems of the present chapter we shall use simplified models, both of a metal and of a semiconductor. Our model of a metal is the following one (see Fig.4.10). We shall assume that the conduction electrons are moving as free, independent, non-interacting particles in a constant negative potential, $-\chi$, produced by the metallic ions. Our problem is thus, first of all, to derive an expression for the energy levels of an electron when its potential energy has the form given in Fig.4.10.[60] Actually, we need only the energy level density, and it turns out that, if we take as zero of energy the value of the potential

---

[58]One sees from Fig.4.9 that the energy levels corresponding to the impurity levels, although often lying very closely — much more closely in fact than is indicated in Fig.4.9 — are different and not necessarily all equal as one might expect from Fig.4.8.

[59]See, for instance, Dekker, *loc.cit.*.

[60]We are not concerned with any detailed features of the potential energy and have therefore assumed U to be constant throughout the metal and to change discontinuously to zero at the metal surface. In actual cases there will be a

energy inside the metal, this density is to a very good approximation the same as that of an electron in a volume V with rigid walls, as long as we are only interested in bound levels. We will therefore take for the number of levels, dZ, with energies between ε and $\varepsilon + d\varepsilon$, where ε is reckoned from the bottom of the potential well, expression (4.604), multiplied by 2 to take the electron spin ito account.

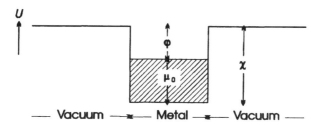

Fig.4.10. The electron Fermi sea in a metal at absolute zero; φ is the work function.

We saw in §4.8 that for a gas of conduction electrons we are dealing with a strongly degenerate gas, so that to a first approximation all levels up to a distance μ_0 from the potential well are occupied. This is again indicated by shading in Fig.4.10. The work function φ is defined by the equation

$$\varphi = \chi - \mu_0. \tag{C}$$

The work function is the energy one needs give to an electron at the top of the "Fermi sea" in order to remove it from the metal.

The model we shall use for an impurity semiconductor is one due to Wilson.[61]   In this model (see Fig.4.11) we assume that we may neglect the influence of the occupied bands. Moreover, we assume that there is a conduction band for $\varepsilon \geqslant 0$ which is completely unoccupied at absolute zero and that there are N_b impurity levels, all situated at $\varepsilon = -\Delta\varepsilon$, which are competely occupied at absolute zero. This can be expressed by writing for the energy level density

$$Z(\varepsilon) = N_b\,\delta(\varepsilon + \Delta\varepsilon), \qquad \varepsilon < 0,$$

where $\delta(\varepsilon)$ is the Dirac δ-function. For the energy level density in the conduction band we use again expression (4.604), multiplied by 2 to take the electron spin into account.

Finally, we shall use for an intrinsic semiconductor a model in which the energy level density in the valence band is given by the expression

$$dZ = 4\pi \left(\frac{2m_h}{h^2}\right)^{3/2} V \sqrt{-\varepsilon}\, d\varepsilon, \qquad \varepsilon < 0, \tag{D}$$

---

smooth transition from $-\chi$ to 0, but the simplified potential of Fig.4.10 is sufficient for our purposes.

[61] A.H.Wilson, *Proc. Roy. Soc. (London)* **A133**, 458 (1931); **A134**, 277 (1931).

where we have assumed the top of the valence band to be at $\varepsilon = 0$ and where m_h is the (effective) hole mass, while the energy level density in the conduction band is given by the expression

$$dZ = 4\pi \left(\frac{2m_{el}}{h^2}\right)^{3/2} V \sqrt{\varepsilon - \Delta}\, d\varepsilon, \qquad \varepsilon > \Delta, \qquad (E)$$

where the bottom of the conduction band is at $\varepsilon = \Delta > 0$ and where m_{el} is the (effective) electron mass.

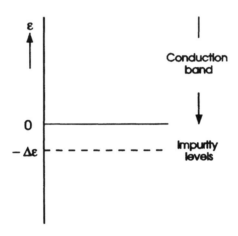

Fig.4.11. Wilson's model of an impurity semiconductor in which there are a number of impurity levels at a depth $\Delta\varepsilon$ below the conduction band.

20. Find for an intrinsic semiconductor (i) the number of electrons in the conduction band as a function of the temperature and (ii) the Fermi level of the semiconductor as a function of m_h/m_{el}, assuming $\beta\Delta \gg 1$.

21. Find for an impurity semiconductor (i) the number of electrons in the conduction band and (ii) the Fermi level of the semiconductor, assuming $\beta\Delta\varepsilon \gg 1$. Find also the distribution of the electrons in the conduction band over the various energies.

22. Consider an impurity semiconductor with a large number of impurity levels and a small gap. Take as an example $N_b/V = 10^{24}$ m^{-3} and $\Delta\varepsilon = 10^{-3}$ eV or 10^{-5} eV. Determine the temperature range in which one cannot treat the gas of conduction electrons as a classical gas, and discuss why at both sufficiently high and at sufficiently low temperatures the gas becomes classical.

23. Find an expression for the Richardson current,[62] that is, the current of the so-called thermionic emission which arises because at finite temperatures

[62] O.W.Richardson, *Phil. Mag.* **23**, 594 (1912).

there will be a certain number of electrons with sufficient energy to leave the metal. Assume that $\beta\varphi \gg 1$ and that the electron gas is strongly degenerate.

24. A metal is illuminated by light of frequency ω. Find the photocurrent, assuming that the components of the electron momentum parallel to the metal surface remain unchanged by the absorption of a photon.

25. Consider the Lorentz model of the transport properties of metals (see Problem 4 in Chapter 2). Taking for f_0 the Fermi-Dirac distribution instead of the Maxwell distribution as was done in Problems 4 to 7 in Chapter 2, evaluate (a) the electrical conductivity, (b) the thermal conductivity, (c) the isothermal Hall coefficient, (d) the Wiedemann-Franz ratio, and (e) the magnetoresistance of a metal.

26. Calculate the spin paramagnetic susceptibility of a system of free electrons (i) at $T = 0$ and (ii) at a finite temperature assuming that $\beta\mu \gg 1$ (Pauli paramagnetism[63]).

27. Given that the energy levels of an electron moving freely in a uniform magnetic field B along the z axis are given by the equation[64]

$$E = \left(n + \tfrac{1}{2}\right)\hbar\omega + \frac{\hbar^2 k_z^2}{2m}, \qquad \omega = \frac{eB}{m},$$

and that the number of levels for a given value of n and with k_z in the range $k_z, k_z + dk_z$ is equal to

$$\frac{V\, eB\, dk_z}{2\pi^2\hbar},$$

find, using the summation formula

$$\sum_{n=0}^{m} g_{n+\frac{1}{2}} \approx \int_0^{m+1} g(x)\, dx - \frac{1}{24}\left[g'(x)\right]_0^{m+1},$$

an expression for the orbital diamagnetism of a strongly degenerate free electron gas in a volume V, assuming that $\beta\mu_B B \ll 1$, where μ_B is the Bohr magneton (Landau diamagnetism[65]).

28. Use Eq.(1.717) as the definition of Boltzmann's H-function and expression (4.306) for ln W; prove that

$$\frac{dH}{dt} \leqslant 0,$$

[63]W.Pauli, Z. Physik 41, 81 (1926).

[64]See, for instance, D.ter Haar, Men of Physics: L.D.Landau, Vol.I, Pergamon, Oxford, 1965, p.31.

[65]L.D.Landau, Z.Physik 64, 629 (1930).

if the average number of transitions $N_{ij \to i'j'}$ per unit time where the groups Z_i and Z_j each lose one particle, while at the same time the groups $Z_{i'}$ and $Z_{j'}$ each gain one particle, is given by the equation

$$N_{ij \to i'j'} = A_{ij \to i'j'} N_i N_j \left(Z_{i'} + \gamma N_{i'} \right) \left(Z_{j'} + \gamma N_{j'} \right),$$

where we assume that

$$A_{ij \to i'j'} = A_{i'j' \to ij}.$$

29. Find the condition that $dH/dt = 0$ from the result of the preceding problem; derive Eqs.(4.314) to (4.316).

Bibliographical Notes

General references for this chapter are
1. G.E.Uhlenbeck, *Over Statistische Methoden in de Theorie der Quanta*, Thesis, Leiden, 1927.
2. P.Jordan, *Statistische Mechanik auf Quantentheoretischer Grundlage*, Vieweg, Braunschweig, 1933.
3. E.Schrödinger, *Statistical Thermodynamics*, Cambridge University Press, 1948.

Section 4.1. The Bose-Einstein statistics were first introduced by Bose using a method which differs only slightly from the elementary method:

4. S.N.Bose, *Z. Physik* **26**, 178 (1924).

This paper had been translated by Einstein, who added the following note: "Boses Ableitung der Planckschen Formel bedeutet nach meiner Meinung einen wichtigen Fortschritt. Die hier benutzte Methode liefert auch die Quantentheorie des idealen Gases, wie ich an anderer Stelle ausführen will."[66]

This application to the case of a perfect gas appeared the same year:

5. A.Einstein, *Berliner Ber.* **1924**, 261; **1925**, 3, 18.

Following the introduction of the exclusion principle by Pauli in 1925,

6. W.Pauli, *Z. Physik* **31**, 776 (1925),

Fermi introduced it into the statistical discussion:

7. E.Fermi, *Z. Physik* **36**, 902 (1926).

The connection between statistics and wave mechanics was investigated by Dirac:

8. P.A.M.Dirac, *Proc. Roy. Soc. (London)* **A112**, 661 (1926).

Section 4.2. The fact that the omission of the factor $N!$ is essential in order that S be an extensive quantity was noted by Gibbs. See, for instance,

---

[66] Bose's derivation is, in my opinion, an important step forward. The method used here also gives a quantum theory of the perfect gas, as I hope to show elsewhere.

9. J.W.Gibbs, *Elementary Principles in Statistical Mechanics* (Vol.II of his Collected Works), Yale University Press, New Haven, Conn., 1948, Chap.15.
Compare also

10. L.Nordheim, *Z. Physik* **27**, 65 (1924).

11. G.S.Rushbrooke, *Introduction to Statistical Mechanics*, Oxford University Press, 1949, pp.36ff.

For the connection between accessible states and probabilities, see Refs.1 and 8. A discussion about the splitting up of the $N!$ possible wavefunctions according to their symmetry properties was given by Wigner:

12. E.Wigner, *Z. Physik* **40**, 492 (1926); 883 (1927).

Sections 4.3 and 4.4. We have used here notes of lectures given by H.A.Kramers in Leiden during the winter of 1944–1945 for a small circle of physicists. In these lectures Kramers introduced the q-potential.

Sections 4.5. The Darwin-Fowler method was first developed in the following papers:

13. C.G.Darwin and R.H.Fowler, *Phil. Mag.* **44**, 450, 823 (1922).

14. C.G.Darwin and R.H.Fowler, *Proc. Cambridge Phil. Soc.* **21**, 262 (1922).

15. R.H.Fowler, *Phil. Mag.* **45**, 1, 497 (1923).

16. C.G.Darwin and R.H.Fowler, *Proc. Cambridge Phil. Soc.* **21**, 391, 730 (1923).

17. R.H.Fowler, *Proc. Cambridge Phil. Soc.* **22**, 861 (1925).

18. R.H.Fowler, *Phil. Mag.* **1**, 845 (1926).

19. R.H.Fowler, *Proc. Roy. Soc.* (*London*) **A113**, 432 (1926).
See also Refs.1 and 3 and

20. R.H.Fowler, *Statistical Mechanics*, Cambridge University Press, 1936.

Sections 4.6 to 4.8. These sections are applications of the formulæ developed in the earlier sections. Compare also Refs.5 and 7 and

21. E.H.Kennard, *Kinetic Theory of Gases*, McGraw-Hill, New York, 1938, pp.397ff.

For a discussion of the Einstein condensation see Ref.5 and

22. D.ter Haar, *Elements of Statistical Mechanics*, Rinehart, New York, 1954 (and the references given there on p.224).

23. F.London, *Phys. Rev.* **54**, 947 (1938).

24. R.H.Fowler and H.Jones, *Proc. Cambridge Phil. Soc.* **34**, 573 (1938).

25. H.Wergeland and K.Hove-Storhoug, *Proc. Norwegian Acad. Sci.*, **15**, 131, 135, 181 (1943).

26. R.Becker, *Z. Physik* **128**, 120 (1950).

27. D.ter Haar, *Proc. Roy. Soc.* (*London*) **212**, 552 (1952).

Section 4.9. Leinaas and Myrheim showed in 1977 that fractional statistics can occur for two-dimensional systems:

28. J.M.Leinaas and J.Myrheim, *Nuovo Cim.* **37B**, 1 (1977).
The term "anyons" was coined by Wilczek

29. F.Wilczek, *Phys. Rev. Lett.* **49**, 957 (1982).

A thorough review of anyon theory can be found in Myrheim's lecture notes:

30. J.Myrheim, *Anyons*, Preprint Theoretical Physics Seminar in Trondheim, No 13 (1993).

    See also

31. I.J.R.Aitchison and N.E.Mavromatos, *Contemp. Phys.* **32**, 219 (1991).

    In this paper the connection between fractional statistics and the Aharanov-Bohm effect is extremely clearly explained and the possible connection between fractional statistics and high-temperature superconductivity and the fractional quantum Hall effect is discussed.

    For a discussion of the applicability of quantum statistics to gases see

32. D.ter Haar, *Am. J. Phys.* **17**, 399 (1949).

    For an application of Bose-Einstein or Fermi-Dirac statistics to the equation of state see

33. J.de Boer, *Contribution to the Theory of Compressed Gases*, Thesis, Amsterdam, 1940.

34. J.de Boer, J.van Kranendonk, and K.Compaan, *Physica* **16**, 545 (1950).

    For the statistical theory of metals see

35. A.Sommerfeld and H.Bethe, *Handb. d. Physik* **24$_2$**, 333 (1933).

    For a discussion of statistics of nuclear particles see

36. H.A.Bethe, *Rev. Mod. Phys.* **9**, 69 (1937).

37. H.Wergeland, *Fra Fysikkens Verden* **1945**, 223.

    For astrophysical applications see

38. S.Chandrasekhar, *Introduction to the Study of Stellar Structure*, University of Chicago Press, 1939, where further references can be found.

39. L.Biermann, *Z. Naturf.* **3a**, 481 (1948).

    We finally refer to part C of Ref.22 for several applications in various fields of physics and also to

40. R.Balian, *Statistical Mechanics*, Springer, Berlin, 1991.

CHAPTER 5

CLASSICAL ENSEMBLES

5.1.   The Γ-Space; Ensembles

In the preceding chapters we have discussed systems that consisted of independent particles and which, moreover, existed isolated from the rest of the universe. We have thus in our treatment made various simplifications. Firstly, except when discussing Boltzmann's H-theorem, where collisions — that is, interactions between particles — play an essential rôle, we neglected interactions within the system. Secondly, we neglected interactions with the surroundings. Thirdly, we neglected any experimental limitations to the amount of information that can be obtained about a physical system. The first point will be dealt with in the present section, where we shall discuss how one can treat systems of interacting particles.

The second point is important in connection with the introduction of the temperature concept. This concept entails necessarily a means of fixing this temperature which, in turn, involves the introduction of a large thermostat in which the system is embedded. However, we then no longer have an isolated system, and we must expect that due to the interaction between system and thermostat, or *temperature bath* as we shall often call the thermostat, the energy of the system is no longer constant. Indeed, there exists a certain *complementarity* between energy and temperature[1] We shall see later (§§ 5.4 and 5.13) that, if we fix the temperature, the energy may have various values which spread around a most probable value — but in most physical cases the actual dispersion is so small as to be negligible — and that, on the other hand, if we are dealing with a really isolated system, we must expect that the temperature measured in the system will be fluctuating.

The third point which has not yet been considered with sufficient care is the fact that there are very serious limitations to the amount of information we can obtain about a physical system that we are considering. We have mentioned, but not sufficiently stressed, that in practice only a few quantities such as the pressure, the total kinetic energy, the total linear mo-

---

[1] It must be emphasised that this complementarity is a classical one and does not involve the quantum of action.

mentum, and the total angular momentum of the system can be measured. In the previous chapters our discussion was such that one might get the impression that it would be possible, if we wished to do so, to determine the distribution function to any degree of detailed accuracy, and hence calculate all average values in which we might be interested. Since only a few data will be available, we are confronted with the problem of determining the distributioon function under those circumstances. This question will be discussed in some detail in §§ 5.13 and 5.14.

Let us discuss first the question of interactions between the particles in the system. We shall briefly consider under what circumstances our neglect of these interactions would be permissible. In the Bibliographical Notes to Chapter 1 we said that a perfect gas, that is, a gas where one can neglect the interactions, would be a system where the total potential energy (or interaction energy) is at all times negligibly small as compared to the total kinetic energy. Using the language of quantum mechanics, we can put this slightly differently. The total energy of an isolated system may be given within certain limits, that is, we may know that the energy of the system lies in the interval $E - \delta, E + \delta$. If it is still to have any sense at all to talk about a system with a fixed energy, δ must be small as compared to E.

> The fact that the energy of a system can never be completrely fixed is, of course, an essential point in quantum mechanics. In order to fix our ideas we may, for instance, take for δ the natural width of the energy of the system. Or, if Δt is the smallest time interval during which observations can be made, we can take δ in accordance with the Heisenberg relations to satisfy the equation
>
> $$\delta \sim \frac{h}{\Delta t}.$$

If, as is mostly the case and as we shall assume to be the case unless we state it not to be so, the system consists of many particles, the spacing ΔE of the energy levels of the particles — treated as independent entities — will be very small (compare the remarks in § 3.5) and we will have $\Delta E \ll \delta$. Now, if the interactions between the particles are so weak that the total interaction energy U is small as compared to δ, we can still treat the system as consisting of independent particles. However, in order that transitions between various states can occur at all, U must be larger than ΔE, since otherwise practically no transitions would occur. We have thus the inequalities

$$\Delta E \ll U \ll \delta \ll E. \tag{5.0101}$$

One way to introduce the interactions between the particles in the system is by treating the whole system as one large molecule. If each particle has s degrees of freedom and if N is the number of particles in the system, the total number of degrees of freedom for the whole system will be sN. The state of the system will be completely determined by $2sN$ coordinates, such as

$$q_1^1, \quad q_2^1, \quad q_3^1, \quad \cdots, \quad q_s^1; \quad p_1^1, \quad p_2^1, \quad p_3^1, \quad \cdots, \quad p_s^1;$$
$$q_1^2, \quad q_2^2, \quad q_3^2, \quad \cdots, \quad q_s^2; \quad p_1^2, \quad p_2^2, \quad p_3^2, \quad \cdots, \quad p_s^2;$$

$$\cdots\cdots\cdots\cdots\cdots\cdots\cdots\cdots\cdots\cdots\cdots\cdots\cdots$$

$$q_1^N, q_2^N, q_3^N, \quad \cdots, \quad q_s^N; \quad p_1^N, p_2^N, p_3^N, \quad \cdots, \quad p_s^N,$$

where the p_k^j and q_k^j are the canonical momenta and coordinates of the j-th particle.[2] Let $\mathcal{T}(p,q)$ be the total kinetic energy and $U(q)$ the total potential energy of the system, where p and q stand, respectively, for all sN p's and all sN q's. We shall only consider conservative systems so that neither $\mathcal{T}$ nor U depend explicitly on the time, and we shall assume that the forces in the system are velocity-independent.[3] The total energy $\mathcal{H}$ is given by the equation

$$\mathcal{H} = \mathcal{T} + U = \mathcal{H}(p,q). \tag{5.0102}$$

It is advantageous to number the p's and q's continuously from 1 to sN, and we shall assume that this has been done, unless stated otherwise. Since the p's and q's are canonically conjugate variables, we have (compare also Eq.(2.207))

$$\dot{p}_i = -\frac{\partial \mathcal{H}}{\partial q_i}, \qquad \dot{q}_i = \frac{\partial \mathcal{H}}{\partial p_i}; \qquad i = 1, 2, \cdots, sN. \tag{5.0103}$$

To each system there corresponds for each value of the time a point, the so-called *representative point*, in the $2sN$-dimensional phase space. Following the Ehrenfests' notation, this space will be called Γ-space where Γ stands for gas.

From theoretical mechanics it is known that for the set of Eqs.(5.0103) there are $2sN$ functions $\varphi_i(p,q)$ with the following properties. Of the $2sN$ functions, one is the total energy ε of the system, one is — apart from a constant — the time t, and the others are constant:

$$\left.\begin{array}{ll}
\varphi_1(p,q) & = \varepsilon(p,q) = c_1, \\
\varphi_2(p,q) & = c_2, \\
\cdots\cdots\cdots\cdots\cdots\cdots\cdots\cdots & \\
\varphi_{2sN-1}(p,q) & = c_{2sN-1}, \\
\varphi_{2sN}(p,q) & = c_{2sN} + t,
\end{array}\right\} \tag{5.0104}$$

---

[2] It should be remarked here that the $2sN$ coordinates need not be the coordinates and momenta of the individual particles. We shall only consider canonical variables so that we have sN q's and sN canonically conjugate p's. The q's can, for instance, be any sN quantities which completely determine the position of all the particles in the system. The p's are then determined from the kinetic energy (compare Eq.(2.205)).

[3] This last assumption is not essential and one can easily generalise to the case where the forces are velocity-dependent.

where the c_i are the $2sN$ integration constants. The c_i $(i = 1, 2, \cdots, 2sN)$ or the corresponding φ_i are called constants of motion or integrals of the motion.

That one can put the integration of Eqs.(5.0103) in the form (5.0104) follows, if one divides the first $2sN - 1$ equations of motion by the last one. The resulting $2sN - 1$ equations do not contain the time and give rise to the first $2sN - 1$ equations of the set (5.0104). One is then left with one equation and from that it is possible, after substitution, to determine t by a quadrature, and the last of the equations (5.0104) results.

As time goes on, the representative point of the system will describe a trajectory in Γ-space. The q_i change their values because of the velocities of the particles, and the p_i because of the external and internal forces acting on the particles. The trajectory will be situated on the hypersurface with equation $\varepsilon(p, q) = c_1$. This hypersurface will be called an *energy surface*.

Instead of considering one system it is often advantageous and even unavoidable — as we shall discuss in § 5.13 — to consider a large number of similar systems that all contain N particles, each of which has s degrees of freedom and which all exist under the same external circumstances. More precisely, all these systems will possess the same Hamiltonian (5.0102), but their representative points will in general not be the same, or, put differently, they will correspond to different sets of the constants of motion c_i. Let us assume that there are so many systems that it is possible to speak of the density $D(p, q)$ with which the representative points are distributed in Γ-space. Such a collection of similar systems was called by Gibbs an *ensemble*.

Since the representative points are moving in phase space, the density will in general be a function of the time. Let the density, that is, the number of representative points per unit volume of Γ-space, at time t be denoted by $D(p, q, t)$. In considering the change of D with time it is important to distinguish between its rate of change at a fixed point in Γ-space and the rate of change at a point moving along a trajectory in Γ-space. The first of these — which we may call the local rate of change — will be denoted by $\partial D/\partial t$, while for the second, or the convected rate of change as it may be called, we use the symbol dD/dt. The relation between the two is given by the equation

$$\frac{dD}{dt} = \frac{\partial D}{\partial t} + \sum_i \left(\frac{\partial D}{\partial p_i} \dot{p}_i + \frac{\partial D}{\partial q_i} \dot{q}_i \right). \qquad (5.0105)$$

Let us consider for a moment a volume element $d\Omega$ in Γ-space corresponding to coordinates in the ranges p_i, $p_i + dp_i$ and q_i, $q_i + dq_i$, so that we have

$$d\Omega = \prod_i dp_i \, dq_i. \qquad (5.0106)$$

Elementary volumes in Γ-space, $d\Omega$, will sometimes be called, following Gibbs, elements of *extension in phase*.

At a later moment, $t + \delta t$, the representative points which at t were inside $d\Omega$ will have moved along their trajectories and their coordinates which were p_i, q_i at t will now be $p_i + \dot{p}_i \,\delta t$, $q_i + \dot{q}_i \,\delta t$. Let $d\Omega'$ be the extension in phase at $t + \delta t$. We shall show presently that

$$\frac{dD}{dt} = 0. \tag{5.0107}$$

This means that the *density in phase*, as D is called by Gibbs, remains constant in the neighbourhood of a point as it moves along its trajectory, or, that we may treat the points in Γ-space as an incompressible fluid. As a result of this we find that

$$d\Omega = d\Omega', \tag{5.0108}$$

expressing the fact that the extension in phase occupied by a collection of representative points is constant in time. Equations (5.0107) and (5.0108) are two different forms of *Liouville's theorem*.

We can prove Eq.(5.0107) as follows by considering the right-hand side of Eq.(5.0105) (compare the derivation of Eq.(2.507)). Consider the extension in phase $d\Omega$ at t and at $t + \delta t$. The number of representative points N in $d\Omega$ at t will be given by the equation

$$N = D\,d\Omega. \tag{5.0109}$$

At $t + \delta t$ we have

$$N + \delta N = \left(D + \frac{\partial D}{\partial t}\,\delta t \right) d\Omega. \tag{5.0110}$$

The difference δN arises from the fact that the number of points entering $d\Omega$ during δt may be different from the number of points leaving $d\Omega$ during this time interval. The number of points entering $d\Omega$ during δt through the "face" at q_i will be given by the expression

$$D\dot{q}_i \frac{d\Omega}{dq_i}\,\delta t, \tag{5.0111}$$

while the number of points leaving $d\Omega$ during δt through the "opposite face" at $q_i + dq_i$ is given by the expression

$$\left(D + \frac{\partial d}{\partial q_i}\,dq_i \right) \left(\dot{q}_i + \frac{\partial \dot{q}_i}{\partial q_i}\,dq_i \right) \frac{d\Omega}{dq_i}\,\delta t. \tag{5.0112}$$

In expressions (5.0111) and (5.0112) $d\Omega/dq_i$ stands for the "area" of the "face", that is, $d\Omega$ divided by dq_i, and not for a derivative.

Combining expressions (5.0111) and (5.0112), summing over all $2sN$ coordinates (q_i and p_i), and neglecting higher-order terms, we get

$$\delta N \; = \; - \sum_i \left[D \left(\frac{\partial \dot{q}_i}{\partial q_i} + \frac{\partial \dot{p}_i}{\partial p_i} \right) + \left(\frac{\partial D}{\partial q_i} \, \dot{q}_i + \frac{\partial D}{\partial p_i} \, \dot{p}_i \right) \right] \, d\Omega \, \delta t,$$

or, using Eqs.(5.0109) and (5.0110),

$$\frac{\partial D}{\partial t} \; = \; - \sum_i \left(\frac{\partial D}{\partial q_i} \, \dot{q}_i + \frac{\partial D}{\partial p_i} \, \dot{p}_i \right), \tag{5.0113}$$

since from Eqs.(5.0103) it follows that

$$\frac{\partial \dot{q}_i}{\partial q_i} + \frac{\partial \dot{p}_i}{\partial p_i} \; = \; 0. \tag{5.0114}$$

Equation (5.0107) follows from Eqs.(5.0105) and (5.0113).

If we use Eqs.(5.0103), we can write Eq.(5.0113) also in the following form:

$$\frac{\partial D}{\partial t} \; = \; \{\mathcal{H}, D\}, \tag{5.0115}$$

which expresses the local rate of change of D in terms of the *Poisson bracket* of D and the Hamiltonian.

The Poisson bracket $\{a, b\}$ is defined by the equation

$$\{a, b\} \; = \; \sum_i \left[\frac{\partial a}{\partial q_i} \frac{\partial b}{\partial p_i} - \frac{\partial a}{\partial p_i} \frac{\partial b}{\partial q_i} \right].$$

An ensemble for which $\partial D / \partial t = 0$ is called a *stationary ensemble* and in the following we shall restrict ourselves mainly to that kind of ensemble. From Eq.(5.0115) and the definition of a stationary ensemble it follows that an ensemble is stationary if

$$\{\mathcal{H}, D\} \; = \; 0. \tag{5.0116}$$

It is shown in theoretical mechanics that Eq.(5.0116) is satisfied if and only if

$$D \; = \; D(\varphi_1, \varphi_2, \cdots, \varphi_{2sN-1}), \tag{5.0117}$$

where the φ_i are the functions defined by Eqs.(5.0114).

Instead of considering stationary ensembles with a density given by the general equation (5.0117), we shall restrict ourselves mainly to those stationary ensembles for which

$$D \; = \; D(\varepsilon), \tag{5.0118}$$

where $\varepsilon = \varphi_1$ is the energy of the system. The reasons for this restriction will be discussed in §5.13.

Instead of working with D we shall mostly use the normalised *coefficient of probability* ϱ, which is given by the equation

$$\varrho = \frac{D}{n}, \tag{5.0119}$$

where n is the total number of systems in the ensemble which is assumed to be large and which is given by the equation

$$n = \int_\Gamma D \, d\Omega, \tag{5.0120}$$

and where in Eq.(5.0120) and henceforth the symbol $\int_\Gamma$ will indicate integration over the whole of the available Γ-space.[4]

We shall often refer to ϱ as the *density of the ensemble*, which should not be confused with the density in phase D.

With Gibbs we also introduce a function η, the *index of probability*, given by the equation

$$\eta = \ln \varrho. \tag{5.0121}$$

As a consequence of Eq.(5.107) we have

$$\frac{d\varrho}{dt} = \frac{d\eta}{dt} = 0, \tag{5.0122}$$

while for stationary ensembles we have

$$\frac{\partial\varrho}{\partial t} = \frac{\partial\eta}{\partial t} = 0. \tag{5.0123}$$

We shall mainly be concerned with ensembles for which Eq.(5.0118) holds, and in that case we have

$$\varrho = \varrho(\varepsilon), \qquad \eta = \eta(\varepsilon). \tag{5.0124}$$

For future reference we remark that ϱ must be a single-valued, real, and non-negative function of the p's and q's which by virtue of Eqs.(5.0119) and (5.0120) satisfies the *normalisation equation*

$$\int_\Gamma \varrho \, d\Omega = 1. \tag{5.0125}$$

---

[4] In most cases, some parts of Γ-space will not be available due to external circumstances or to our choice of coordinates. For instance, in the case of a gas enclosed in a finite volume, the x_i, y_i, z_i can range only over a finite range, while a q_i which is an angle is restricted to the interval 0 to 2π.

Any function $G(p, q)$ of the p_i and q_i is called a *phase function* and its average value $\overline{G}$, where the average is taken over the ensemble, is defined by the equation

$$\overline{G} = \int_{\Gamma} G\varrho \, d\Omega. \tag{5.0126}$$

If we are dealing with a stationary ensemble, $\overline{G}$ will be a constant, that is, independent of the time, provided G does not contain the time explicitly.

We want to indicate briefly, and very incompletely, how one can derive the Boltzmann transport equation (2.507) from the Liouville equation in the form

$$\frac{d\varrho^{(N)}}{dt} = 0, \tag{5.0127}$$

where we have emphasised that we are dealing with a system consisting of N particles — all of which we assume to be identical — by the addition of a superscript on ϱ.

We now consider the case where the only degrees of freedom are those of the centre of mass; that is, we consider the case of point particles without internal degrees of freedom. In that case we can write Eq.(5.0127) in the form

$$\frac{\partial \varrho^{(N)}}{\partial t} + \sum_{j} \left(\nabla_{pj} \mathcal{H} \cdot \nabla_j \varrho^{(N)} \right) - \sum_{j} \left(\nabla_j \mathcal{H} \cdot \nabla_{pj} \varrho^{(N)} \right) = 0, \tag{5.0128}$$

or

$$\frac{\partial \varrho^{(N)}}{\partial t} + \sum_{j} \left[\frac{1}{m} \left(p_j \cdot \nabla_j \varrho^{(N)} \right) + \left(\left\{ F_j^{\text{ext}} + F_j^{\text{int}} \right\} \cdot \nabla_{pj} \varrho^{(N)} \right) \right] = 0, \tag{5.0129}$$

where we have used the canonical equations of motion (5.0103) and the fact that for point particles the kinetic energy is a sum of terms $p_j^2/2m$, while the potential energy consists of two parts, one part corresponding to external forces, F_j^{ext}, and the other part corresponding to the interparticle forces, F_j^{int}. To simplify our notation we have indicated by ∇_p the vector operator with components $\partial/\partial p_x$, $\partial/\partial p_y$, and $\partial/\partial p_z$.

We now introduce the single-particle distribution function $f(r, p)$ by the equation

$$f(r, p) = \int \varrho^{(N)} \, d^3r_2 \cdots d^3r_N \, d^3p_2 \cdots d^3p_N. \tag{5.0130}$$

We first of all note that it is immaterial over which $N - 1$ of the N particles we integrate, since all particles are identical. Secondly, we note that we are here, in contradistinction to § 2.3, considering f as a function of r and p, rather than of r and c. Finally, we note that it follows from Eqs.(5.0125) and (5.0130), where the integration is over all allowable values of $r_2, \cdots, p_N$, that $f(r, p)$ is normalised.

To find the equation to be satisfied by $f(r, p)$ we must integrate Eq.(5.0129) over $N - 1$ particles. Before doing that, we shall introduce one further assumption — that the forces between the particles are two-particle forces so that

$$F_j^{\text{int}} = \sum_{k(\neq j)} F_{jk}^{\text{int}}.$$

Because of the identical nature of the particle and the fact that f will vanish as $|p| \to \infty$ and at the walls of the vessel containing the system, we find after integrating

$$\frac{\partial f_1}{\partial t} + \frac{1}{m}\left(p_1 \cdot \nabla_1 f_1\right) + \left(F_1^{\text{ext}} \cdot \nabla_{p1} f_1\right) = -\int \left(F_{12}^{\text{int}} \cdot \nabla_{p1} f_{12}^{(2)}\right) d^3r_2\, d^3p_2,$$

$$(5.0131)$$

where $f_{12}^{(2)}$ is the two-particle distribution function defined by the equation

$$f_{12}^{(2)} = \int \varrho^{(N)}\, d^3r_3 \cdots d^3r_N\, d^3p_3 \cdots d^3p_N. \qquad (5.0132)$$

We note that Eq.(5.0131) already begins to look like the Boltzmann equation (2.507). Kirkwood has shown how a time-averaging procedure can reduce the right-hand side of Eq.(5.0131) to the collisional integral of Eq.(2.507).[5] Since the analysis is too complicated to be reproduced here, we merely draw attention to the fact that integrating over time will produce something like the difference between $f_{12}^{(2)}$ before and after a collision, while if there were no correlations between the particles we would expect that $f_{12}^{(2)}$ would simply be the product $f_1 f_2$.

5.2.   Stationary Ensembles

In the present section we shall briefly introduce three kinds of stationary ensembles which we shall use in the subsequent discussion.

We mentioned that we would restrict ourselves to ensembles with densities given by Eq.(5.0124), that is, densities depending on the energy only. Since the density must satisfy Eq.(5.0125), we cannot use the following two expressions for ϱ, which are the two simplest possibilities,[6]

$$\varrho = \text{const}, \quad \text{or} \quad \varrho = \text{const} \times \varepsilon. \qquad (5.0201)$$

However, we can try a linear dependence of η on ε,

$$\eta = \beta(\psi - \varepsilon), \qquad (5.0202)$$

---

[5] J.G.Kirkwood, *J. Chem. Phys.* **15**, 72 (1947). See also: N.N.Bogolyubov, *Studies Stat. Mech.* **1**, 1 (1962); E.G.D.Cohen and T.H.Berlin, *Physica* **26**, 717 (1960).

[6] It is, of course, possible that ϱ satisfies Eq.(5.0201) for a restricted range of energy values; examples are given by the microcanonical and by the F-ensembles to be introduced presently.

and we then have for the density

$$\varrho = e^{\beta(\psi-\varepsilon)}, \tag{5.0203}$$

where ψ and β are constants. Gibbs called ensembles the density of which is given by Eq.(5.0203) *canonical ensembles*. In order to distinguish them from the microcanonical ensembles to be introduced presently, we shall call them *macrocanonical ensembles*. The quantity β is related to the temperature, as we shall see in the next section. The inverse of β which Gibbs denoted by Θ is called the *modulus* of the macrocanonical ensemble. The quantity ψ is independent of the p_i and q_i, but depends on β and the external parameters a_i — which Gibbs called the *external coordinates* — on which the enery ε depends, since by virtue of the normalisation condition (5.0125) it is determined by the equation

$$e^{-\beta\psi} = \int_{\Gamma} e^{-\beta\varepsilon} \, d\Omega. \tag{5.0204}$$

We may remark here that there is a strong resemblance between the density of a macrocanonical ensemble, on the one hand, and the distribution function (2.107) of the Maxwell-Boltzmann distribution, on the other hand. It looks as if a macrocanonical ensemble can be considered to be a system of "molecules" that interact weakly and therefore tend toward the Maxwell-Boltzmann distribution. Indeed, it can be shown (see § 5.13) that a system in a thermostat will show the average behaviour of a system belonging to a macrocanonical ensemble.

Another reason why macrocanonical ensembles play a special rôle among the stationary ensembles is that if two of them with the same value of β are coupled together, the resulting ensemble is again a macrocanonical ensemble. We shall return to this point in § 5.6.

Apart from the macrocanonical ensembles we shall introduce two other stationary ensembles, namely, the microcanonical and the F-ensembles.

A *microcanonical ensemble* — sometimes called an *ergodic ensemble* — consists of systems that all have the same energy ε_0; its density is given by the equation

$$\varrho = A \, \delta(\varepsilon - \varepsilon_0), \tag{5.0205}$$

where A is a constant to be determined from the normalisation condition and $\delta(x)$ is Dirac's δ-function.

The microcanonical ensemble corresponds to a surface density $\sigma(p,q)$ which is given by the equation

$$\sigma(p,q) = \frac{1}{Q(p,q)}, \tag{5.0206}$$

where $Q(p,q)$ is given by the equation

$$Q = \sqrt{\sum_{i=1}^{sN} \left[\left(\frac{\partial \mathcal{H}}{\partial p_i}\right)^2 + \left(\frac{\partial \mathcal{H}}{\partial q_i}\right)^2 \right]}. \tag{5.0207}$$

The proof of Eq.(5.0206) is as follows. According to Liouville's theorem the density in Γ-space is invariant if one follows the repesentative points along their trajectories in Γ-space. Consider now a volume element lying between two neighbouring energy surfaces — corresponding to E_0 and $E_0 + \delta E$. The volume element will retain its extension if we follow its points along their trajectories. Since the extension is given by the expression $dS \cdot \delta N$, where dS is a surface element on the energy surface and δN the distance between the two energy surfaces, and since δN will be proportional to $1/Q$, we see that dS will vary as Q. In this way the number of points inside a fixed surface element ($= \sigma \cdot dS$) will remain constant in time if we choose σ according to Eq.(5.0206).

Sometimes it is convenient to define the microcanonical ensemble by the equations

$$\left. \begin{array}{ll} \varrho = A, & \varepsilon_0 \leqslant \varepsilon \leqslant \varepsilon_0 + \Delta, \\ \varrho = 0, & \varepsilon < \varepsilon_0 \quad \text{or} \quad \varepsilon > \varepsilon_0 + \Delta, \end{array} \right\} \tag{5.0208}$$

instead of by Eq.(5.0205). In this case the ensemble is sometimes called an *energy-shell ensemble*. The microcanonical ensemble is really a limiting case of an energy-shell ensemble.

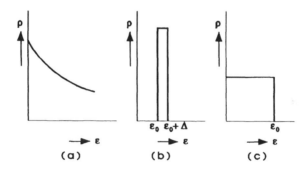

Fig.5.1. Ensemble densities: (a) macro-canonical ensemble; (b) energy-shell ensemble; (c) F-ensemble.

The density of the F-*ensemble* is given by the equation

$$\varrho = B\,F(\varepsilon), \tag{5.0209}$$

where B is to be determined by the normalisation condition and where $F(x)$ is a function defined by the equations

$$\left. \begin{array}{ll} F(x) = 1, & x \leqslant \varepsilon_0; \\ F(x) = 0, & x > \varepsilon_0. \end{array} \right\} \tag{5.0210}$$

In Fig.5.1 we have drawn ϱ as a function of ε for the three ensembles mentioned in this section.

5.3.   The Macrocanonical Ensemble

We mentioned in the preceding section the following theorem. *The macroscopic behaviour of a system at equilibrium at a temperature T will be described correctly by taking the average behaviour of a system in a macrocanonical ensemble with a β that is related to T by the equation*

$$\frac{1}{\beta} = kT. \tag{5.0301}$$

That this is a sensible way of describing the behaviour of a system in temperature equilibrium is due to the following two facts:

(a) The fluctuations of most quantities from the average values are negligibly small — provided the number of particles, N, in each system of the ensemble is large, as we shall assume to be the case.

(b) Those fluctuations which are not small are so fast that they cannot be measured with our instruments.

Fluctuations, that is, deviations from the average value, will be discussed in the next section, when we shall have an opportunity to see how far conditions (a) and (b) are satisfied.

Let us now prove our basic theorem. From the normalisation of ϱ it follows that

$$e^{-\beta\psi} = \int_{\Gamma} e^{-\beta\varepsilon} \, d\Omega. \tag{5.0302}$$

Remembering that ε depends on the external parameters a_i we shall, as in § 2.6, introduce the generalised forces A_i through the equations

$$A_i = -\frac{\partial\varepsilon}{\partial a_i}. \tag{5.0303}$$

If the external parameters a_i are varied from a_i to $a_i + \delta a_i$ we have for the variation $\delta\varepsilon$ of the energy

$$\delta\varepsilon = -\sum_i A_i \, \delta a_i. \tag{5.0304}$$

Let us now consider a new macrocanonical ensemble, with the same value of the number of systems in the ensemble and the same Hamiltonian for the systems in the ensemble, but for which the values of the external parameters in the Hamiltonian as well as the value of β are changed. Since the normalisation condition is the same for this new ensemble, we have

$$-e^{-\beta\psi} \, \delta(\beta\psi) = \int_{\Gamma} e^{-\beta\varepsilon} \left[-\varepsilon \, \delta\beta - \beta \, \delta\varepsilon \right] d\Omega,$$

that is,

$$\delta(\beta\psi) = \int_{\Gamma} e^{\beta(\psi-\varepsilon)} \left[\varepsilon \, \delta\beta + \beta \, \delta\varepsilon \right] d\Omega,$$

or

$$\delta(\beta\psi) \; = \; \bar{\varepsilon}\,\delta\beta + \beta\,\overline{\delta\varepsilon} \; = \; \bar{\varepsilon}\,\delta\beta - \beta \sum_i \overline{A}_i\,\delta a_i, \tag{5.0305}$$

where we have used Eq.(5.0126) in the form

$$\overline{G} \; = \; \int_\Gamma G\,e^{\beta(\psi-\varepsilon)}\,d\Omega. \tag{5.0306}$$

Since

$$\delta\beta\bar{\varepsilon} \; = \; \bar{\varepsilon}\,\delta\beta + \beta\,\delta\bar{\varepsilon}, \tag{5.0307}$$

we can write Eq.(5.0305) in the form

$$\delta\left[\beta\left(\bar{\varepsilon} - \psi\right)\right] \; = \; -\delta\bar{\eta} \; = \; \beta\left(\delta\bar{\varepsilon} - \overline{\delta\varepsilon}\right). \tag{5.0308}$$

Using arguments similar to those of § 4.4 (compare also § 2.7) we find that

(a) $\overline{\delta\varepsilon}$ corresponds to the work done on the system when the external parameters are changed from a_i to $a_i + \delta a_i$;

(b) $\delta\bar{\varepsilon} - \overline{\delta\varepsilon}$ is thus the heat, δQ, added to the system, that is, the increase in energy not accounted for by macroscopically measurable work, this last quantity, δA, being equal to $-\overline{\delta\varepsilon}$;

(c) hence, β, being an integrating factor of δQ, must be inversely proportional to the absolute temperature T, or

$$\frac{1}{\beta} \; = \; kT,$$

which proves Eq.(5.0301);

(d) $-\bar{\eta}$, apart possibly from an additive constant, corresponds to the entropy S divided by k, or

$$S \; = \; k\beta\left(\bar{\varepsilon} - \psi\right) + \text{const} \; = \; -k\bar{\eta} + \text{const}; \tag{5.0309}$$

(e) and, finally, ψ, apart possibly from an additive multiple of the temperature is the (Helmholtz) free energy F.

We shall fix the additive constants by putting

$$e^{\beta F} \; = \; N!\,h^{sN}\,e^{\beta\psi}, \tag{5.0310}$$

where h is Planck's constant. From Eq.(5.0302) we then have

$$e^{-\beta F} \; = \; \frac{1}{N!}\int_\Gamma e^{-\beta\varepsilon}\,\frac{d\Omega}{h^{sN}}, \tag{5.0311}$$

and we see that both sides of Eq.(5.0311) are dimensionless.

The factor $N!$ enters in order that the free energy and the entropy will be extensive quantities and the factor h^{sN} ensures the correct dimensions; together they ensure that the formulæ given here are, indeed, the classical limits of the corresponding quantum-mechanical equations.

From this discussion it follows that once ψ is given as a function of β and the a_i, or β is given and ε is given as a function of the external parameters — in which case ψ can be evaluated from Eq.(5.0302) — we have obtained an expression for the free energy of a system at equilibrium at temperature T for the given values of the a_i. It is well known that once the free energy is given as a function of T and the external parameters, the thermodynamical state of the system is determined. Most thermodynamical quantities can be derived from the free energy by differentiation. As δF and $\delta \psi$ are the same, we should expect relations between the average values of phase functions and partial derivatives of ψ. Indeed, from Eq.(5.0305) we easily have

$$\bar{\varepsilon} \;=\; \frac{\partial(\beta\psi)}{\partial\beta}, \tag{5.0312}$$

which is the Gibbs-Helmholtz equation, and

$$\bar{A}_i \;=\; -\,\frac{\partial\psi}{\partial a_i}. \tag{5.0313}$$

For instance, if a_i corresponds to the volume V, Eq.(5.0313) gives

$$P \;=\; -\,\frac{\partial\psi}{\partial V}, \tag{5.0314}$$

from which the equation of state can be derived once ψ is calculated.

One often denotes the right-hand side of Eq.(5.0311) by Z_Γ and this *partition function* plays a similar rôle to that of Z_μ of Chapter 2. Expressing $\bar{\varepsilon}$, $\bar{A}_i$, and P in terms of Z_Γ we have, instead of Eqs.(5.0312) to (5.0314)

$$\bar{\varepsilon} \;=\; -\,\frac{\partial\ln Z_\Gamma}{\partial\beta}, \tag{5.0315}$$

$$\bar{A}_i \;=\; \frac{1}{\beta}\,\frac{\partial\ln Z_\Gamma}{\partial a_i}, \tag{5.0316}$$

$$P \;=\; \frac{1}{\beta}\,\frac{\partial\ln Z_\Gamma}{\partial V}. \tag{5.0317}$$

Other averages can be calculated by straightforward integration using Eq. (5.0306). In that way we can, for instance, evaluate the average value of the kinetic energy $\mathcal{T}$. This kinetic energy is a homogeneous quadratic expression in the p_i and hence we have (compare Eq.(2.211))

$$\mathcal{T} \;=\; \frac{1}{2}\sum_{i=1}^{sN} p_i\,\frac{\partial\mathcal{T}}{\partial p_i}. \tag{5.0318}$$

By using the same method that led to Eq.(2.212) we find

$$\overline{\mathcal{T}} = \frac{sN}{2\beta} = \frac{1}{2} sNkT, \tag{5.0319}$$

expressing the *equipartition* of kinetic energy. Each degree of freedom contributes $1/2\beta = \frac{1}{2}kT$ to the average kinetic energy.

Let us now consider an ensemble of systems, each containing N independent particles in a volume V. Let ε_j be the total energy of the j-th particle $(j = 1, 2, \cdots, N)$ and p_k^j and q_k^j the k-th generalised momentum and coordinate pertaining to the j-th particle. We then have

$$\varepsilon = \sum_{j=1}^{N} \varepsilon_j \tag{5.0320}$$

and

$$d\Omega = \prod_j d\omega_j, \tag{5.0321}$$

where

$$d\omega_j = \prod_{k=1}^{s} dp_k^j \, dq_k^j. \tag{5.0322}$$

We can now evaluate Z_Γ as

$$Z_\Gamma = \frac{1}{N!} \int_\Gamma e^{-\beta \sum \varepsilon_j} \prod_j \frac{d\omega_j}{h_s} = \left[\frac{e}{N} \int e^{-\beta \varepsilon_j} \frac{d\omega_j}{h^s} \right]^N = Z_\mu^N, \tag{5.0323}$$

where Z_μ is given by Eq.(2.701) with $K = e/Nh^s$ and where we have used Stirling's formula for the factorial (Eq.(1.708)).

If the system consists of free monatomic particles, Z_Γ can be evaluated directly. The energy is given by the equation

$$\varepsilon = \sum_j \frac{p_j^2}{2m} + \sum_j F(x_j, y_j, z_j), \tag{5.0324}$$

where $F(x, y, z)$ is the function defined by Eq.(2.721),

$$\left.\begin{array}{ll} F(x, y, z) = 0, & \text{if } x, y, z \text{ lies inside } V; \\ F(x, y, z) = \infty, & \text{if } x, y, z \text{ lies outside } V. \end{array}\right\} \tag{5.0325}$$

The integration over Γ-space is elementary and leads to the result (compare Eqs.(2.722) and (5.0323))

$$Z_\Gamma = \frac{1}{N!} V^N \left(\frac{2\pi m}{\beta h^2} \right)^{3N/2}. \tag{5.0326}$$

Using Eqs.(5.0315) and (5.0317) we find

$$\bar{\varepsilon} \,(= \,\bar{\mathcal{T}}) \,=\, -\,\frac{\partial \ln Z_\Gamma}{\partial \beta} \,=\, \frac{3N}{2\beta}, \tag{5.0327}$$

$$P \,=\, \frac{1}{\beta}\,\frac{\partial \ln Z_\Gamma}{\partial V} \,=\, \frac{N}{\beta V}. \tag{5.0328}$$

The last equation leads once more, as expected, to the perfect gas law.

5.4.   Fluctuations in a Macrocanonical Ensemble

In the preceding section we derived formulæ for the average values of the total energy and the generalised forces. If we wish to ascertain what the chances are of finding these average values realised when we pick a system at random from the macrocanonical ensemble, we must calculate the fluctuations of these quantities. Doing this we shall at the same time also find formulæ for the correlations between the various quantities.

The easiest way of obtaining formulæ for the fluctuations is to start from the normalising equation in the form

$$\int_\Gamma e^{\beta(\psi-\varepsilon)}\,d\Omega \,=\, 1. \tag{5.0401}$$

Taking the derivative with respect to β we get

$$0 \,=\, \int_\Gamma \frac{\partial}{\partial \beta}\,[\beta(\psi - \varepsilon)]\,e^{\beta(\psi-\varepsilon)}\,d\Omega \,=\, \int_\Gamma \left[\frac{\partial \beta\psi}{\partial \beta} - \varepsilon\right]e^{\beta(\psi-\varepsilon)}\,d\Omega, \tag{5.0402}$$

or

$$\bar{\varepsilon} \,=\, \frac{\partial(\beta\psi)}{\partial \beta}, \tag{5.0403}$$

in accordance with Eq.(5.0312).

Similarly, by taking the derivative with respect to a_i in Eq.(5.0401) we get

$$0 \,=\, \int_\Gamma \frac{\partial}{\partial a_i}\,[\beta(\psi - \varepsilon)]\,e^{\beta(\psi-\varepsilon)}\,d\Omega \,=\, \beta\int_\Gamma \left(\frac{\partial \psi}{\partial a_i} - \frac{\partial \varepsilon}{\partial a_i}\right)e^{\beta(\psi-\varepsilon)}\,d\Omega, \tag{5.0404}$$

or, using Eq.(5.0303),

$$\bar{A}_i \,=\, -\,\frac{\partial \psi}{\partial a_i}. \tag{5.0405}$$

Differentiating Eq.(5.0402) with respect to β gives us

$$\int_\Gamma \left[\frac{\partial^2(\beta\psi)}{\partial \beta^2} + \left\{\frac{\partial(\beta\psi)}{\partial \beta} - \varepsilon\right\}^2\right]e^{\beta(\psi-\varepsilon)}\,d\Omega \,=\, 0, \tag{5.0406}$$

or, using Eq.(5.0403),

$$\overline{(\varepsilon - \bar{\varepsilon})^2} = - \frac{\partial^2 (\beta \psi)}{\partial \beta^2} = - \frac{\partial \bar{\varepsilon}}{\partial \beta}. \tag{5.0407}$$

Similarly, differentiating Eq.(5.0404) with respect to a_j we get

$$\int_\Gamma \left[\frac{\partial^2 \psi}{\partial a_i\, \partial a_j} - \frac{\partial^2 \varepsilon}{\partial a_i\, \partial a_j} + \beta \left(\frac{\partial \psi}{\partial a_i} - \frac{\partial \varepsilon}{\partial a_i} \right) \left(\frac{\partial \psi}{\partial a_j} - \frac{\partial \varepsilon}{\partial a_j} \right) \right]$$
$$\times\ e^{\beta(\psi - \varepsilon)}\ d\Omega = 0, \tag{5.0408}$$

or, using Eqs.(5.0405) and (5.0303),

$$\overline{(A_i - \overline{A}_i))(A_j - \overline{A}_j))} = \frac{1}{\beta} \left[\overline{\frac{\partial^2 \varepsilon}{\partial a_i\, \partial a_j}} - \frac{\partial^2 \psi}{\partial a_i\, \partial a_j} \right], \tag{5.0409}$$

which is valid both for $i = j$ and for $i \neq j$.

Finally by differentiating either Eq.(5.0402) with respect to a_i or Eq. (5.0404) with respect to β we get

$$\overline{(A_i - \overline{A}_i))(\varepsilon - \bar{\varepsilon})} = \frac{\partial^2 \psi}{\partial a_i\, \partial \beta}. \tag{5.0410}$$

We shall see later on that in many cases ψ and its derivatives, and thus also $\bar{\varepsilon}$ and $\overline{A}_i$, will be proportional to the number of particles, N, in the system. In that case it follows from Eqs.(5.0407), (5.0409), and (5.0410) that all quantities such as

$$\frac{\overline{(\varepsilon - \bar{\varepsilon})^2}}{\bar{\varepsilon}^2}, \qquad \frac{\overline{(A_i - \overline{A}_i))(A_j - \overline{A}_j))}}{\overline{A}_i\, \overline{A}_j}, \qquad \frac{\overline{(A_i - \overline{A}_i))(\varepsilon - \bar{\varepsilon})}}{\overline{A}_i\, \bar{\varepsilon}}$$

will be of the order of $1/N$ and hence negligibly small if N is sufficiently large.

It sometimes happens that fluctuations can be very large. At an ideal wall, for instance, the force $(-\partial U/\partial q)$ will be infinite (compare Fig.1.5 where we have a similarly idealised potential energy) and the pressure will fluctuate violently. In those cases it is practically always possible to show that the fluctuations of the time average B_i, defined by the equation

$$B_i = \frac{1}{\tau} \int_t^{t+\tau} A_i\, dt, \tag{5.0411}$$

are small, that is,

$$\frac{\overline{(B_i - \overline{B}_i)^2}}{\overline{B}_i^2} \sim \frac{1}{N} \ll 1. \tag{5.0412}$$

The fact that any actual measurement of the pressure will take a finite time means that it is more in accordance with physical reality to consider the B_i rather than the A_i (compare also the discussion in §5.8).

We must, however, mention that there are circumstances where the fluctuations are, indeed, large. This is, for instance the case when we are dealing with the neighbourhood of a phase transition point or at a critical point.

5.5.   The Entropy in a Macrocanonical Ensemble

We saw in §5.3 that the entropy S of a system is related to the average value $\bar{\eta}$ of the index of probability through Eq.(5.0309). We shall now show that there are also other ways to correlate the entropy with quantities relating to the ensemble.

Let the smallest possible value of the energy for given values of the external parameters be $\varepsilon_{\min}$ which in general will be a function of the a_i. Since in general it will always be possible to choose p_i to be equal to zero, independent of the values of the q_i, so that $\mathcal{T}_{\min} = 0$, we have

$$\varepsilon_{\min}(a_i) = U_{\min}(a_i). \tag{5.0501}$$

We now introduce a quantity Ω by the equation

$$\Omega = \int_{\varepsilon_{\min}(a_i)}^{\varepsilon} d\Omega. \tag{5.0502}$$

From this equation we see, firstly, that Ω is the extension in phase between the two energy surfaces corresponding, respectively, to $\varepsilon_{\min}$ and ε and, secondly, that Ω depends both on ε and on the a_i, the latter through $\varepsilon_{\min}$, that is,

$$\Omega = \Omega(\varepsilon, a_i). \tag{5.0503}$$

We also introduce a function ϕ through the equation

$$e^{\phi} = \frac{\partial \Omega}{\partial \varepsilon}, \tag{5.0504}$$

or

$$\phi(\varepsilon, a_i) = \ln \frac{\partial \Omega}{\partial \varepsilon}. \tag{5.0505}$$

We shall need presently the value of e^{ϕ} for $\varepsilon = \varepsilon_{\min}$. We shall for the sake of simplicity assume that the kinetic energy $\mathcal{T}$ depends on the p_i only. We now introduce the following two elementary volumes:

$$d\Omega_p = \prod_{i=1}^{sN} dp_i, \qquad d\Omega_q = \prod_{i=1}^{sN} dq_i, \tag{5.0506}$$

which in the simplified case considered here are, apart from a multiplying constant, what Gibbs calls, respectively, an element of *extension in velocity* and an element of *extension in configuration*. (For a discussion of the general case, see Problem 3 of the present chapter.)

Furthermore, we write

$$\Omega_p = \int_0^{\mathcal{T}} d\Omega_p, \qquad \Omega_q = \int_{U_{\min}}^{U} d\Omega_q, \qquad (5.0507)$$

and we introduce two quantities ϕ_p and ϕ_q by the equations

$$e^{\phi_p} = \frac{\partial \Omega_p}{\partial \mathcal{T}}, \qquad e^{\phi_q} = \frac{\partial \Omega_q}{\partial U}. \qquad (5.0508)$$

From Eqs.(5.0502) and (5.0507) it follows that

$$\Omega = \int\int d\Omega_p \, d\Omega_q = \int_{U_{\min}}^{U=\varepsilon} \Omega_p \, d\Omega_q, \qquad (5.0509)$$

where Ω_p depends on the q_i through the fact that the upper limit $\mathcal{T}$ of the integral in Eq.(5.0507) must satisfy the relation

$$\mathcal{T} + U = \varepsilon. \qquad (5.0510)$$

Taking the derivative with respect to ε in Eq.(5.0509) and using the fact that

$$\frac{\partial \Omega_p}{\partial \varepsilon} = \frac{\partial \Omega_p}{\partial \mathcal{T}},$$

we get

$$e^{\phi} = \frac{\partial \Omega}{\partial \varepsilon} = \int_{U_{\min}}^{U=\varepsilon} e^{\phi_p} \, d\Omega_q + \Omega_p \, e^{\phi_q} \Big|_{U=\varepsilon}, \qquad (5.0511)$$

and hence by virtue of Eq.(5.0501) and the fact that $\Omega_p = 0$ for $U = \varepsilon$

$$e^{\phi}\Big|_{\varepsilon=\varepsilon_{\min}} = 0. \qquad (5.0512)$$

If a phase function G is a function of the energy only, we have for its average value

$$\overline{G} = \int_\Gamma G(\varepsilon)\, \varrho(\varepsilon)\, d\Omega, \qquad (5.0513)$$

or

$$\overline{G} = \int_{\varepsilon=\varepsilon_{\min}}^{\infty} G(\varepsilon)\, \varrho(\varepsilon)\, e^{\phi}\, d\varepsilon. \qquad (5.0514)$$

We must here draw attention to the different ways through which the right-hand sides of Eqs.(5.0513) and (5.0514) depend on the a_i. In the first equation, $\overline{G}$ depends on the a_i through ε, but in the second equation $\overline{G}$ depends on the a_i through ϕ and the lower limit of integration, while ε is an integration variable.

From the normalisation equation we get in this way

$$e^{-\beta\psi} = \int e^{-\beta\varepsilon+\phi} \, d\varepsilon. \tag{5.0515}$$

In the preceding section we saw that the fluctuations of the energy in a macrocanonical ensemble are in general very small. This means that the great majority of the members of the ensemble will possess energies that are practically equal to $\overline{\varepsilon}$, and thus Eq.(5.0515) can be written in the form

$$e^{-\beta\psi} \approx \left[e^{-\beta\varepsilon+\phi}\right]_{\varepsilon=\overline{\varepsilon}} \cdot \Delta, \tag{5.0516}$$

where Δ is the "half-width" of the energy fluctuations, which is of the order of magnitude of $\overline{\varepsilon}/\sqrt{N} \sim \sqrt{N}$ (compare the discussion at the end of the preceding section),

$$\Delta^2 = C \, \overline{(\varepsilon - \overline{\varepsilon})^2}. \tag{5.0517}$$

In order to derive Eq.(5.0516) we first note that, since the energy fluctuations are small, $\overline{\varepsilon}$ will be the energy possessed by the largest number of systems in the ensemble. Since the fraction of systems, $f(\varepsilon) \, d\varepsilon$, with energies between ε and $\varepsilon + d\varepsilon$ is given by the equation

$$f(\varepsilon) \, d\varepsilon = e^{\beta(\psi-\varepsilon)+\phi} \, d\varepsilon, \tag{5.0518}$$

as follows from the definition of ϱ and ϕ, $\overline{\varepsilon}$ will also be the energy for which $e^{-\beta\varepsilon+\phi}$ will be maximum, or

$$\beta = \left.\frac{\partial\phi}{\partial\varepsilon}\right|_{\varepsilon=\overline{\varepsilon}}. \tag{5.0519}$$

Developing $\phi - \beta\varepsilon$ in a Taylor series, we get

$$\phi - \beta\varepsilon = \phi|_{\varepsilon=\overline{\varepsilon}} - \beta\overline{\varepsilon} + \frac{1}{2}\left.\frac{\partial^2\phi}{\partial\varepsilon^2}\right|_{\varepsilon=\overline{\varepsilon}} \left(\varepsilon - \overline{\varepsilon}\right)^2 + \cdots, \tag{5.0520}$$

where we have used Eq.(5.0519). From Eq.(5.0520) it follows that to a first approximation

$$e^{\beta\varepsilon+\phi} = \left[e^{-\beta\varepsilon+\phi}\right]_{\varepsilon=\overline{\varepsilon}} \cdot \exp\left\{\frac{1}{2}\left.\frac{\partial^2\phi}{\partial\varepsilon^2}\right|_{\varepsilon=\overline{\varepsilon}} \left(\varepsilon - \overline{\varepsilon}\right)^2\right\}. \tag{5.0521}$$

Substituting Eq.(5.0521) into Eq.(5.0515) and integrating, we get approximately (by integrating from $-\infty$ to ∞ instead of from $\varepsilon_{\min}$ to ∞)

$$e^{-\beta\psi} = \left[e^{-\beta\varepsilon+\phi}\right]_{\varepsilon=\overline{\varepsilon}} \cdot \left(-\frac{1}{2\pi}\left.\frac{\partial^2\phi}{\partial\varepsilon^2}\right|_{\varepsilon=\overline{\varepsilon}}\right)^{-1/2}. \tag{5.0522}$$

On the other hand, substituting Eq.(5.0521) into the expression for $(\varepsilon - \bar{\varepsilon})^2$, we get

$$\overline{(\varepsilon - \bar{\varepsilon})^2} = \frac{\int e^{\beta(\psi - \varepsilon) + \phi} (\varepsilon - \bar{\varepsilon})^2 \, d\varepsilon}{\int e^{\beta(\psi - \varepsilon) + \phi} \, d\varepsilon}$$

$$\cong \frac{\int (\varepsilon - \bar{\varepsilon})^2 \exp\left\{ \frac{1}{2} \frac{\partial^2 \phi}{\partial \varepsilon^2} \Big|_{\varepsilon = \bar{\varepsilon}} (\varepsilon - \bar{\varepsilon})^2 \right\} d\varepsilon}{\int \exp\left\{ \frac{1}{2} \frac{\partial^2 \phi}{\partial \varepsilon^2} \Big|_{\varepsilon = \bar{\varepsilon}} (\varepsilon - \bar{\varepsilon})^2 \right\} d\varepsilon}$$

$$\cong \frac{3}{2} \cdot \left(-\frac{\partial^2 \phi}{\partial \varepsilon^2} \Big|_{\varepsilon = \bar{\varepsilon}} \right)^{-1}. \tag{5.0523}$$

Combining Eqs.(5.0522) and (5.0523), we obtain Eq.(5.0516).

We may remark here that Eq.(5.0521) shows that to a first approximation the distribution in energy is a Gaussian distribution.

From Eq.(5.0516) it follows that

$$\beta (\bar{\varepsilon} - \psi) = \phi(\bar{\varepsilon}, a_i) + \ln \Delta. \tag{5.0524}$$

Since $\ln \Delta$ is in general of the order of magnitude of $\ln N$ while $\bar{\varepsilon}$ and ψ are of the order of magnitude N, we may neglect the last term on the right-hand side of Eq.(5.0524). Using Eq.(5.0309) for the entropy we have, apart possibly from an additive constant,

$$\frac{S}{k} = \phi(\bar{\varepsilon}, a_i) = \ln \frac{\partial \Omega}{\partial \varepsilon} \Big|_{\varepsilon = \bar{\varepsilon}}. \tag{5.0525}$$

From the well-known relation between entropy and temperature (compare Eq.(5.0308)),

$$\frac{1}{T} = \frac{\partial S}{\partial \varepsilon} \Big|_{\varepsilon = \bar{\varepsilon}}, \tag{5.0526}$$

we get, using Eq.(5.0525)

$$\beta = \frac{1}{kT} = \frac{\partial S/k}{\partial \varepsilon} \Big|_{\varepsilon = \bar{\varepsilon}} = \frac{\partial \phi}{\partial \varepsilon} \Big|_{\varepsilon = \bar{\varepsilon}}, \tag{5.0527}$$

which we derived a moment ago (Eq.(5.0519)) by slightly different arguments.

If we calculate the average value of $\partial \phi / \partial \varepsilon$ over the ensemble, we get

$$\overline{\frac{\partial \phi}{\partial \varepsilon}} = \int \frac{\partial \phi}{\partial \varepsilon} e^{\beta(\psi - \varepsilon) + \phi} \, d\varepsilon = \left[e^{\beta(\psi - \varepsilon) + \phi} \right]_{\varepsilon_{\min}}^{\infty} + \beta \int e^{\beta(\psi - \varepsilon) + \phi} \, d\varepsilon,$$

or

$$\overline{\frac{\partial \phi}{\partial \varepsilon}} = \beta, \tag{5.0528}$$

where we have used Eq.(5.0512), to show that the integrated part is zero, and Eq.(5.0515).

Comparing Eqs.(5.0527) and (5.0528), we see that the average value of $\partial\phi/\partial\varepsilon$ is the same as its value when the energy has its most probable value. This property is equally true for other quantities and is a consequence of the fact that the dispersion in energy is extremely small.

We mentioned before that by far the largest fraction of systems in the ensemble have an energy in the neighbourhood of $\bar{\varepsilon}$. From Eq.(5.0518) we see that this means that e^{ϕ} must increase steeply when ε increases from $\varepsilon_{\min}$ to $\bar{\varepsilon}$. Let us compare a macrocanonical enesemble with $\eta = \beta(\psi - \varepsilon)$ and an average energy $\bar{\varepsilon}$ with an F-ensemble with density given by

$$\varrho_F = a\, e^{\beta(\psi-\bar{\varepsilon})}\, F(\varepsilon), \qquad (5.0529)$$

where $F(\varepsilon)$ is given by Eq.(5.0210), where a is a constant, and where we have put $\varepsilon_0 = \bar{\varepsilon}$.

If G is a phase function that depends only on the energy, we have for its average value $\overline{G}_F$, taken over the F-ensemble,

$$\overline{G}_F = \int_{\Gamma} \varrho_F\, G(\varepsilon)\, d\Omega = \int_{\varepsilon_{\min}}^{\bar{\varepsilon}} a\, e^{\beta(\psi-\bar{\varepsilon})}\, G(\varepsilon)\, e^{\phi}\, d\varepsilon, \qquad (5.0530)$$

or, since e^{ϕ} has a steep maximum for $\varepsilon \cong \bar{\varepsilon}$,

$$\overline{G}_F \approx a\, \Delta\, e^{\beta(\psi-\bar{\varepsilon})+\phi(\bar{\varepsilon})}\, G(\bar{\varepsilon}). \qquad (5.0531)$$

On the other hand, the average taken over the macrocanonical ensemble is given by the equation

$$\overline{G} = \int_{\Gamma} e^{\beta(\psi-\bar{\varepsilon})}\, G(\varepsilon)\, d\Omega \approx e^{\beta(\psi-\bar{\varepsilon})+\phi(\bar{\varepsilon})}\, G(\bar{\varepsilon})\, \Delta. \qquad (5.0532)$$

We note that in deriving Eqs.(5.0531) and (5.0532) we have used Eqs. (5.0520) and (5.0523) and neglected numerical factors of order unity, as we also did in Eq.(5.0516). From Eqs.(5.0531) and (5.0532) we see that the two ensembles will lead to roughly the same average values and thus to be roughly equivalent, if

$$a \approx 1. \qquad (5.0533)$$

The normalisation of the F-ensemble gives

$$1 = \int_{\varepsilon_{\min}}^{\bar{\varepsilon}} a\, e^{\beta(\psi-\bar{\varepsilon})}\, d\Omega = a\, e^{\beta(\psi-\bar{\varepsilon})} \int_{\varepsilon_{\min}}^{\bar{\varepsilon}} d\Omega = a\, e^{\beta(\psi-\bar{\varepsilon})}\, \Omega(\bar{\varepsilon}).$$
$$(5.0534)$$

Taking the logarithm, neglecting $\ln a$ in comparison with the other terms, and using Eq.(5.0309) for the entropy, we get, apart possibly from an additive constant,

$$\frac{S}{k} = \ln\, \Omega(\bar{\varepsilon}); \qquad (5.0535)$$

for β we get the equation

$$\beta = \left.\frac{\partial S/k}{\partial \varepsilon}\right|_{\varepsilon=\bar{\varepsilon}} = \left.\frac{\partial \ln \Omega}{\partial \varepsilon}\right|_{\varepsilon=\bar{\varepsilon}} = \left.\frac{e^{\phi}}{\Omega}\right|_{\varepsilon=\bar{\varepsilon}}. \qquad (5.0536)$$

We note from Eqs.(5.0525) and (5.0535) that, once we know $\Omega(\varepsilon)$, we can evaluate the entropy. As S is one of the thermodynamic potentials, it follows that a knowledge of $\Omega(\varepsilon)$ enables us to evaluate the thermodynamic properties of the system represented by the macrocanonical ensemble.

5.6.   The Coupling of Two Macrocanonical Ensembles

The coupling of two ensembles consists of forming a new ensemble with systems obtained by combining one system of the first ensemble with one system from the second ensemble. The coupling is supposed to be a physical one so that energy can be exchanged between the two systems that are combined to form a system of the new ensemble.

In § 5.2 we mentioned that one way to couple two stationary ensembles in such a way that the resulting ensemble is again a stationary ensemble is to couple two macrocanonical ensembles with the same value of β. We shall now justify that statement. Once more our discussion will follow closely Gibbs's considerations.

Let us consider two ensembles. The first one may have a density ϱ_1, index of probability η_1, energy ε_1, and if it is macrocanonical, modulus $\Theta_1 = \beta_1^{-1}$, while the corresponding quantities in the second ensemble will have indices 2. Let us couple these two ensembles at $t = t'$. The density, index of probability, and energy of the new ensemble will be denoted by ϱ_{12}, η_{12}, and ε_{12}. After a while, at $t = t''$, the two systems are separated and we have again two separate ensembles. At t' we have clearly

$$\varrho_{12}' = \varrho_1' \cdot \varrho_2', \qquad (5.0601)$$

and we must show that, if

$$\varrho_1 = e^{\beta_1(\psi_1 - \varepsilon_1)}, \qquad \varrho_2 = e^{\beta_2(\psi_2 - \varepsilon_2)}, \qquad (5.0602)$$

the equation

$$\varrho_{12}' = \varrho_{12}'' \qquad (5.0603)$$

will be satisfies only if $\beta_1 = \beta_2$. Henceforth, as in Eqs.(5.0601) and (5.0603), primes and double primes will indicate values at $t = t'$ and $t = t''$, respectively.

That Eq.(5.0603) is satisfied if $\beta_1 = \beta_2 = \beta_{12}$ follows easily from Eqs. (5.0601), (5.0602), and the equation

$$\varepsilon_{12} = \varepsilon_1 + \varepsilon_2, \qquad (5.0604)$$

whence we have

$$\varrho_{12} = e^{\beta_1(\psi_1 - \varepsilon_1)} \cdot e^{\beta_2(\psi_2 - \varepsilon_2)} = e^{\beta_{12}(\psi_{12} - \varepsilon_{12})} \qquad (\psi_{12} = \psi_1 + \psi_2),$$
$$(5.0605)$$

which is a stationary macrocanonical distribution.

Let us now consider what would happen if we coupled two macrocanonical ensembles with different moduli. We see immediately that the resulting ensemble is not a stationary one, since the density is not a function of ε_{12}, but contains ε_1 and ε_2 separately. We shall prove that, if

$$\beta_1 < \beta_2, \qquad (5.0606)$$

a system from the first ensemble will on average have lost energy to the system from the second ensemble, with which it was combined between t' and t''. Since β is inversely proportional to the temperature, this corresponds to the thermodynamical fact that energy will pass from a system of higher temperature to a system of lower temperature.

We must remark here that, in order that energy can be transported from a system of the first ensemble to a system from the second ensemble, it is necessary that there exists an interaction energy, ε_{int} so that instead of Eq.(5.0604) we in fact have

$$\varepsilon_{12} = \varepsilon_1 + \varepsilon_2 + \varepsilon_{int}. \qquad (5.0607)$$

However, we shall assume, firstly, that $\varepsilon_{int} \ll \varepsilon_{12}$ and, secondly, that the coupling takes place in such a way that the systems are brought together and pulled apart so gradually that both at t' and at t'' Eq. (5.0604) is satisfied, that is, that

$$\varepsilon'_{12} = \varepsilon'_1 + \varepsilon'_2 \qquad \text{and} \qquad \varepsilon''_{12} = \varepsilon''_1 + \varepsilon''_2, \qquad (5.0608)$$

Before proving that, on average, energy will go from the first ensemble to the second one if Eq.(5.0606) is satisfied, we shall prove a few necessary lemmas.

Consider an ensemble which is such that all its systems consist of two parts, denoted by indices "a" and "b" while an index "ab" indicates the whole system. We then have

$$\int_{\Gamma_{ab}} \rho_{ab} \, d\Omega_{ab} = \int_{\Gamma_{ab}} e^{\eta_{ab}} \, d\Omega_{ab} = 1, \qquad (5.0609)$$

where

$$d\Omega_{ab} = d\Omega_a \, d\Omega_b. \qquad (5.0610)$$

Introducing

$$\varrho_a = e^{\eta_a} = \int_{\Gamma_b} e^{\eta_{ab}} \, d\Omega_b, \qquad (5.0611)$$

$$\varrho_b = e^{\eta_b} = \int_{\Gamma_a} e^{\eta_{ab}} \, d\Omega_a, \qquad (5.0612)$$

we see that from Eq.(5.0609) it follows that

$$\int_{\Gamma_a} e^{\eta_a}\, d\Omega_a \;=\; \int_{\Gamma_b} e^{\eta_b}\, d\Omega_b \;=\; 1. \tag{5.0613}$$

We also see that ϱ_a and ϱ_b evidently denote the densities of the subensembles made up out of part a and part b, respectively.

Lemma I. In the case of an ensemble such as the one just discussed we have

$$\overline{\eta}_{ab} \;\geqslant\; \overline{\eta}_a + \overline{\eta}_b, \tag{5.0614}$$

where the averages are taken over the corresponding ensembles or subensembles.

Proof. First let us note that the expression $u \cdot e^u + 1 - e^u$ is positive for $u \neq 0$ and is zero for $u = 0$.

Secondly, let us evaluate $\overline{\eta}_{ab} - \overline{\eta}_a - \overline{\eta}_b$. Using Eqs.(5.0611) and (5.0612), we have

$$\overline{\eta}_{ab} - \overline{\eta}_a - \overline{\eta}_b \;=\; \int_{\Gamma_{ab}} \eta_{ab}\, e^{\eta_{ab}}\, d\Omega_{ab} - \int_{\Gamma_a} \eta_a\, e^{\eta_a}\, d\Omega_a - \int_{\Gamma_b} \eta_b\, e^{\eta_b}\, d\Omega_b$$

$$= \int_{\Gamma_{ab}} (\eta_{ab} - \eta_a - \eta_b)\, e^{\eta_{ab}}\, d\Omega_{ab} \tag{5.0615}$$

From Eq.(5.0613) we have

$$\int_{\Gamma_a} e^{\eta_a}\, d\Omega_a \;\cdot\; \int_{\Gamma_b} e^{\eta_b}\, d\Omega_b \;=\; \int_{\Gamma_{ab}} e^{\eta_a + \eta_b}\, d\Omega_{ab} \;=\; 1, \tag{5.0616}$$

and using Eqs.(5.0615), (5.0616), and (5.0609), we get finally

$$\overline{\eta}_{ab} - \overline{\eta}_a - \overline{\eta}_b \;=\; \int_{\Gamma_{ab}} (x e^x + 1 - e^x)\, e^{\eta_a + \eta_b}\, d\Omega_{ab} \;\geqslant\; 0, \tag{5.0617}$$

$(x = \eta_{ab} - \eta_a - \eta_b)$ by virtue of the fact that the integrand is non-negative. This proves our lemma.

Lemma II. Consider two ensembles with indices of probability given by the equations

$$\eta_1 \;=\; \beta(\psi - \varepsilon), \qquad \eta_2 \;=\; \beta(\psi - \varepsilon) + \Delta\eta, \tag{5.0618}$$

that is, one macrocanonical and one arbitrary ensemble. We then have

$$\overline{(\eta + \beta\varepsilon)_1} \;\leqslant\; \overline{(\eta + \beta\varepsilon)_2}. \tag{5.0619}$$

Proof. We have from Eq.(5.0618)

$$\overline{(\eta + \beta\varepsilon)_2} - \overline{(\eta + \beta\varepsilon)_1} \;=\; \int_{\Gamma} \left[(\beta\psi + \Delta\eta)\, e^{\eta_1 + \Delta\eta} - \beta\psi\, e^{\eta_1} \right]\, d\Omega. \tag{5.0620}$$

Moreover, from the normalisation condition we have

$$\int_\Gamma e^{\eta_1 + \Delta\eta}\, d\Omega \;=\; \int_\Gamma e^{\eta_1}\, d\Omega, \tag{5.0621}$$

and combining Eqs.(5.0620) and (5.0621), we find that the right-hand side of Eq.(5.0620) is equal to

$$\int_\Gamma \Delta\eta\, e^{\eta_1 + \Delta\eta}\, d\Omega \;=\; \int_\Gamma \left[\Delta\eta\, e^{\Delta\eta} + 1 - e^{\Delta\eta}\right] e^{\eta_1}\, d\Omega \;\geqslant\; 0, \tag{5.0622}$$

from which our second lemma follows.

We can now consider the joining together and separating of the two ensembles. From Eq.(5.0601) it follows that

$$\eta_{12}' \;=\; \eta_1' + \eta_2', \tag{5.0623}$$

and thus that

$$\overline{\eta}_{12}' \;=\; \overline{\eta}_1' + \overline{\eta}_2', \tag{5.0624}$$

where in going over from Eq.(5.0623) to Eq.(5.0624), we have used the fact that ϱ_1 and ϱ_2 are normalised.[7]

At t'' we have (see Eq.(5.0614))

$$\overline{\eta}_{12}'' \;\geqslant\; \overline{\eta}_1'' + \overline{\eta}_2''. \tag{5.0625}$$

We saw in §5.3 that for a macrocanonical ensemble $-\overline{\eta}$, possibly apart from additive and multiplying constants, corresponded to the entropy. We can also easily see that, in general, $\overline{\eta}$ should correspond to a generalisation of Boltzmann's H, as defined by Eq.(2.301), since

$$\overline{\eta} \;=\; \int_\Gamma \eta e^\eta\, d\Omega \;=\; \int_\Gamma \varrho \ln \varrho\, d\Omega, \tag{5.0626}$$

while ϱ can be considered to be a generalisation of the distribution function f.

From the second law of thermodynamics, or from a generalised H-theorem (see §5.14), we should then expect that

$$\overline{\eta}_{12}'' \;\leqslant\; \overline{\eta}_{12}'. \tag{5.0627}$$

---

[7] From Eq.(5.0623) one has

$$\int_{\Gamma_{12}} \eta_{12}' \varrho_{12}\, d\Omega_{12} \;=\; \int_{\Gamma_{12}} \eta_1' \varrho_{12}\, d\Omega_{12} + \int_{\Gamma_{12}} \eta_2' \varrho_{12}\, d\Omega_{12}$$

$$=\; \int_{\Gamma_1} \eta_1' \varrho_1\, d\Omega_1 \cdot \int_{\Gamma_2} \varrho_2\, d\Omega_2 + \int_{\Gamma_1} \varrho_1\, d\Omega_1 \cdot \int_{\Gamma_2} \eta_2' \varrho_2\, d\Omega_2.$$

Since at t' the two ensembles were macrocanonical, while this is no longer necessarily so at t'', the conditions under which Lemma II is valid are satisfied and we have from inequality (5.0619)

$$\overline{\eta}_1' + \beta_1 \overline{\varepsilon}_1' \;\leqslant\; \overline{\eta}_1'' + \beta_1 \overline{\varepsilon}_1'' \tag{5.0628}$$

and

$$\overline{\eta}_2' + \beta_2 \overline{\varepsilon}_2' \;\leqslant\; \overline{\eta}_2'' + \beta_2 \overline{\varepsilon}_2''. \tag{5.0629}$$

Combining Eqs.(5.0624), (5.0625), and (5.0627) to (5.0629), we get

$$\beta_1 \big(\overline{\varepsilon}_1'' - \overline{\varepsilon}_1'\big) + \beta_2 \big(\overline{\varepsilon}_2'' - \overline{\varepsilon}_2'\big) \;\geqslant\; 0. \tag{5.0630}$$

From Eqs.(5.0608) and the energy conservation principle we have

$$\overline{\varepsilon}_1'' - \overline{\varepsilon}_1' + \overline{\varepsilon}_2'' - \overline{\varepsilon}_2' \;=\; 0, \tag{5.0631}$$

and hence, combining Eqs.(5.0630) and (5.0631),

$$\big(\overline{\varepsilon}_1'' - \overline{\varepsilon}_1'\big)\big(\beta_1 - \beta_2\big) \;\geqslant\; 0. \tag{5.0632}$$

From Eqs.(5.0606) and (5.0632) we then have

$$\overline{\varepsilon}_1' \;\geqslant\; \overline{\varepsilon}_1'', \tag{5.0633}$$

or, the system with the smaller β will, on average, have lost energy to the other system.

We have just seen that, if two macrocanonical ensembles are physically coupled, they will form a stationary ensemble if the values of their β are equal, and that if the values of their β are not equal the one with the smaller β will on average lose energy to the other one. If we couple two microcanonical ensembles, the resulting ensemble will also be a microcanonical ensemble; hence one might ask whether there is any reason to prefer the macrocanonical to the microcanonical ensembles as we have done. The reason is that while in both cases the resulting ensemble is again a stationary one, separation at t'' reproduces the situation before t' in the case of macrocanonical, but not in the case of microcanonical ensembles.

To see this we first consider the case of the coupling of two microcanonical ensembles. Consider two systems A and B which together form one system of the combined ensemble. If ε_1 is the energy of the first and ε_2 that of the second microcanonical ensemble, system A had originally at t' an energy ε_1 and system B an energy ε_2. During the interval between t' and t'' the two systems were coupled and could exchange energy. When they are separated at t'' there is no reason to assume that all systems A still have the same energy, and hence they will no longer form a microcanonical ensemble.

On the other hand, the situation is different when two macrocanonical ensembles with the same β are coupled. Since, according to our assump-, tions, the interaction between the two systems from the coupled ensembles which together make up one system of the combined ensemble is neglegibly small, we can at any time — also between t' and t'' — calculate the average values of the powers of the energy ε_1 of the system from the first of the two coupled ensembles. We then find that they correspond at any time to the average values in a macrocanonical ensemble with the given β so that, indeed, after t'' we will again have two macrocanonical ensembles.[8]

The average value of ε_1^n can be calculated as follows. Let the indices "A", "B", and "tot" indicate, respectively, quantities pertaining to the first, the second, and the combined ensembles, and let p_A, q_A stand for the coordinates pertaining to the first and p_B, q_B for those of the second ensemble. We have then

$$\varepsilon_1 = \varepsilon_1(p_A, q_A), \qquad \varepsilon_2 = \varepsilon_2(p_B, q_B), \qquad \varepsilon_{tot} = \varepsilon_1(p_A, q_A) + \varepsilon_2(p_B, q_B)$$

and

$$d\Omega_{tot} = d\Omega_A \, d\Omega_B.$$

The average value of ε_1^n in the combined ensemble then satisfies the equations

$$
\begin{aligned}
\langle \varepsilon_1^n \rangle_{\text{av tot}} &= \int_{\Gamma_{tot}} \varepsilon_1^n(p_A, q_A) \, \varrho_{tot} \, d\Omega_{tot} \\
&= \int_{\Gamma_A} \varepsilon_1^n(p_A, q_A) \, \varrho_A \, d\Omega_A \cdot \int_{\Gamma_B} \varrho_B \, d\Omega_B \\
&= \int_{\Gamma_A} \varepsilon_1^n(p_A, q_A) \, \varrho_A \, d\Omega_A = \langle \varepsilon_1^n \rangle_{\text{av A}} .
\end{aligned}
$$

5.7.   Microcanonical Ensembles

In this section we shall consider microcanonical ensembles. We saw in § 5.1 and know from theoretical mechanics that the energy of an (isolated) system is one of its integrals. We might therefore be tempted to prefer microcanonical ensembles to macrocanonical ones, especially as long as our considerations are classical so that in principle we could hope to determine the energy of the system under observation as accurately as we want to.

A different reason why microcanonical ensembles are often used is that one can show that — under conditions to be discussed briefly in § 5.14 — the time average of a phase function is equal to the average of that phase function taken over a suitable microcanonical ensemble. We can derive this result as follows. Consider a system of a given energy ε, the representative point of which moves over the appropriate energy surface. The time average $\tilde{G}$ of a phase function G is defined by the equation

$$\tilde{G} = \lim_{T \to \infty} \frac{1}{2T} \int_{-T}^{+T} G(p, q) \, dt, \tag{5.0701}$$

---

[8] We have assumed here without proof that the ensemble density is uniquely determined by the average values of all powers of the energy.

where p, q denote the values of the coordinates of the representative point at the time t.

Consider now the orbit of the system on the energy surface (see Fig.5.2). Let P_0 be the position of the representative point at $t = t_0$, P_1 that at $t_0 + \tau$, P_2 that at $t_0 + 2\tau$, ..., while P_{-1}, P_{-2}, $\cdots$ were the positions at $t_0 - \tau$, $t_0 - 2\tau$, ..., in general, P_i at $t_0 + i\tau$ ($i = 0, \pm 1, \pm 2, \ldots$). The points P_i form a stationary ensemble: any collection of points in Γ-space forms an ensemble and since the P_i at a later time, say $t_0 + n\tau$, can be obtained from the P_i at t_0 by putting

$$P_i(t = t_0 + n\tau) = P_{i-n}(t = t_0), \tag{5.0702}$$

the ensemble is stationary.[9]

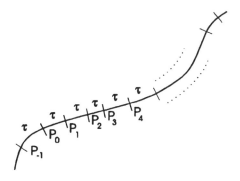

Fig.5.2. The path of a representative point on the energy surface.

Let us now consider the average $\langle G \rangle_{\mathrm{av}}$ of a phase function $G(p, q)$ over the P_i,

$$\langle G \rangle_{\mathrm{av}} = \frac{\cdots + G_{-2} + G_{-1} + G_0 + G_1 + G_2 + \cdots}{\cdots + 1 + 1 + 1 + 1 + 1 + \cdots}, \tag{5.0703}$$

where G_i is the value of $G(p, q)$ at P_i. If we now take the limit as $\tau \to 0$, the right-hand side of Eq.(5.0703) will become the time average $\tilde{G}$. On the other hand, we see that the average on the right-hand side of Eq.(5.0703) resembles strongly an average taken over the whole of the energy surface, that is, an average over a microcanonical ensemble. If we accept that identification for a moment, and if we denote here and henceforth the average over a microcanonical ensemble by a double bar, we have

$$\overline{\overline{G}} = \tilde{G}. \tag{5.0704}$$

---

[9] Strictly speaking, only in the limit as $\tau \to 0$.

From the fact that it can be proved (see § 5.14) that Eq.(5.0704) is valid for most physical systems and the fact that, while one measures time averages, ensemble averages can nearly always be much more easily computed, it follows that it might be advantageous to consider the microcanonical ensemble in slightly more detail. We shall then see that averages over a microcanonical ensemble will be the same as averages taken over a suitably chosen macrocanonical ensemble. Thus, we see that for actual calculations we can practically always confine ourselves to macrocanonical ensembles, which is rather gratifying, since they are more amenable to mathematical treatment and since there are also other arguments that point to macrocanonical ensembles, as we have seen and shall see again in § 5.13.

We now introduce the microcanonical ensemble by Eq.(5.0205). Let us consider the identity

$$\int_\Gamma \delta(\varepsilon - \varepsilon_0)\, d\Omega \;=\; \int_{\varepsilon_{\min}(a_i)}^{\infty} \delta(\varepsilon - \varepsilon_0)\, e^{\phi(\varepsilon, a_i)}\, d\varepsilon. \tag{5.0705}$$

If we take the derivative with respect to a_j of both sides of this equation, the result is

$$\left[\overline{\overline{A}}_j(\varepsilon)\, \delta(\varepsilon - \varepsilon_0)\, e^\phi \right]_{\varepsilon_{\min}}^{\infty} + \int_{\varepsilon_{\min}}^{\infty} \left(\frac{\partial \overline{\overline{A}}_j}{\partial \varepsilon} + \overline{\overline{A}}_j \frac{\partial \phi}{\partial \varepsilon} \right) \delta(\varepsilon - \varepsilon_0)\, e^\phi\, d\varepsilon$$

$$= \int_{\varepsilon_{\min}}^{\infty} \frac{\partial \phi}{\partial a_j} \delta(\varepsilon - \varepsilon_0)\, e^\phi\, d\varepsilon - \left[\delta(\varepsilon - \varepsilon_0)\, e^\phi \frac{\partial \varepsilon}{\partial a_j} \right]_{\varepsilon_{\min}}^{\infty}. \tag{5.0706}$$

Equation (5.0706) can be derived as follows. We note, first of all, that while the left-hand side of Eq.(5.0705) depends on a_j because ε depends on it, the right-hand side depends on a_j through ϕ and $\varepsilon_{\min}$. Secondly, we used the identity

$$\int_\Gamma G\, \delta(\varepsilon - \varepsilon_0)\, d\Omega \;=\; \int_{\varepsilon_{\min}}^{\infty} \overline{\overline{G}}(\varepsilon)\, \delta(\varepsilon - \varepsilon_0)\, e^\phi\, d\varepsilon, \tag{5.0707}$$

where $\overline{\overline{G}}(\varepsilon)$ is the average of G over a microcanonical ensemble with energy ε.

As the integrated parts clearly vanish, we get from Eq.(5.0706)

$$e^\phi \frac{\partial \overline{\overline{A}}_j}{\partial \varepsilon} + \overline{\overline{A}}_j\, e^\phi \frac{\partial \phi}{\partial \varepsilon} \;=\; e^\phi \frac{\partial \phi}{\partial a_j} \left(= \frac{\partial^2 \Omega}{\partial a_j\, \partial \varepsilon} \right). \tag{5.0708}$$

This equation can be integrated and the result is[10]

$$\overline{\overline{A}}_j\, e^\phi \;=\; \frac{\partial \Omega}{\partial a_j}, \tag{5.0709}$$

---

[10]There should be an integration constant $K_j(a_i)$ on the right-hand side of Eq.(5.0709). However, by writing down the variation of Ω (compare Eq. (5.0711)), considering such variations of the a_i and of ε that $\delta\varepsilon_{\min} = 0$, and using Eq.(5.0512), one can prove that all the K_j are equal to zero.

or

$$\overline{\overline{A}}_j = e^{-\phi} \frac{\partial \Omega}{\partial a_j}.$$ (5.0710)

The variation of Ω is now given by the equation

$$\delta \Omega = e^{\phi} \left(\delta \varepsilon + \sum_j \overline{\overline{A}}_j \, \delta a_j \right).$$ (5.0711)

This last equation can be written in the form

$$\delta \ln \Omega = \frac{\delta \varepsilon + \sum_j \overline{\overline{A}}_j \, \delta a_j}{e^{-\phi} \, \Omega},$$ (5.0712)

and comparison with the thermodynamical equation for the variation of the entropy shows that we can take $\ln \Omega$ as being proportional to the entropy of the system and $e^{-\phi} \Omega$ as a measure for the temperature. Since that gives us the same as expressions (5.0535) and (5.0536), which were derived for macrocanonical ensembles, we have found the result that the averages over a microcanonical ensemble with energy ε_0 are the same as those taken over a macrocanonical ensemble for which $\overline{\varepsilon} = \varepsilon_0$.[11] We shall consider the relationship between micro- and macrocanonical ensembles in somewhat more detail in § 5.13.

5.8.   Application: the Perfect Gas

We shall briefly discuss here the simple case of a perfect gas in order to illustrate the theory developed in the preceding sections.

The discussion was started in § 5.3 where an expression for $e^{\beta \psi}$, or rather for the partition function Z_Γ, was derived. We had there (Eqs.(5.0325) and (5.0310))

$$e^{-\beta \psi} = V^N \left(\frac{2\pi m}{\beta} \right)^{3N/2}.$$ (5.0801)

From this equation we can calculate, for instance, the fluctuations in the energy, using Eq.(5.0407), which gives

$$\overline{(\varepsilon - \overline{\varepsilon})^2} = -\frac{\partial^2 (\beta \psi)}{\partial \beta^2} = \frac{3N}{2\beta^2},$$ (5.0802)

and combining this with the result for the average energy,

$$\overline{\varepsilon} = \frac{3N}{2\beta},$$ (5.0803)

---

[11]It must be borne in mind that, for instance, in Eq.(5.0710) all expressions on the right-hand side must be taken for $\varepsilon = \varepsilon_0$.

we have

$$\frac{\overline{(\varepsilon - \bar{\varepsilon})^2}}{\bar{\varepsilon}^2} = \frac{2}{3N}, \tag{5.0804}$$

showing that the dispersion is indeed very small as soon as N is large.

Let us now consider the fluctuations in the pressure. We mentioned in §5.4 that in this case we cannot use formula (5.0409) as long as we use ideal walls, that is, as long as we use the function $F(x_i, y_i, z_i)$ of Eq.(5.0325) for the potential energy function. We must now consider the time average of the pressure and its fluctuations. We can do this in two steps. The first step consists in reminding ourselves of the fact that, for a system of many degrees of freedom, we found in §5.7 that we obtain the same results whether we consider a macro- or a microcanonical ensemble. The second step then follows immediately, since we also saw that time averages and averages taken over the corresponding microcanonical ensemble are the same. We can now use Eq.(5.0710) for an average in a microcanonical ensemble; thus we have

$$\tilde{P} = \overline{\overline{P}} = e^{-\phi} \frac{\partial \Omega}{\partial V}, \tag{5.0805}$$

and since

$$e^\phi = \frac{\partial \Omega}{\partial \varepsilon}, \tag{5.0806}$$

we have

$$\tilde{P} = \frac{\frac{\partial \Omega}{\partial V}}{\frac{\partial \Omega}{\partial \varepsilon}}, \tag{5.0807}$$

where all quantities are evaluated for $\varepsilon = \bar{\varepsilon}$, since we are dealing with a microcanonical ensemble.

The quantity Ω can easily be computed, in the case of a perfect gas, from the equations

$$\Omega(\varepsilon, V) = \int_{\varepsilon=0}^{\varepsilon} d\Omega = \int_{\varepsilon=0}^{\varepsilon} \prod_i d^3 p_i \prod_i d^3 r_i, \tag{5.0808}$$

$$\varepsilon = \sum_i \left(\frac{p_i^2}{2m} + F(x_i, y_i, z_i) \right). \tag{5.0809}$$

The integration over the r_i gives a factor V^N and the integral over the p_i is the expression for the volume of a $3N$-dimensional sphere of radius $\sqrt{2m\varepsilon}$. This expression is different for even or odd N, but as we are only interested in the dependence of Ω on ε and V we can write

$$\Omega(\varepsilon, V) = K \varepsilon^{3N/2} V^N, \tag{5.0810}$$

where K is a constant.

From Eqs.(5.0807) and (5.0810) we get the familiar equation (see Eq. (4.715))

$$\widetilde{P} = \frac{2}{3} \frac{\bar{\varepsilon}}{V}. \tag{5.0811}$$

From Eqs.(5.0811) and (5.0804) we now get for the fluctuations in the time average of the pressure

$$\frac{\overline{\left(\widetilde{P} - \overline{\widetilde{P}}\right)^2}}{\overline{\widetilde{P}}^2} = \frac{\overline{\left(\varepsilon - \bar{\varepsilon}\right)^2}}{\bar{\varepsilon}^2} = \frac{2}{3N}, \tag{5.0812}$$

which proves that, in this case at least, the fluctuations are negligible for large N.

5.9.   Grand Ensembles

In the last chapter of his famous monograph Gibbs considers systems in which chemical reactions can take place. This means that the number of particles of a given kind is no longer constant, and we must thus abandon the condition of a constant number of particles, just as we abandoned the condition of a constant total energy when we went from microcanonical to macrocanonical ensembles. Gibbs called ensembles with a constant number of particles *petit ensembles* and those without a restriction on the total number of particles *grand ensembles*. The ensembles discussed so far in this chapter were all petit ensembles. Apart from systems in which chemical reactions can take place, the need of grand ensembles also arises when we are dealing with systems at very high temperatures in which pair creation and annihilation processes may occur. In § 5.13 we shall discuss other, more fundamental, reasons for introducing grand ensembles.

Just as we imagined the temperature of a system to be fixed through the interaction between the system and a "temperature bath", we can now imagine an interaction between the system and a "particle bath" or "particle bank" with which it can exchange particles, just as it could exchange energy with the temperature bath. To retain as close an analogy as possible with the term temperature bath, such a particle bath should be called "activity bath", "fugacity bath", or "chemical potential bath".

We shall start with developing the theory of the grand ensembles in a rather general way and assume that our system contains r different kinds of particles. The number of particles of the i-th kind will be denoted by ν_i, and the density of our representative grand ensemble will now be a function not only of the appropriate p's and q's, but also of the ν_i,

$$\varrho = \varrho(\nu_1, \nu_2, \cdots, \nu_r; p, q), \tag{5.0901}.$$

where, of course, the number of degrees of freedom, that is the number of the p's and q's, depends on the actual values of the ν_i. If each particle of the i-th kind possesses s_i degrees of freedom, the total number, s, of degrees of freedom of a system characterised by $\nu_1, \nu_2, \cdots, \nu_r$ will be given by the equation

$$s = \sum_i \nu_i s_i. \tag{5.0902}$$

The density of a grand ensemble is defined in such a way that

$$\varrho(\nu_1, \nu_2, \cdots, \nu_r; p_1, \cdots, p_s, q_1, \cdots, q_s)\, dp_1 \cdots dp_s\, dq_1 \cdots dq_s$$

is the fraction of systems in the ensemble with ν_i particles of the i-th kind and with the p_k and q_k within the phase-space volume element $dp_1 \cdots dp_s$ $dq_1 \cdots dq_s$. We shall write for the sake of brevity

$$d\Omega = dp_1 \cdots dp_s\, dq_1 \cdots dq_s, \tag{5.0903}$$

where we have to remember that $d\Omega$ depends on the ν_i, since s depends on them.

From the above definition of the density it follows that ϱ satisfies the normalisation condition

$$\sum_{\nu_1=0}^{\infty} \sum_{\nu_2=0}^{\infty} \cdots \sum_{\nu_r=0}^{\infty} \int_{\Gamma} \varrho(\nu_1, \nu_2, \cdots, \nu_r; p, q)\, d\Omega = 1. \tag{5.0904}$$

Let us now discuss briefly the implications of the fact that the ν_i atoms of the i-th kind are identical. Two situations differing only in the interchange of two particles of the same kind can be considered to correspond either to different phases of the system, or to the same phase. If they are considered to give us the same phase, we are dealing with *generic phases*, as Gibbs calls them, while if they are considered to correspond to different phases, we call the phase defined in that way a *specific phase*, another term due to Gibbs. It is easily seen that for given values of the ν_i each generic phase is made up out of $\prod \nu_i!$ different specific phases.

We shall only be interested in stationary grand ensembles and, moreover, in *canonical grand ensembles* which are defined by the following specific density:

$$\varrho_{\text{spec}} = \frac{1}{\nu_1! \nu_2! \cdots \nu_r} e^{-q + \alpha_1 \nu_1 + \alpha_2 \nu_2 + \cdots + \alpha_r \nu_r - \beta\varepsilon}, \tag{5.0905}$$

where $\varepsilon = \varepsilon(\nu_i; p, q; a_k)$ is the total energy of the system. Gibbs had, instead of β, a parameter Θ, the *modulus* of the system, which is the reciprocal of β. The quantity q does not depend on the p's and the q's, but will depend on β, the α_i, and the external parameters a_k. We shall presently see the physical meaning of q, β, and the α_i (see Eqs.(5.0915), (5.0918) and (5.0920), and (5.0926) or (5.0927))

From Eq.(50905) it follows that the generic density ϱ_{gen} is given by the equation

$$\varrho_{\text{gen}} = \sum \varrho_{\text{spec}} = e^{-q+\alpha_1\nu_1+\alpha_2\nu_2+\cdots+\alpha_r\nu_r-\beta\varepsilon}, \qquad (5.0906)$$

where the summation is over all different specific phases making up one generic phase and where we have used the fact that the energies of all the specific phases corresponding to the same generic phase will be the same.

Just as a macrocanonical (petit) ensemble can be considered to be built up out of microcanonical ensembles, each canonical grand ensemble is the aggregate of a large number of macrocanonical petit ensembles, as can be seen from Eqs.(5.0905) or (5.0906), since for each of the petit ensembles the ν_i are fixed and the formula reduces to the familiar formula for the density of a macrocanonical ensemble.

The quantity q, which is called the *grand potential* by Kramers, is determined by the normalisation condition (5.0904), and we have

$$e^q = \sum_{\nu_1} \cdots \sum_{\nu_r} \frac{e^{\alpha_1\nu_1+\cdots+\alpha_r\nu_r}}{\nu_1! \cdots \nu_r!} \int_\Gamma e^{-\beta\varepsilon} \, d\Omega. \qquad (5.0907)$$

The right-hand side of Eq.(5.0907) is called the *grand partition function*.

Using Eq.(5.0204) from the theory of the macrocanonical ensembles for the integral in Eq.(5.0907), we can write this last equation in the form

$$e^q = \sum_{\nu_1} \cdots \sum_{\nu_r} \frac{e^{\alpha_1\nu_1+\cdots+\alpha_r\nu_r-\beta\psi}}{\nu_1! \cdots \nu_r!}. \qquad (5.0908)$$

From Eqs.(5.0907) or (5.0908) we see that

$$q = q(\beta, \alpha_i, a_k). \qquad (5.0909)$$

We shall denote the average over a canonical grand ensemble by pointed brackets: $\langle \cdots \rangle$. The average of a phase function $G(p, q; \nu_i)$, which now will in general also be a function of the ν_i, is given by the equation

$$\langle G \rangle = \sum_{\nu_i} \int_{\text{spec}} G(p, q; \nu_i) \varrho_{\text{spec}} \, d\Omega = \sum_{\nu_i} \int_{\text{gen}} G(p, q; \nu_i) \varrho_{\text{gen}} \, d\Omega, \qquad (5.0910)$$

where $\int_{\text{spec}}$ indicates integration over all different specific phases while $\int_{\text{gen}}$ indicates integration over all different generic phases only.

Let us now consider the variation of q as a function of variations in β, the α_i, and the a_k. From Eqs.(5.0907), (5.0910), and (5.0905) we get, after a straightforward calculation,

$$\delta q = \sum_i \langle \nu_i \rangle \, \delta\alpha_i - \langle \varepsilon \rangle \, \delta\beta - \beta \, \delta\langle\varepsilon\rangle, \qquad (5.0911)$$

where

$$\langle \varepsilon \rangle \;=\; -\sum_k \langle A_k \rangle \, \delta a_k, \tag{5.0912}$$

$$A_k \;=\; -\frac{\partial \varepsilon}{\partial a_k}. \tag{5.0913}$$

We can write Eq.(5.0911) in the form

$$\delta \left[-q + \sum_i \langle \nu_i \rangle \, \alpha_i - \beta \, \langle \varepsilon \rangle \right] \;=\; \sum_i \alpha \, \delta \langle \nu_i \rangle - \beta \big(\delta \langle \varepsilon \rangle - \langle \delta \varepsilon \rangle \big), \tag{5.0914}$$

and by arguments similar to thos used in §§ 2.7, 4,4, or 5.3 we find that

$$\beta^{-1} \;=\; kT, \tag{5.0915}$$

$$\frac{S}{k} \;=\; q - \sum_i \alpha_i \, \langle \nu_i \rangle + \beta \, \langle \varepsilon \rangle, \tag{5.0916}$$

where T is the temperature and S the entropy.[12]

Comparing Eqs.(5.0916) and (5.0906), we see that, apart from a multiplying constant, the entropy is equal to the average of the logarithm of the generic density,

$$S \;=\; -k \, \langle \ln \varrho_{\text{gen}} \rangle.$$

We can compare this result with the relationship found in Chapter 1 between Boltzmann's H and the entropy. There we saw that H, which is the average of the logarithm of the density function, and $-S/k$ were the same.

From Eq.(5.0916) it follows that the Helmholtz free energy F, expressed in terms of q, β, the α_i, and the $\langle \nu_i \rangle$, is given by the equation

$$F \;=\; \langle \varepsilon \rangle - ST \;=\; -\frac{q}{\beta} + \sum_i \mu_i \, \langle \nu_i \rangle, \tag{5.0917}$$

where

$$\mu_i \;=\; \frac{\alpha_i}{\beta}. \tag{5.0918}$$

The variation of βF is given by the equation

$$\delta \, \beta F \;=\; \sum_i \alpha_i \, \delta \langle \nu_i \rangle + \langle \varepsilon \rangle \, \delta \beta + \beta \, \langle \delta \varepsilon \rangle, \tag{5.0919}$$

---

[12]It is perhaps well to remind ourselves that when we talk glibly about "the entropy", we really mean "the entropy of the system represented by the canonical grand ensemble under discussion", and similarly for other quantities.

and we see that the μ_i are the partial (Helmholtz) free energies,

$$\mu_i = \left(\frac{\partial F}{\partial \langle \nu_i \rangle} \right)_{\beta, a_k}. \tag{5.0920}$$

Let us assume for a moment that the volume V is one of the external parameters and, furthermore, that it is the only one. Introducing the thermal potential, or Gibbs free energy, G,

$$G = F + PV = -\frac{q}{\beta} + \sum_i \mu_i \langle \nu_i \rangle + PV, \tag{5.0921}$$

and considering the variation of βG,

$$\delta \beta G = \sum_i \alpha_i \, \delta \langle \nu_i \rangle + \left(\langle \varepsilon \rangle + PV \right) \delta \beta + \beta V \, \delta P, \tag{5.0922}$$

we see that the μ_i are also the partial thermal potentials or the partial (Gibbs) free energies,

$$\mu_i = \left(\frac{\partial G}{\partial \langle \nu_i \rangle} \right)_{\beta, P}. \tag{5.0920}$$

It may be mentioned in passing that the α_i are related to the absolute *activities*, z_i, which are sometimes called the *fugacities*[13] or the active number density[14] and are also sometimes denoted by λ_i, by the equation

$$z_i = e^{\alpha_i}. \tag{5.0924}$$

If the system under consideration is homogeneous in all the kinds of particles in it, G will be a homogeneous function of the first degree in the $\langle \nu_i \rangle$ and we have

$$G = \sum_i \langle \nu_i \rangle \frac{\partial G}{\partial \langle \nu_i \rangle} = \sum_i \mu_i \langle \nu_i \rangle. \tag{5.0925}$$

Combining Eqs.(5.0921) and (5.0925), we get in this case[15]

$$q = \frac{pV}{kT}. \tag{5.0926}$$

It will have been noticed that the grand potential resembles in many ways the q-potential discussed in Chapter 4. Indeed, Eq.(5.0926) is the same

---

[13] J.E. and M.G.Mayer, *Statistical Mechanics*, Wiley, New York (1940).

[14] J.O.Hirschfelder, C.F.Curtiss, and R.B.Bird, *Molecular Theory of Gases and Liquids*, Wiley, New York (1954).

[15] The grand potential q is related to the quantity Ω introduced by Gibbs through the equation $\Omega = -q/\beta$.

as Eq.(4.426), while Eqs.(5.0911) and (5.0916) have as their counterparts
Eqs.(4.421) and (4.420). In order to see the analogy between Eqs.(5.0916)
and (4.420) more clearly, we write Eq.(5.0916) in the form

$$q = \beta \left(ST - \sum_i \mu_i \langle \nu_i \rangle - \langle \varepsilon \rangle \right). \tag{5.0927}$$

From Eq.(5.0911) it follows that the grand potential can be used to derive
various averages, as follows (compare Eqs.(4.422) to (4.424)):

$$\langle \nu_i \rangle = \frac{\partial q}{\partial \alpha_i}, \tag{5.0928}$$

$$\langle \varepsilon \rangle = -\frac{\partial q}{\partial \beta}, \tag{5.0929}$$

$$\langle A_k \rangle = \frac{1}{\beta} \frac{\partial q}{\partial a_k}. \tag{5.0930}$$

In the special case $a_k \equiv V$ we have from Eq.(5.0930)

$$P = kT \frac{\partial q}{\partial V}, \tag{5.0931}$$

which will give us the equation of state, once q has been found. Of course,
in the case of a homogeneous system, the equation of state follows directly
from Eq.(5.0926), which is then the same as Eq.(5.0931), since in this
particular case q will be directly proportional to V.

5.10.   Fluctuations in a Canonical Grand Ensemble

In order to show that canonical grand ensembles can be used to describe
systems in statistical equilibrium, we must consider the fluctuations of
the various quantities around their average values and show that these
in general will be small, at least if the total number of particles involved
is large. Let us immediately remark here that sometimes subsystems of a
larger system will be considered, and then the fluctuations may not be small
but quite substantial, for instance, when we consider density fluctuations in
a gas at a temperature near the critical temperature. In that case, however,
these fluctuations are real; it is thus a virtue rather than a drawback of the
canonical grand ensembles that they contain the possibility of describing
real fluctuations.

However, before deriving expressions for the fluctuations in a canonical
grand ensemble and discussing some special cases, we shall briefly consider
an approach to fluctuations which was first developed by Einstein in his
considerations of critical opalescence. Although Einstein based his considerations upon the relation between entropy and probability (compare
Eq.(4.413) and see also Problem 7 at the end of the present chapter), we
shall use ensemble theory.

We first of all note that in a grand canonical ensemble the quantity

$$P(\nu; p, q; a_k)\, d\Omega \;=\; e^{-q + \alpha\nu - \beta\varepsilon}\, d\Omega \tag{5.1001}$$

gives us the probability that there are in the ensemble — for which we have taken, to simplify the discussion, a one-component ensemble — systems with ν particles and with p- and q-values lying within the extension in phase $d\Omega$ (compare Eq.(5.0906)). Let us now look at the fluctuations in some variable, λ say. To find the probability $P(\lambda)$ that a certain value of λ is realised within a range $d\lambda$, we must take the sum of all $P(\nu; p, q, a_k)$ compatible with those values of λ. We thus find

$$P(\lambda)\, d\lambda \;=\; \int_{\lambda}^{\lambda + d\lambda} d\lambda \sum_{\nu} \int d\Omega\, e^{-q + \alpha\nu - \beta\varepsilon}, \tag{5.1002}$$

where the summation over ν and the integration over phase space is only over those regions that are compatible with λ lying between λ and $\lambda + d\lambda$. If we introduce a quantity $q(\lambda)$ by the equation

$$e^{q(\lambda)} \;=\; \left[\sum_{\nu} \int d\Omega\, e^{\alpha\nu - \beta\varepsilon} \right]_{\lambda < \lambda < \lambda + d\lambda}, \tag{5.1003}$$

we can write Eq.(5.1002) in the form

$$P(\lambda) \;=\; e^{-\Delta q}, \qquad \Delta q = q - q(\lambda), \tag{5.1004}$$

where q is the equilibrium value of q.[16] If $\langle \lambda \rangle$ is the equilibrium value of λ and we define $\Delta\lambda$ as $\Delta\lambda = \lambda - \langle \lambda \rangle$, it follows from the thermodynamic meaning of q that $\Delta q / \beta$ is equal to the minimum work ΔA required to produce the fluctuation $\Delta\lambda$. We can therefore rewrite Eq.(5.1004) in the form

$$P(\lambda) \;=\; e^{-\beta \Delta A}, \tag{5.1005}$$

and one can easily convince oneself that this formula is valid, independent of which kind of ensemble we are considering. If we are considering a macrocanonical rather than a grand canonical ensemble, ΔA is equal to the change in the Helmholtz free energy.[17]

---

[16]We should note that in our derivation we have neglected a possible factor in $P(\lambda)$ depending on λ, which follows from the fact that we are integrating over a range of λ-values. For a discussion of this factor see, for instance, R.C.Tolman, *The Principles of Statistical Mechanics*, Oxford University Press (1938), § 141.

[17]Although there is a superficial resemblance between expression (5.1005) and the Maxwell-Boltzmann distribution, it must be emphasised that δA is here a free energy and not an internal energy (compare also Problem 5 at the end of this chapter).

In most cases of interest — the main exception being the case of critical fluctuations (*vide infra*) — we have $\beta \Delta A \gg 1$ as soon as λ becomes appreciably different from $\langle \lambda \rangle$; this is, of course, stating in a different way that almost all fluctuations are negligibly small, a fact we have appealed to repeatedly. Moreover, if q corresponds to the equilibrium situation it follows that we have $[\partial q / \partial \lambda]_{\lambda = \langle \lambda \rangle} = 0$, so that we can to a fair approximation write

$$\Delta q = \frac{1}{2} \frac{\partial^2 q}{\partial \lambda^2} (\lambda - \langle \lambda \rangle)^2, \tag{5.1006}$$

and we see that to this approximation the probability distribution (5.1004) is a Gaussian one. It also easily follows that the mean square deviation in λ satisfies the equation

$$\langle (\lambda - \langle \lambda \rangle)^2 \rangle = \left[\frac{\partial^2 q}{\partial \lambda^2} \right]^{-1}. \tag{5.1007}$$

Although this formula looks simple, it is often difficult to apply, since it is not at all obvious, in general, which variables must be kept constant when the derivative is taken with respect to λ.

For many applications it is advantageous to use the following thermodynamic expression for the minimum work ΔA:[18]

$$\Delta A = \frac{1}{2} (\Delta S \Delta T - \Delta P \Delta V), \tag{5.1008}$$

and Eq.(5.1005) becomes

$$P(\lambda) = e^{-\beta(\Delta S \Delta T - \Delta P \Delta V)/2}. \tag{5.1009}$$

Let us now return to specific cases and for the moment generalise our discussion to that of an ensemble of systems with several components. The calculation of the fluctuations in the ν_i, the A_k, and ε is completely analogous to the calculation of the fluctuations in §5.4, and we shall therefore only indicate the way in which these fluctuations can be calculated and give the results.

Taking the derivative of the normalising condition with respect to the α_i, the a_k, or β, respectively, we obtain Eqs.(5.0928) to (5.0930). Once more taking the derivative with respect to the same variables, we get, respectively,

$$\left\langle (\nu_i - \langle \nu_i \rangle)^2 \right\rangle = \frac{\partial^2 q}{\partial \alpha_i^2} \left(= \frac{\partial \langle \nu_i \rangle}{\partial \alpha_i} \right), \tag{5.1010}$$

$$\left\langle (A_k - \langle A_k \rangle)^2 \right\rangle = \frac{1}{\beta^2} \frac{\partial^2 q}{\partial a_k^2} + \frac{1}{\beta} \left\langle \frac{\partial^2 \varepsilon}{\partial a_k^2} \right\rangle, \tag{5.1011}$$

$$\left\langle (\varepsilon - \langle \varepsilon \rangle)^2 \right\rangle = \frac{\partial^2 q}{\partial \beta^2} \left(= -\frac{\partial \langle \varepsilon \rangle}{\partial \beta} \right). \tag{5.1012}$$

---

[18]See, for instance, L.D.Landau and E.M.Lifshitz, *Statistical Physics*, Oxford University Press (1958), p.351.

To derive Eqs.(5.1010) to (5.1012), we write the normalising condition in the form

$$\sum_i \int_\Gamma \left[\prod_i \frac{e^{\alpha_i \nu_i}}{\nu_i!} \right] e^{-q-\beta\varepsilon} \, d\Omega \; = \; 1.$$

Taking the derivative with respect to, for example, a_k, we find

$$\sum_i \int_\Gamma \left[\prod_i \frac{e^{\alpha_i \nu_i}}{\nu_i!} \right] e^{-q-\beta\varepsilon} \left\{ -\frac{\partial q}{\partial a_k} - \beta \frac{\partial \varepsilon}{\partial a_k} \right\} d\Omega \; = \; 0,$$

and, using Eqs.(5.0910), (5.0905), and (5.0913), we find Eq.(5.0930). Taking once more the derivative with respect to a_k we get

$$\sum_i \int_\Gamma \left[\prod_i \frac{e^{\alpha_i \nu_i}}{\nu_i!} \right] e^{-q-\beta\varepsilon}$$
$$\left\{ \left(-\frac{\partial q}{\partial a_k} - \beta \frac{\partial \varepsilon}{\partial a_k} \right)^2 - \frac{\partial^2 q}{\partial a_k^2} - \beta \frac{\partial^2 \varepsilon}{\partial a_k^2} \right\} d\Omega \; = \; 0,$$

from which Eq.(5.1011) follows. Equations (5.1010) and (5.1012) can be derived in a similar manner.

In general q will be proportional to the total number of particles, and

$$\frac{\left\langle (\varepsilon - \langle\varepsilon\rangle)^2 \right\rangle}{\langle\varepsilon\rangle^2} \qquad \text{and} \qquad \frac{\left\langle (A_k - \langle A_k\rangle)^2 \right\rangle}{\langle A_k\rangle^2}$$

will be of the order of magnitude of $1/N$, where $N = \sum_i \langle\nu_i\rangle$, while

$$\frac{\left\langle (\nu_i - \langle\nu_i\rangle)^2 \right\rangle}{\langle\nu_i\rangle^2}$$

will be of the order of $\langle\nu_i\rangle^{-1}$.

The correlations between different quantities can be evaluated in the same way as the fluctuations, as in § 5.4.

We shall apply some of the results we have just obtained to the case of a perfect gas in § 5.12, but here we wish to consider a few special cases.

First of all, we see that it follows from Eq.(5.1012) that we have

$$\left\langle (\varepsilon - \langle\varepsilon\rangle)^2 \right\rangle \; = \; -\frac{\partial\langle\varepsilon\rangle}{\partial\beta} \; = \; kT^2 C_v, \tag{5.1013}$$

where C_v is the specific heat at constant volume. Usually, C_v will be of the order of magnitude of the number of particles in the system, and as $\langle\varepsilon\rangle$

is of the same order of magnitude, it follows that the relative fluctuations in energy will be of the order of $\langle \nu \rangle^{-1}$, and thus negligibly small (compare Eq.(5.1210) for the case of a perfect gas). An exception will occur when we have a specific heat anomaly, and we see that we may expect large energy fluctuations at a phase transition point where one has such a specific heat anomaly.

Another case where large energy fluctuations may occur was suggested by Gibbs. Consider a box containing a liquid with over it saturated vapour and on top of it all a piston (see Fig.5.3). If this system is placed in a thermostat, it will have an infinite specific heat, since any amount of heat taken up or given off by the system will result in evaporation or condensation of vapour, but not in a change in the temperature. Energy fluctuations may thus become arbitrarily large, provided not all of the substance is either evaporated or condensed.

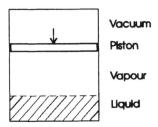

Fig.5.3. A system in which very large energy fluctuations will occur.

Let us now turn to density fluctuations and once again consider an ensemble of one-component systems. From Eq.(5.1010) we have

$$\frac{\left\langle (\nu - \langle \nu \rangle)^2 \right\rangle}{\langle \nu \rangle^2} = \frac{1}{\langle \nu \rangle^2} \frac{\partial \langle \nu \rangle}{\partial \alpha}. \tag{5.1014}$$

From Eq.(5.0922) and the fact that for a homogeneous system $\beta G = \alpha \langle \nu \rangle$ it follows that

$$\langle \nu \rangle \, \delta \alpha = \beta V \, \delta P, \qquad (\beta : \text{const}) \tag{5.1015}$$

and thus that[19]

$$\langle \nu \rangle \left(\frac{\partial \alpha}{\partial \langle \nu \rangle} \right)_{\beta, V} = -\frac{\beta V^2}{\langle \nu \rangle} \left(\frac{\partial P}{\partial V} \right)_{\beta, \langle \nu \rangle}. \tag{5.1016}$$

---

[19]In deriving Eq.(5.1016) we have used the relation

$$\left(\frac{\partial y}{\partial x} \right)_z = -\left(\frac{\partial y}{\partial z} \right)_x \left(\frac{\partial z}{\partial x} \right)_y,$$

We can thus rewrite Eq.(5.1014) in the form

$$\frac{\left\langle (\nu - \langle\nu\rangle)^2 \right\rangle}{\langle\nu\rangle^2} = -\frac{1}{\beta V^2}\frac{\partial V}{\partial P} = \frac{\kappa}{\beta V}, \qquad (5.1017)$$

where κ is the compressibility,

$$\kappa = -\frac{1}{V}\frac{\partial V}{\partial P}. \qquad (5.1018)$$

In a critical point $\partial P/\partial V$ vanishes and we should therefore expect large density fluctuations. Smoluchowski suggested that this was the reason for the *critical opalescence* effect, but, in fact, in the scattering of light, different regions of a system take part and it is necessary that there are long-range correlations between the density at different points in the system — as was first shown by Ornstein and Zernike. We shall in a moment show that there exists a relation between this long-range correlation and the fluctuations in the number of particles ΔN contained in a small, but macroscopic volume ΔV of the system.

To find how Eq.(5.1017) for the density fluctuations must be changed in the case when $\partial P/\partial V$ vanishes, we use the considerations of § 5.9 to write the expression for the density fluctuations in the form

$$\langle (\nu - \langle\nu\rangle)^2 \rangle = \sum_\nu \frac{e^{-q+\alpha\nu}}{\nu!}(\nu - \langle\nu\rangle)^2 \int e^{-\beta\varepsilon}\,d\Omega. \qquad (5.1019)$$

Changing the sum over ν into an integral, we can write Eq.(5.1019) in the form

$$\langle (\nu - \langle\nu\rangle)^2 \rangle = \int d\varepsilon \int d\nu\,(\nu - \langle\nu\rangle)^2\,e^{-\Phi(\nu,\varepsilon)-\beta\varepsilon}, \qquad (5.1020)$$

and the normalisation condition for the ensemble density in the form

$$1 = \int d\nu \int d\varepsilon\,e^{-\Phi(\nu,\varepsilon)-\beta\varepsilon}; \qquad (5.1021)$$

---

which follows from writing

$$\delta y = \left(\frac{\partial y}{\partial x}\right)_z \delta x + \left(\frac{\partial y}{\partial z}\right)_x \delta z,$$

and substituting in it a similar relation for δz in terms of δy and δx, as well as the fact that for a homogeneous system we have

$$\left(\frac{\partial \langle\nu\rangle}{\partial V}\right)_P = \frac{\langle\nu\rangle}{V}.$$

the function $\Phi(\nu, \varepsilon)$ is a generalisation of the function $\phi(\varepsilon)$ from §5.5.

In order to find the replacement for Eq.(5.1017) we first of all expand Φ as a function of ν in the neighbourhood of the value $\nu = \langle \nu \rangle$, bearing in mind that the integrands in Eqs.(5.1020) and (5.1021) are steeply peaked around that value of ν. The partial derivatives occurring in the expansion of Φ can be expressed in terms of partial derivatives with respect to V, and hence in terms of partial derivatives of P with respect to V, if we use the relation

$$\beta P = -\left\langle \left(\frac{\partial \Phi}{\partial V} \right)_\beta \right\rangle ; \qquad (5.1022)$$

this equation follows from the normalisation equation (5.1021) (compare Eq.(5.0930)). We now use the fact that $\partial \Phi / \partial \nu$ vanishes for $\nu = \langle \nu \rangle$ and that in the critical point both $\partial P / \partial V$ and $\partial^2 P / \partial V^2$ vanish, and after a simple integration we get, instead of Eq.(5.1017), the relation

$$\frac{\left\langle (\nu - \langle \nu \rangle)^2 \right\rangle}{\langle \nu \rangle^2} = \frac{B}{\sqrt{\langle \nu \rangle}} \left[\beta n^2 \frac{\partial^3 P}{\partial n^3} \right]^{-1/2}, \qquad (5.1023)$$

where we have used the fact that $\langle \nu \rangle$ is proportional to V, where the constant B is of the order of magnitude unity, and where n is the particle density,

$$n = \frac{\langle \nu \rangle}{V}. \qquad (5.1024)$$

We note that, indeed, the fluctuations are anomalously large.

Let us now briefly consider the relation between density fluctuations and the so-called *radial distribution function*, a relation which plays an important role in the discussion of critical opalescence. As we mentioned earlier, the scattering of electromagnetic waves by a fluid is the result of the density fluctuations in the fluid. In fact, the intensity of the scattered light can be shown to be proportional to[20]

$$1 + n \int [g(r) - 1] d^3 r, \qquad (5.1025)$$

where $g(r)$ is the radial distribution function which has the meaning that $ng(r) d^3 r$ is the number of particles in the volume element $d^3 r$ around r, if there is a particle at the origin.

Let now $\Delta(r)$ be the function which is unity, if r lies within the volume ΔV and which vanishes elsewhere. For the number ΔN contained in a small, but macroscopic volume ΔV we can then write

$$\Delta N = \sum_{i=1}^{\nu} \Delta(r_i). \qquad (5.1026)$$

---

[20] R.Balescu, *Equilibrium and Nonequilibrium Statistical Mechanics*, Wiley, New York (1975), §8.1.

One can easily prove that

$$\langle \Delta N \rangle = n\Delta V., \tag{5.1027}$$

and since $[\Delta(\mathbf{r})]^2 = \Delta(\mathbf{r})$ it follows that

$$\langle (\Delta N - \langle \Delta N \rangle)^2 \rangle = \langle \Delta N \rangle \left[1 + n \int \{g(\mathbf{r}) - 1\} d^3\mathbf{r} \right], \tag{5.1028}$$

which proves the relation between density fluctuations and the radial distribution, and because of Eq.(5.1025) the relation between the density fluctuations and the intensity of scattered light.

We finally want to consider pressure fluctuations. We discussed these briefly in §5.8 for the case of a perfect gas when we could invoke Eq. (5.0811). In that discussion we emphasised, as we had done in §5.4, that the physical meaning of a pressure naturally involves the idea that one is dealing with a time-averaged quantity. The question of pressure fluctuations in a perfect gas has been the subject of an extensive discussion between Klein, Münster, and Wergeland.[21] As this topic provides us both with an instructive example of the application of fluctuation theory and with an occasion to clear up a few difficult conceptual points, we shall discuss it here in some detail.

To begin with, we remind ourselves that according to Eq.(5.1011) the fluctuations in the pressure are given by the formula

$$\left\langle (P - \langle P \rangle)^2 \right\rangle = \frac{1}{\beta} \frac{\partial \langle P \rangle}{\partial V} + \frac{1}{\beta} \left\langle \frac{\partial^2 \varepsilon}{\partial V^2} \right\rangle, \tag{5.1029}$$

where, of course, here P is not a truly macroscopic quantity — as was pointed out by Münster — but the microscopic quantity $-\partial \varepsilon / \partial V$. We did point out in §5.4 that the second term on the right-hand side of Eq.(5.1029) becomes infinite for an infinitely steep wall potential. This is not surprising, since it corresponds to the instantaneous infinite forces exerted by the wall on the gas particles when they collide with the wall. Of course, time-averaging these instantaneous forces will lead to a finite result. The fact that the fluctuations depend crucially upon the nature of the potential field near the wall by itself shows that we must tread extremely warily. For some special choices of the wall potential one can evaluate the right-hand side of Eq.(5.1029). One example was worked out by Fowler,[22] who used a potential varying as an inverse power of the distance of the particles from the wall and found that in that case we have

$$\frac{\left\langle (P - \langle P \rangle)^2 \right\rangle}{\langle P \rangle^2} = A \cdot N^{-2/3}, \tag{5.1030}$$

[21] H.Wergeland, *Proc. Norwegian Acad. Sci.* **28**, 106 (1955); M.J.Klein, *Physica* **26**, 1073 (1960); A.Münster, *Physica* **26**, 1111 (1960); H.Wergeland, in *Fundamental Problems in Statistical Mechanics* (Ed. E.G.D.Cohen), North-Holland, Amsterdam (1962).

[22] R.H.Fowler, *Statistical Mechanics*, Cambridge University Press (1936), p.756.

that is, small fluctuations, but fluctuations that are larger than normal ones, which are proportional to N^{-1}. Fowler also found that the constant A tended to infinity when the potential became infinitely steep. A different potential was investigated by Wergeland (see Problem 17 at the end of this chapter); he found normal fluctuations, but again fluctuations tending to infinity as the steepness of the wall potential increased.

To see why it is reasonable to expect Eq.(5.1029) to give non-physical results, we must make clear to ourselves what kind of fluctuations we are discussing. Strictly speaking, Eq.(5.1029) tells us the spread in the values of P we can expect when we are looking successively at the different systems which together make up our ensemble. As $\langle P \rangle$ itself is finite, we know that the chance of picking out of the ensemble a system for which we get an infinite pressure is infinitesimal. On the other hand, in an actual physical situation we are looking at the fluctuations in the pressure observed *in one system*. If we treat an ensemble as representing the time evolution of a single system, that is, as mathematically equivalent to a microcanonical ensemble and hence, through the ergodic theorem, as determining the time averages of phase functions, we are back to the approach used in §5.8 — and incidentally to the approach favoured by Münster. However, there is a different way of looking at Eq.(5.1029). When we are considering fluctuations in a system represented by an ensemble, and when we are then using the formalism of ensemble theory to derive Eq.(5.1029), we are comparing two situations with slightly different values of V (compare the discussion in §5.9). That means that we are investigating how small reversible changes in the a_k affect the system represented by the ensemble. In order that we can throughout use the ensemble formalism, it is necessary — as was emphasised by Wergeland — that the change $a_k \rightarrow a_k + \delta a_k$ must be quasistatic as well as infinitesimal, which means that it must be an *adiabatic* change. This means, in turn, that we should have specified when writing down $\partial^2 \varepsilon / \partial V^2$ *not* that this derivative was taken for constant p's and q's — which, as shown by Wergeland, corresponds to infinitely fast changes in V — but for constant entropy. This means that, strictly speaking, Eq.(5.1011) should be written in the form

$$\left\langle (A_k - \langle A_k \rangle)^2 \right\rangle = \frac{1}{\beta} \left. \frac{\partial \langle A_k \rangle}{\partial a_k^2} \right|_{\alpha, \beta} + \frac{1}{\beta} \left\langle \left. \frac{\partial A_k}{\partial a_k} \right|_S \right\rangle. \qquad (5.1031)$$

If in the case of a perfect gas we apply the corresponding equation for the macrocanonical ensemble, we can use thermodynamic relations to derive again Eq.(5.0812). On the other hand, if we consider a perfect gas in a cubical vessel we can use the well known equation from classical mechanics[23]

$$J_i = \oint p_i \, dq_i, \qquad (5.1032)$$

---

[23]See, for instance, D.ter Haar, *Elements of Hamiltonian Mechanics*, Pergamon, Oxford (1971), §6.2.

for the action integrals J_i, which are also the adiabatic invariants. For a particle moving with a kinetic energy $\mathcal{T}_i$ corresponding to q_i Eq.(5.1032) gives us

$$J_i = 2V^{1/3} \sqrt{2m\mathcal{T}_i},\tag{5.1033}$$

so that the energy of the system, expressed in terms of the J_i is given by the equation

$$\varepsilon = \sum_{i=1}^{3N} \mathcal{T}_i = \sum_i \frac{J_i^2}{8mV^{2/3}},\tag{5.1034}$$

and since constant entropy now means that the adiabatic invariants J_i must be kept constant so that we find

$$\left(\frac{\partial^2 \varepsilon}{\partial V^2} \right)_S = \left(\frac{\partial^2 \varepsilon}{\partial V^2} \right)_{J_i} = \frac{10\varepsilon}{9V^2},\tag{5.1035}$$

and this together with expression (5.0803) for the average value of the energy in a macrocanonical ensemble and the relation $P = 2\bar{\varepsilon}/3V$ again leads to Eq.(5.0812).

5.11.   The Coupling of Two Canonical Grand Ensembles

In § 5.6 we discussed in some detail the coupling of two petit ensembles. Since the situation is much the same in the case of grand ensembles, we shall discuss only those points where new features are appearing. Consider two canonical grand ensembles, A and B, and assume for the sake of simplicity that in each of the two systems only one kind of particle is present (the generalisation to the case where there are more kinds of particles present is straightforward). Their specific densities are

$$\varrho_{\text{spec}}^{A} = \frac{1}{\nu_A!} e^{-q_A + \alpha_A \nu_A - \beta_A \varepsilon_A}, \qquad \varrho_{\text{spec}}^{B} = \frac{1}{\nu_B!} e^{-q_B + \alpha_B \nu_B - \beta_B \varepsilon_B},\tag{5.1101}$$

and their generic densities are

$$\varrho_{\text{gen}}^{A} = e^{-q_A + \alpha_A \nu_A - \beta_A \varepsilon_A}, \qquad \varrho_{\text{gen}}^{B} = e^{-q_B + \alpha_B \nu_B - \beta_B \varepsilon_B},\tag{5.1102}$$

If the two ensembles are physically coupled to form an ensemble AB and if we may neglect the interaction energy, we see that the specific density of the AB ensemble is given by the equation

$$\varrho_{\text{spec}}^{AB} = \varrho_{\text{spec}}^{A} \varrho_{\text{spec}}^{B} = \frac{1}{\nu_A! \nu_B!} e^{-q_A + \alpha_A \nu_A - \beta_A \varepsilon_A} e^{-q_B + \alpha_B \nu_B - \beta_B \varepsilon_B},\tag{5.1103}$$

and its generic density by the equation

$$\varrho_{\text{gen}}^{AB} = \varrho_{\text{gen}}^{A} \varrho_{\text{gen}}^{B} = e^{-q_A + \alpha_A \nu_A - \beta_A \varepsilon_A} e^{-q_B + \alpha_B \nu_B - \beta_B \varepsilon_B}.\tag{5.1104}$$

We see from Eqs.(5.1103) and (5.1104) that as long as the particles in system A are different from those in system B, we have for the AB ensemble the correct relation between specific and generic densities. Moreover, if $\beta_A = \beta_B$, the AB ensemble will again be a canonical grand ensemble. However, if the systems A and B contain the same kind of particles, we see that the AB ensemble is only canonically distributed as far as the generic phases are concerned — provided $\beta_A = \beta_B$ and $\alpha_A = \alpha_B$, the two familiar conditions of equality of temperature and thermal potential — but not as far as specific phases are concerned. We see here the importance of the generic phases which was clearly recognised by Gibbs. We can also understand now why $\langle \ln \varrho_{gen} \rangle$ rather than $\langle \ln \varrho_{spec} \rangle$ might play the role of Boltzmann's H. Once this point of the pre-eminence of the generic phases is accepted, we must start from Eq.(5.1104) and if the particles in the A and B systems are the same, we must write for the specific density, instead of Eq.(5.1103),

$$\varrho_{spec}^{AB} = \frac{1}{\nu_{AB}!} \varrho_{gen}^{AB} = \frac{1}{\nu_{AB}!} e^{-q_{AB}+\alpha_{AB}\nu_{AB}-\beta_{AB}\varepsilon_{AB}}, \tag{5.1105}$$

where[24]

$$\left. \begin{array}{l} \nu_{AB} = \nu_A + \nu_B, \quad q_{AB} = q_A + q_B, \quad \varepsilon_{AB} = \varepsilon_A + \varepsilon_B, \\ \alpha_{AB} = \alpha_A = \alpha_B, \quad \beta_{AB} = \beta_A = \beta_B, \end{array} \right\} \tag{5.1106}$$

The choice of

$$S = -k \langle \ln \varrho_{gen} \rangle \tag{5.1107}$$

entails the introduction of the factor $1/N!$ in Eq.(5.0311); this can be seen as follows. From Eq.(5.1107) and the equation $F = \langle \varepsilon \rangle - ST$ we get

$$e^{-\beta F} = e^{q - \sum \alpha_i \langle \nu_i \rangle}. \tag{5.1108}$$

If now the $\langle \nu_i \rangle$ are sufficiently large so that their dispersion is negligible, the canonical grand ensemble will practically be a macrocanonical ensemble and we can in all equations substitute $\nu_i \cong \langle \nu_i \rangle$. This corresponds to the equivalence of the macro- and microcanonical ensembles which we discussed in the precediung chapter. In that way, using Eq.(5.0907), we get from Eq.(5.1108)

$$e^{-\beta F} = \frac{1}{\prod_i \langle \nu_i \rangle!} \left[\int_\Gamma e^{-\beta \varepsilon} \, d\Omega \right]_{\nu_i = \langle \nu_i \rangle}, \tag{5.1109}$$

which reduces to Eq.(5.0311) in the case of only one kind of particle — apart from the factor h^{-sN}, which we should have introduced throughout in

---

[24]Note that the indices "A", "B", and "AB" denote different ensembles here and not, as in Eq.(5.1114), different particles taking part in a reaction.

order to get formulæ that are the limiting cases of the quantum-mechanical formulæ to be derived in the next chapter — and also in oirder that the right-hand side of equations such as Eq.(5.1109) are dimensionless.

We may remark here that, although we see from this discussion the plausibility of the introduction of the factor $1/N!$ in Z_Γ, the only decisive argument is the one given by considering quantum statistics and looking at the classical limit of the quantum-mechanical expressions. Since these lead to the factor $1/N!$, we can be sure of the correctness instead of feeling rather safe in including them.

In § 5.6 we discussed at some length the coupling of two petit ensembles. We noted then that the only way in which we could at time t' couple two systems physically and at time t'' decouple them again in such a way that the situation after t'' is the same as that before t' was to couple two macrocanonical ensembles with the same value of β. In proving that and also in proving that, if we couple two macrocanonical ensembles with different values of β, energy will flow from systems in the ensemble with the lower β-value to those in the ensemble with the larger β-value, we used certain lemmas as well as the consequences of a generalized H-theorem.

We can now proceed in a similar manner. Using Eq.(5.0121) to introduce the index of probability η, we see that Eq.(5.1107) can be written in the form

$$S = -k \langle \eta_{\text{gen}} \rangle, \tag{5.1110}$$

which is the counterpart of Eq.(5.0309) for the case of petit ensembles. From the generalised H-theorem we now get instead of Eq.(5.0627)

$$\langle \eta_{\text{gen}} \rangle''_{12} \leqslant \langle \eta_{\text{gen}} \rangle'_{12}, \tag{5.1111}$$

where the index "12" again indicates a combined ensemble. From Eq.(5.1111) we now, by calculations completely similar to those in § 5.6, get instead of Eq.(5.0630) the inequality

$$\beta_1 \left(\langle \varepsilon \rangle''_1 - \langle \varepsilon \rangle'_1 \right) + \beta_2 \left(\langle \varepsilon \rangle''_2 - \langle \varepsilon \rangle'_2 \right)$$
$$+ \sum_i \left[\alpha_{i1} \left(\langle \nu_i \rangle'_1 - \langle \nu_i \rangle''_1 \right) + \alpha_{i2} \left(\langle \nu_i \rangle'_2 - \langle \nu_i \rangle''_2 \right) \right] \geqslant 0, \tag{5.1112}$$

where the equal sign holds, if the two ensembles before the coupling were canonical with $\beta_1 = \beta_2$ and $\alpha_{i1} = \alpha_{i2}$ (for all i).

If we consider for a moment ensembles with only one kind of particle, and if we consider the coupling of two canonical grand ensembles for which $\beta_1 = \beta_2$, but $\alpha_1 \neq \alpha_2$, it follows from Eq.(5.1112) and the conservation of particle number that

$$\langle \nu \rangle'_1 < \langle \nu \rangle''_1, \qquad \text{if } \alpha_1 < \alpha_2, \tag{5.1113}$$

that is, the system with the larger value of α will, on the average, have lost particles to the other system.

Let us now consider the case in which we have three kinds of particles in the systems in the grand ensemble — particles A, B, and AB — and let us assume that the following reaction can take place:

$$A + B \; \rightleftarrows \; AB. \tag{5.1114}$$

Equation (5.1112) can in the case where we couple two ensembles with $\beta_1 = \beta_2$ be written in the form

$$(\alpha_A + \alpha_B - \alpha_{AB}) \left(\langle \nu_A \rangle' - \langle \nu_A \rangle'' \right) \geqslant 0, \tag{5.1115}$$

where we have used the relations

$$\langle \nu_A \rangle' - \langle \nu_A \rangle'' \; = \; \langle \nu_B \rangle' - \langle \nu_B \rangle'' \; = \; \langle \nu_{AB} \rangle'' - \langle \nu_{AB} \rangle', \tag{5.1116}$$

which follow from Eq.(5.1114).

Just as from Eq.(5.0630) we find the condition $\beta_1 = \beta_2$ for equilibrium, so we find from Eq.(5.1115) the well known relation

$$\alpha_A + \alpha_B = \alpha_{AB}, \qquad \text{or} \qquad \mu_A + \mu_B = \mu_{AB}, \tag{5.1117}$$

as an equilibrium condition.

5.12.   Application of the Theory of Classical Grand Ensembles to a Perfect Gas

Although the results we shall obtain are not new and were derived previously in Chapters 1, 2, and 4, and in § 8 of the present chapter, we shall once more illustrate the theory by considering the case of a perfect gas enclosed in a volume V. The energy of the system is given by the equation

$$\varepsilon \; = \; \sum_{i=1}^{\nu} \left(\frac{p_i^2}{2m} + F_i \right), \tag{5.1201}$$

where

$$F_i \; \equiv \; F(x_i, y_i, z_i) \; = \; 0, \qquad \text{if } x_i, y_i, z_i \text{ lies inside } V,$$

and

$$F(x_i, y_i, z_i) \; = \; \infty, \qquad \text{if } x_i, y_i, z_i \text{ lies outside } V.$$

From Eq.(5.0907) we get for the grand potential q the equation

$$e^q \; = \; \sum_{\nu=0}^{\infty} \frac{e^{\alpha \nu}}{\nu!} \int_\Gamma \exp \left[-\beta \sum_i \left(\frac{p_i^2}{2m} + F_i \right) \right] d\Omega. \tag{5.1202}$$

Integration over the x_i, y_i, and z_i gives a factor V^ν, and we get

$$e^q \; = \; \sum_\nu \frac{e^{\alpha \nu} V^\nu}{\nu!} \left[\int_{-\infty}^{+\infty} e^{-\beta u^2 / 2m} \, du \right]^{3\nu},$$

or

$$e^q = \sum_\nu \frac{1}{\nu!} \left[e^\alpha V (2\pi mkT)^{3/2} \right]^\nu,$$

which leads to the equation (compare Eq.(4.610))[25]

$$q = e^\alpha V (2\pi mkT)^{3/2}. \tag{5.1203}$$

Using Eq. (5.0928), we get for $\langle \nu \rangle$

$$\langle \nu \rangle = \frac{\partial q}{\partial \alpha} = q, \tag{5.1204}$$

and combining this with Eq.(5.0926), we have

$$\langle \nu \rangle = \frac{pV}{kT}, \quad \text{or} \quad pV = \langle \nu \rangle kT, \tag{5.1205}$$

the perfect gas law.

We find from Eq.(5.1001) for the fluctuations in the number of particles

$$\left\langle (\nu - \langle \nu \rangle)^2 \right\rangle = \frac{\partial^2 q}{\partial \alpha^2} = \langle \nu \rangle, \tag{5.1206}$$

which shows that, indeed,

$$\frac{\left\langle (\nu - \langle \nu \rangle)^2 \right\rangle}{\langle \nu \rangle^2} = \frac{1}{\langle \nu \rangle}. \tag{5.1207}$$

The average energy and its fluctuations satisfy the equations

$$\langle \varepsilon \rangle = -\frac{\partial q}{\partial \beta} = \frac{3q}{2\beta} = \frac{3}{2} \langle \nu \rangle kT = \left(= \frac{3}{2} PV \right), \tag{5.1208}$$

$$\left\langle (\varepsilon - \langle \varepsilon \rangle)^2 \right\rangle = \frac{\partial^2 q}{\partial \beta^2} = \frac{15 \langle \nu \rangle}{4\beta^2}, \tag{5.1209}$$

$$\frac{\left\langle (\varepsilon - \langle \varepsilon \rangle)^2 \right\rangle}{\langle \varepsilon \rangle^2} = \frac{5}{3 \langle \nu \rangle}. \tag{5.1210}$$

The free energy and the entropy of the system are given by the equations

$$F = -qkT + \alpha \langle \nu \rangle = \langle \nu \rangle kT \left[\ln P - \tfrac{5}{2} \ln T - \tfrac{3}{2} \ln M + \text{const} \right], \tag{5.1211}$$

---

[25]The difference lies in the absence of a factor h^3. We should have introduced a factor $h^{3\nu}$ in Eq.(5.1202) in order to make the right-hand side dimensionless and to make sure that Eq.(5.1202) is the classical limit of the corresponding quantum-mechanical equation. However, these extra factors h do not affect the remaining results of this section and we shall therefore continue to omit them.

and

$$S = qk - k\alpha\langle\nu\rangle - \frac{\langle\varepsilon\rangle}{T} = \langle\nu\rangle k \left[\tfrac{5}{2}\ln T - \ln P + \tfrac{3}{2}\ln M + \text{const}\right],$$
$$(5.1212)$$

where M is the molecular weight of the gas and where we have used for α the equation

$$e^\alpha = \langle\nu\rangle V^{-1} \left(2\pi mkT\right)^{-3/2},$$
$$(5.1213)$$

which follows from Eqs.(5.1203) and (5.1204).

We shall conclude this section by mentioning the paradox that bears Gibbs's name, which is briefly mentioned in the last paragraph of his famous monograph on the equilibrium of heterogeneous substances.[26]   To see that there is a paradox we consider two systems, one consisting of particles of the kind A in a volume V_A at a temperature T, the other of particles of the kind B in a volume V_B at the same temperature T. We now join these two systems together and remove any walls dividing them, so that both systems now have a volume $V_A + V_B$ at their disposal. We then ask what the value of the entropy of the combined system will be, assuming throughout that we can treat both the separate systems and the combined system as perfect gases. From Eq.(5.1107) and calculations similar to those leading to Eq.(5.1212) we find — provided the particles of kind A are different from those of kind B — that

$$S_{AB} = S_A + S_B + \langle\nu_A\rangle k \ln \frac{V_A + V_B}{V_A} + \langle\nu_B\rangle k \ln \frac{V_A + V_B}{V_B}, \quad (5.1214)$$

where S_A and S_B are given by Eq.(5.1212) with the appropriate subscripts. We see that the entropy is increased by the removal of any dividing walls. This is as was to be expected, since the removal of a dividing wall is an irreversible process and the mixing entropy, as the extra term is called, is a consequence of the diffusion of the two gases into one another.

If, however, the particles of kind A and of kind B were identical, and if to simplify the calculations we assume that $V_A = V_B (= V)$ and $\langle\nu_A\rangle = \langle\nu_B\rangle(= N)$, we are dealing with a system of $2N$ particles in a volume $2V$ and we find now, as should be the case,

$$S_{AB} = S_A + S_B. \quad (5.1215)$$

Equation (5.1215) is, of course, not the limit of Eq.(5.1214), but differs from it in the limit by a term $2Nk \ln 2$. The paradoxical result is that somehow or other the particles must have sensed that they were identical.

It is at this point that we can once again demonstrate the reason why the generic rather than the specific density occurs in the definition of the entropy. If we had used instead of Eq.(5.1107) the equation

$$S' = -k \langle\ln \varrho_{spec}\rangle, \quad (5.1216)$$

---

[26]See *The Collected Works of J. Willard Gibbs*, Vol.I, New Haven, Conn. (1948), p.165.

and evaluated the value of S'_{AB} for the case where the two kinds of particles were identical, we would have found instead of Eq.(5.1215) the equation

$$S'_{AB} = S'_A + S'_B + 2kN \ln 2. \quad .$$

$$(5.1217)$$

This shows that we must, indeed, use the generic density.

5.13.   The Relationship between Ensembles and Actually Observed Systems

In this and the next section we shall be concerned with some of the fundamental problems of statistical mechanics referring to the foundations of the subject. We cannot enter very deeply here into a discussion of these problems for which we refer to the literature mentioned in the bibliographical notes.

One could argue that the fact that statistical mechanics has been able to predict accurately and successfully the behaviour of physical systems under equilibrium and near-equilibrium conditions — and even often deal successfully with some non-equilibrium situations — should be a sufficient justification for the methods used. However, a great deal of time and effort has been spent in attempts to show that the statistical formalism follows in a straightforward manner from classical or quantum mechanics. The two following questions are the main ones with which we shall be concerned here:

(A) "Why is it possible to describe the behaviour of almost all physical systems by considering only equilibrium situations?" with the corrolary (A1) "How can one define an equilibrium situation?"

(B) "Why is it possible to describe the behaviour of an actual physical system by considering a large number of identical systems (in ensemble theory) and identifying the average behaviour of this group of systems with the physical system in which we are interested?" with the corollary (B1) "How can we construct an ensemble such that it will "represent" an actual given physical system?"

Question A had occupied Boltzmann in his "Gastheorie" who showed by his H-theorem (see §§ 1.5 and 2.3) that any non-equilibrium situation would develop in the direction of an equilibrium situation. The answer to Question A1 was in that case that the equilibrium situation is the most probable situation, compatible with a few restricting conditions — which still left open the question of how to define the probability of a situation. In its original, unrestricted form the H-theorem proved that any system will tend toward equilibrium, if equilibrium does not originally exist. This would mean that, provided one waited a sufficiently long time, one would find the system in an equilibrium situation and, moreover, this equilibrium would persist forever afterwards. From this follows that equilibrium would be the rule and non-equilibrium the exception, so that Question A is answered.

We shall not consider here the various problems concerning the limitations of the unrestricted H-theorem and the necessity — stressed especially by the Ehrenfests — of the statistical nature of the H-theorem in kinetic theory.

Apart from introducing the H-theorem to explain A, Boltzmann also tried to prove that the average behaviour of a system is the same as its equilibrium behaviour. This statement is meant in the following way: The time average of any phase function taken over a sufficiently long period should be equal to the value of the phase function at equilibrium. To prove this, one has to prove Eq.(5.0704); this is done by the so-called ergodic theory, which we shall briefly discuss in the next section.

We now come to Questions B and B1. One might ask what reason there is to introduce ensembles at all since in actual cases one is always dealing with a single system. Indeed, Gibbs introduced ensembles in order to use them for statistical considerations rather than to illustrate the behaviour of physical systems, even though he did not adhere strictly to his original intentions as set out in the introduction to his monograph. Einstein considered the possibility that the average behaviour of a system in a macrocanonical ensemble would represent the actual behaviour of a system in thermodynamical equilibrium. However, the Ehrenfests, Ornstein,[27] and Uhlenbeck[28] considered ensemble theory still mainly as a mathematical trick to lighten the calculation of average values of phase functions for a system with a given total energy and a given total number of particles, much in the same way as we did in our discussion of Eq.(5.0705) in preceding sections. Uhlenbeck therefore considered Gibbs's method and the Darwin-Fowler method (see § 4.5) to be more or less equivalent. However, Tolman in his monograph succeeded in showing the importance of ensemble theory from a physical point of view. We shall give some of his arguments in the present section, although referring the reader to his monograph for a thorough discussion of this point.

We have mentioned before that there are very severe limitations to the possibilities of obtaining accurate knowledge about the initial condition, say, of the phusical system in which we are interested. One mole of a monatomic gas contains 6×10^{23} atoms, and instead of having available the values of the 36×10^{23} coordinates and momenta which we need to know in order completely to characterise the situation, we can only obtain a few relations between them, for instance, by measuring the pressure, the total linear momentum, and the total angular momentum of the system. Moreover, these quantities will be known to us only within certain margins, just as we will not known the exact total number of particles in the system: Avogadro's number is only known to a few parts in 10^7. Having obtained this scant information, we now want to make predictions about the future behaviour of the system, starting from this information. Clearly, the only way to do so is by using probability theory and statistical methods. If we could have obtained the values of all the coordinates and momenta at one

[27]L.S.Ornstein, *Toepassing der Statistische Mechanica van Gibbs op Molekulair-theoretische Vraagstukken*, Thesis, Leiden (1908).

[28]G.E.Uhlenbeck, *Over Statistische Methoden in de Theorie der Quanta*, Thesis, Leiden (1927).

moment — and if we could have solved the equations of motion — the behaviour of the system would be completely determined, as we would then be dealing with a problem of classical mechanics with a sufficient number of boundary conditions to determine the orbit of the representative point of the system in phase space. However, as we cannot obtain these values, we are reduced to making statements about the most probable behaviour of the system. In order to determine this most probable behaviour, we compare various systems possessing the same values of those quantities that have been measured, but which otherwise may differ widely. This means that we construct an ensemble to represent our system. The question remains as to how we can construct such a representative ensemble. This introduces the problem of a *priori* probabilities. In constructing a classical ensemble, we use the assumption of equal a *priori* probabilities for equal volumes in Γ-space. We shall briefly discuss the plausibility or justification of this choice at the end of this section.

To see what kind of ensemble will best represent a system, we shall state some results, but only prove some of them (proofs of the other statements are asked for as problems at the end of this chapter). We find that

(1) If a system is completely isolated so that it can exchange neither energy nor particles with its surroundings, it will be represented by a microcanonical ensemble;

(2) If a system can exchange energy, but cannot exchange particles with its surroundings, it will be represented by a macrocanonical ensemble;

(3) If a system can exchange both energy and particles with its surroundings, it will be represented by a canonical grand ensemble.

We shall only prove Statement 3 here. Consider the system together with its surroundings and let this large system be isolated so that the total number of particles (for the sake of simplicity we shall assume that only one kind of particles is present) and the total energy of the large system are constant. Let $\varrho\,(\varrho')$, $\varepsilon\,(\varepsilon')$, and $\nu\,(\nu')$ denote, respectively, the (generic) ensemble density, the energy, and the number of particles in the system (surroundings). We then have the following equations to be satisfied by ϱ and ϱ':

$$\sum \int_{\Gamma} \varrho \, d\Omega \; = \; 1, \tag{5.1301}$$

$$\sum \int_{\Gamma'} \varrho' \, d\Omega' \; = \; 1, \tag{5.1302}$$

$$\sum \int_{\Gamma_{\text{tot}}} (\varepsilon + \varepsilon') \, \varrho\varrho' \, d\Omega \, d\Omega' \; = \; \text{const}, \tag{5.1303}$$

$$\sum (\nu + \nu') \int_{\Gamma_{\text{tot}}} \varrho\varrho' \, d\Omega \, d\Omega' \; = \; \text{const}, \tag{5.1304}$$

Furthermore we assume that an equilibrium situation has been reached. This means, as we shall see in the next section, that the average value of the index of probability is a minimum for the large system,[29] or

$$\langle \eta_{\text{gen}} \rangle = \sum \int_{\Gamma_{\text{tot}}} (\varrho \varrho') \ln(\varrho \varrho') \, d\Omega \, d\Omega' = \text{minimum}. \tag{5.1305}$$

Using Eqs.(5.1301) and (5.1302), we can rewrite Eqs.(5.1303) and (5.1304) as

$$\sum \int_{\Gamma} \epsilon \varrho \, d\Omega + \sum \int_{\Gamma'} \epsilon' \varrho' \, d\Omega' = \text{const}, \tag{5.1306}$$

$$\sum \nu \int_{\Gamma} \varrho \, d\Omega + \sum \nu' \int_{\Gamma'} \varrho' \, d\Omega' = \text{const}. \tag{5.1307}$$

From Eqs.(5.1301), (5.1302), and (5.1305) to (5.1307) we now find in the usual way, by the method of Lagrangian multipliers, that the equilibrium expression for ϱ will be

$$\varrho = e^{-q + \alpha \nu - \beta \epsilon}, \tag{5.1308}$$

which is the same as Eq.(5.0906). This concludes our proof.

We should mention here that we have tacitly introduced our assumption about equal a priori probabilities for equal volumes in Γ-space by not introducing any weight function into Eqs.(5.1301) to (5.1304).

We note that it follows from our analysis that any part of an isolated system will be represented by a canonical ensemble. The analogy between Eq.(5.1308) and the expression for the Maxwell distribution is not accidental, as we are there considering an example of a subsystem (one atom) in an isolated system (the gas).

Before considering the question of a priori probabilities, let us briefly discuss the idea that energy and temperature are complementary quantities. The term "complementarity" is, of course, used here in a slightly different manner from its use in the discussion of the Heisenberg relations. While in quantum mechanics we cannot define two canonically conjugate quantities simultaneously with arbitrary accuracy, the point here is that we cannot reduce the fluctuations in both temperature and energy arbitrarily. We discussed one side of this question, namely, that *in a system in temperature equilibrium energy fluctuations occur.* On the other hand, if we consider an isolated system, say a volume of helium gas, its total energy is fixed. Is it possible to speak about the temperature of such a system? In order to do this, we must envisage a temperature measurement, for instance, by considering a small part of the gas and measuring its mean kinetic energy per atom.[30] Since we have just seen that such a subsystem will behave as the average member

---

[29] Many authors (see, for instance, R.Balian, *From Microphysics to Macrophysics*, Springer Verlag, Berlin (1991), Vol.1, p.144) use the so-called "maximum statistical entropy principle" to define the ensemble which represents a system.

[30] We cannot introduce an external thermometer, because this would destroy the isolation of the system.

of a macrocanonical ensemble, we shall find fluctuations in the value of the mean kinetic energy per atom, and hence in the temperature. *In an isolated system we must expect temperature fluctuations.*

Let us now consider the problem of a priori probabilities. For the sake of simplicity we restrict ourselves to petit ensembles. Consider two regions Ω_0 and Ω_0' of Γ-space. Let those two regions at t_0 be filled with representative points of an ensemble, and let $p(\Omega_0)$ and $p(\Omega_0')$ be the a priori probabilities for the two regions. At t_1, t_2, $\cdots$ the representative points, which at t_0 filled up Ω_0 and Ω_0', will fill up regions Ω_1 and Ω_1', Ω_2 and Ω_2', $\cdots$ with a priori probabilities $p(\Omega_1)$ and $p(\Omega_1')$, $p(\Omega_2)$ and $p(\Omega_2')$, $\cdots$. One can clearly tolerate only such a choice of a priori probabilities that

$$\frac{p(\Omega_0)}{p(\Omega_0')} = \frac{p(\Omega_1)}{p(\Omega_1')} = \frac{p(\Omega_2)}{p(\Omega_2')} = \cdots , \tag{5.1309}$$

and these relations are satisfied, if we take

$$p(\Omega) = W(\Omega), \qquad p(\Omega') = W(\Omega'), \tag{5.1310}$$

where $W(\Omega)$ and $W(\Omega')$ are the volumes of the regions Ω and Ω' in Γ-space. We remark here in passing that Eq.(5.1310) will only satisfy Eq.(5.1309), if we are dealing with a phase space made up of canonically conjugate p's and q's (compare the remark in § 2.3).

Instead of Eq.(5.1310) one could choose the general form

$$p(\Omega) = \int_\Omega F(\varepsilon, \varphi_i) \, d\Omega, \tag{5.1311}$$

where the integration extends over the region Ω and where $F(\varepsilon, \varphi_i)$ is a once and for all chosen function of the energy ε and the other constants of motion φ_i of Eqs.(5.0104).

The reason for choosing $F(\varepsilon, \varphi_i)$ equal to 1, which means choosing Eq.(5.1310) rather than the more general form of Eq.(5.1311) was originally historical and based upon the assumption that ergodic systems (see next section) existed. Since nowadays we are convinced that most mechanical systems are quasi-ergodic (again, see § 5.14), which means essentially that the various possible values of φ_i all occur with practically equal frequency in each part of phase space, we can immediately restrict ourselves to what the Ehrenfests call the ergodic choice,

$$p(\Omega) = \int_\Omega F(\varepsilon) \, d\Omega. \tag{5.1312}$$

However, since the density ϱ of the representative ensembles of equilibrium systems are functions of the energy only, the choice of Eq. (5.1312) instead of Eq.(5.1310) would correspond to a density

$$\varrho'(\varepsilon) = \varrho(\varepsilon)/F(\varepsilon) \tag{5.1313}$$

instead of the $\varrho(\varepsilon)$ given, for instance, by Eq.(5.1308). The $F(\varepsilon)$ would occur in all integrals; but the final results would be the same, as the $F(\varepsilon)$ in the denominator of ϱ' would be cancelled by the same function now occurring as a weight factor in all integrals.

5.14.   Ergodic Theory and the H-Theorem in Ensemble Theory

In the preceding section, we based the proof that the canonical grand ensemble is an equilibrium ensemble upon Eq.(5.1305). To derive that equation we need a generalised H-theorem. Before discussing this, we want to dwell very briefly on Eq.(5.0704), stating the equivalence of time- and ensemble averages.

Boltzmann expected the existence of so-called *ergodic* [31] systems. An ergodic system is one such that the orbit in Γ-space will go through every point of its energy surface. Once the existence of ergodic systems is accepted, one can easily see the advantage of considering time averages, since the proof of Eq.(5.0704) is now complete. One can, however, easily see that ergodic systems do not exist, as was pointed out by Rosenthal and Plancherel. Plancherel's argument was that the points of an orbit form a set of measure zero on the energy surface, the measure of which itself is not zero.[32]

We are thus left to find a different proof of Eq.(5.0704). This is done in ergodic theory, which is in fact a branch of pure mathematics rather than one of physics. We shall therefore only just mention the main results of ergodic theory. In 1931 Birkhoff proved (1) that for bounded mechanical motion and for "reasonable" phase functions G the time average $\widetilde{G}$ almost always[33] exists and (2) that for metrically transitive systems this time average is equal to the (microcanonical) ensemble average. A mechanical system is metrically transitive, if the energy surface cannot be divided into two finite regions such that orbits starting from points in one of the regions always remain in that region. The question then remained whether or not the systems considered in physics are metrically transitive, and in 1963 Sinai proved that a system of hard sphere molecules is metrically transitive. We refer to the literature given in the Bibliographical Notes at the end of this chapter for discussions of ergodic theory.

---

[31] From $\grave{\varepsilon}\rho\gamma o\nu$ = work (used here as "energy"; compare the erg as unit of energy) and $o\delta\acute{o}\varsigma$ = path: the orbit passes through every point of the energy surface. This name was introduced by Boltzmann (*J. f. Math.* **100**, 201 (1887).

[32] The measure of a point set on a surface can be defined as follows (see, for example, P.R.Halmos, *Measure Theory*, New York (1950)) Let $f(P)$ be a function that is 1, if P belongs to the point set, and that is 0 otherwise. The Lebesgue integral $\int f \, d\sigma$, which extends over the whole of the surface, is then called the Lebesgue measure of the point set on the surface.

[33] The expression "almost always" must be interpreted as by meaning "in by far the most cases", and can only be made into a more rigorous statement by using the language of measure theory.

With the arrival of powerful computers it became feasible to investigate whether physical systems with a modest number of degrees of freedom are, indeed, ergodic. In 1955 Fermi, Pasta, and Ulam[34] found to their surprise that this was not the case for a linear chain of points interacting non-linearly. Since then there has been great activity in studying non-linear systems with a modest number of degrees of freedom, many of which show deterministic rather than chaotic (ergodic) behaviour. Unfortunately it would take up too much space to do justice to these developments and we must refer to the literature for more details.[35]

We now come to the *H*-theorem in ensemble theory. First of all, we shall give three statements and prove the last of them, leaving the proof of the other two as problems.

1. If our ensemble is such that all systems have the same number of particles N (for the sake of simplicity we restrict the discussion to systems containing only one kind of particles) and that the energy ε of all systems lies in the interval $E \leqslant \varepsilon \leqslant E + \delta E$, the average $\langle \eta \rangle$ of the generic index of probability, given by the equation

$$\langle \eta \rangle = \sum_{\nu} \int_{\Gamma} \varrho_{\text{gen}} \ln \varrho_{\text{gen}} \, d\Omega, \tag{5.1401}$$

will be a minimum, if ϱ (we drop here and henceforth the index "gen") is given by the equation

$$\left. \begin{array}{ll} \varrho = \text{const} \times \delta(\nu - N), & E \leqslant \varepsilon \leqslant E + \delta E; \\[2mm] \varrho = 0, & \varepsilon < E, \quad E + \delta E < \varepsilon, \end{array} \right\} \tag{5.1402}$$

where $\delta(x)$ is the Dirac δ-function.

2. If our ensemble is such that all systems have the same number of particles, but only the average energy is given, that is, ϱ must satisfy the condition[36]

$$\sum_{\nu} \int_{\Gamma} \varepsilon \varrho \, d\Omega = E, \tag{5.1403}$$

---

[34] E.Fermi, J.R.Pasta, and S.Ulam, Los Alamos Report LA-1940; see *Collected Works of Enrico Fermi* Vol.2 (1965), p.978.

[35] We may refer to the following books where further extensive bibliographies can be found: *Dynamical Systems, Theory and Applications* (Ed. J.Moser), Springer, Berlin (1975); *Topics in Nonlinear Dynamics* (Ed. S.Jorna), American Institute of Physics, New York (1978); *Nonlinear Dynamics and the Beam-Beam Interaction* (Eds. M.Month and J.C.Herrera), American Institute of Physics, New York (1979); A.J.Lichtenberg and M.A.Lieberman, *Regular and Stochastic Motion*, Springer, New York (1983); P.Cvitanović, *Universality in Chaos*, Adam Hilger, Bristol (1984); R.S.MacKay and J.D.Meiss, Hamiltonian Dynamical Systems, Adam Hilger, Bristol (1987); H.G.Schuster, *Deterministic Chaos*, VCH Verlag, Weinheim (1988).

[36] The summation over ν is here trivial as only one term of the sum contributes.

then $\langle \eta \rangle$ will be a minimum, if ϱ satisfies the equation

$$\varrho = \delta(\nu - N)\, e^{\beta(\psi - \varepsilon)}. \tag{5.1404}$$

3. If only the average number of particles and the average energy of the systems in the ensemble are given so that ϱ must satisfy Eq.(5.1403) and the condition

$$\sum_\nu \nu \int_\Gamma \varrho\, d\Omega = N, \tag{5.1405}$$

$\langle \eta \rangle$ will be a minimum, if ϱ satisfies the equation

$$\varrho = e^{q + \alpha\nu - \beta\varepsilon}. \tag{5.1406}$$

Before proving Statement 3, we mention that the proof of the minimum properties of $\langle \eta \rangle$ rests mainly on the inequality

$$x\, e^x - e^x + 1 \geqslant 0, \tag{5.1407}$$

for all x, the equal sign holding only when $x = 0$, which was used to prove Lemma I in § 5.6.

The proof of Statement 3 now proceeds as follows. Consider two densities ϱ_1 and ϱ_2, where ϱ_1 is given by Eq.(5.1406), while ϱ_2 is given by

$$\varrho_2 = \varrho_1 e^{\Delta\eta}, \tag{5.1408}$$

where $\Delta\eta$ is an unspecified function of the p's, of the q's, and of ν. Both ϱ_1 and ϱ_2 satisfy Eqs.(5.1403) and (5.1405), as well as the normalisation condition

$$\sum_\nu \int_\Gamma \varrho\, d\Omega = 1. \tag{5.1409}$$

Now consider the equation

$$\langle \eta \rangle_2 - \langle \eta \rangle_1 = \sum_\nu \int_\Gamma (\varrho_2 \ln \varrho_2 - \varrho_1 \ln \varrho_1)\, d\Omega. \tag{5.1410}$$

Adding to the right-hand side of Eq.(5.1410) the expression

$$\sum_\nu \int_\Gamma [q - \alpha\nu + \beta\varepsilon - 1](\varrho_2 - \varrho_1)\, d\Omega,$$

which vanishes by virtue of conditions (5.1403), (5.1405), and (5.1409), we find after some straightforward transformations

$$\langle \eta \rangle_2 - \langle \eta \rangle_1 = \sum_\nu \int_\Gamma \varrho_1 \left[\Delta\eta\, e^{\Delta\eta} - e^{\Delta\eta} + 1 \right] d\Omega \geqslant 0, \tag{5.1411}$$

which concludes the proof.

Comparing expression (5.1401) with the definition of Boltzmann's H, we might hope to be able to prove that $d\langle\eta\rangle/dt$ is non-positive, and thus that there is always a tendency to establish equilibrium ensembles. However, if we compare the values of $\langle\eta\rangle$ at times t' and t'', we find

$$\langle\eta\rangle'' = \sum_\nu \int_\Gamma \varrho'' \ln \varrho'' \, d\Omega'' = \sum_\nu \int_\Gamma \varrho' \ln \varrho' \, J \, d\Omega'$$

$$= \sum_\nu \int_\Gamma \varrho' \ln \varrho' \, d\Omega' = \langle\eta\rangle', \qquad (5.1412)$$

which shows that in fact $d\langle\eta\rangle/dt = 0$. In Eq.(5.1412) ϱ' (ϱ'') is a shorthand notation for $\varrho(p',q';t')$ $(\varrho(p'',q'';t''))$, where the relation between p'',q'' and p',q' is such that a representative point p',q' at time t' will have moved to p'',q'' by t'', while J is the Jacobian of the transformation from p'',q'' to p',q' which by virtue of Liouville's theorem is equal to 1.

Although we have just seen that $\langle\eta\rangle$ is constant, there is still in a certain sense an approach to a stationary state. Let us illustrate this by an example introduced by Gibbs. Consider a container with a liquid, say water, in which is put some colouring material; let us assume that this colouring material is non-diffusible and consists of colloidal particles. It is a well known empirical fact that, if we start from a state where the colouring material is unevenly distributed, practically any kind of stirring will produce a situation where the colour distribution is, as far as our eye can see, uniform. That means that stirring will produce an "equilibrium" state. However, if we look at the system very closely, we will still find that, in microscopic volumes, part of the space is occupied by the water and part occupied by the colloidal particles. Although the coarse distribution is uniform, the finer distribution is still uneven.

From this example it follows that it might be advantageous to introduce apart from the *fine-grained density* ϱ a *coarse-grained density* P (read "capital ϱ") defined as follows: Divide Γ-space into finite, but small, cells Ω_i and let P_i be the average of ϱ taken over Ω_i, or

$$P_i = \frac{1}{W^\nu(\Omega_i)} \int_{\Omega_i} \varrho \, d\Omega, \qquad (5.1413)$$

where $W^\nu(\Omega_i)$ is the volume of the cell Ω_i corresponding to a system with ν particles.[37]

From the normalisation of ϱ we then get

$$\sum_\nu \sum_i P_i \, W^\nu(\Omega_i) = 1. \qquad (5.1414)$$

---

[37] We should really also have lumped together a range of ν-values, but since this would not have added anything new to the discussion we have left this complication for the reader to carry out.

We now introduce the coarse-grained density $P(p, q)$ by putting it constant in each cell Ω_i and equal to P_i. Equation (5.1414) can then be written in the form

$$\sum_\nu \int_\Gamma P(p, q) \, d\Omega = 1. \tag{5.1415}$$

Instead of $\langle \eta \rangle$ we can now introduce a function $\langle H \rangle$ by the equation[38]

$$\langle H \rangle = \sum_\nu \sum_i P_i \ln P_i \, W^\nu(\Omega_i). \tag{5.1416}$$

Using the coarse-grained density P, Eq.(1.416) takes the form

$$\langle H \rangle = \sum_\nu \int_\Gamma P \ln P \, d\Omega. \tag{5.1417}$$

Since $\ln P$ is constant in each cell Ω_i and

$$\int_{\Omega_i} P \, d\Omega = P_i \, W^\nu(\Omega_i) = \int_{\Omega_i} \varrho \, d\Omega,$$

we can also write Eq.(5.1417) in the form

$$\langle H \rangle = \sum_\nu \int_\Gamma \varrho \ln P \, d\Omega, \tag{5.1418}$$

and we see that

$$\langle H \rangle = \langle \ln P \rangle. \tag{5.1419}$$

The conclusions we reached at the beginning of this section about $\langle \eta \rangle$ will also hold for $\langle H \rangle$, and one can, indeed, easily show, for instance, that in an ensemble of which the constituent systems all have energies within the range $E, E + \delta E$, $\langle H \rangle$ will be a minimum if P is constant in the region between the two corresponding energy surfaces. And when there is a slight interaction between the different members of the ensemble, $\langle H \rangle$ will be a minimum if P corresponds to a coarse-grained macrocanonical distribution, and so on.

We can now consider the change of $\langle H \rangle$ with time. Let us assume that we have made some observations about a physical system at $t = t'$. Since these observations will never give us the maximum possible information, we can construct an ensemble, the average properties of which at t' correspond to the observed properties of the system at observation at t'. Because of experimental limitations we shall at most be able to give $\varrho(p, q)$ changing from one cell Ω_i to another, provided we have chosen the cells of the cells Ω_i in accordance with the experimental limitations, as we shall assume that we have done. In accordance with

---

[38]The symbol $\langle H \rangle$ can be read either as the ensemble average of Boltzmann's H or rather as the average of a function "capital η".

the principle of equal a *priori* probabilities for equal volumes in Γ-space, the fine-grained density will be chosen constant in each Ω_i and at t' we have thus

$$\varrho' = P', \tag{5.1420}$$

where the primes indicate values at time t'. For $\langle H \rangle'$ we have thus

$$\langle H \rangle' = \sum_{\nu} \int_{\Gamma} P' \ln P' \, d\Omega = \sum_{\nu} \int_{\Gamma} \varrho' \ln \varrho' \, d\Omega.$$

If the situation at t' already corresponds to an equilibrium situation — either microcanonical, macrocanonical, or grand canonical, according to whether or not interactions between the members of the ensemble are excluded from our considerations — the situation will be the same at a later time, since it is a stationary distribution. Let us therefore assume that ϱ' is not a stationary distribution. It can then easily be seen that at a later moment, t'',[39] Eq.(5.1420) will no longer hold and we have

$$\varrho'' \neq P''. \tag{5.1421}$$

This inequality is due to the fact that, although ϱ will stay constant in extensions in phase of unchanging magnitude Ω_i, the shape of these extensions will change and at a later moment each of the Ω_i will, in general, be covered by points which at t' belonged to many different cells.

We have now for $\langle H \rangle''$

$$\langle H \rangle'' = \sum_{\nu} \int_{\Gamma} P'' \ln P'' \, d\Omega, \tag{5.1422}$$

and we can no longer replace P'' by ϱ''. Comparing $\langle H \rangle'$ and $\langle H \rangle''$, we have

$$\langle H \rangle' - \langle H \rangle'' = \sum_{\nu} \int_{\Gamma} \left(\varrho' \ln \varrho' - P'' \ln P'' \right) d\Omega$$

$$= \sum_{\nu} \int_{\Gamma} \left(\varrho'' \ln \varrho'' - \varrho'' \ln P'' \right) d\Omega$$

$$= \sum_{\nu} \int_{\Gamma} \left[\varrho'' \ln \varrho'' - \varrho'' \ln P'' + P'' - \varrho'' \right] d\Omega, \tag{5.1423}$$

or

$$\langle H \rangle' - \langle H \rangle'' = \sum_{\nu} \int_{\Gamma} P'' \left[\Delta e^{\Delta} + 1 - e^{\Delta} \right] d\Omega > 0, \tag{5.1424}$$

---

[39]It would be more general to say at *another* moment, leaving open the question whether t'' is later or earlier than t'. However, since we have set up our ensemble to represent a system that was observed at t', we shall be interested only in predictions that can actually be verified, that is, in times for which $t'' > t'$.

where Δ is defined by the equation

$$\varrho'' = P'' e^{\Delta}.$$

The second equality in Eq.(5.1423) follows from Eqs.(5.1412), (5.1417), and (5.1418), and the last one from Eqs.(5.1409) and (5.1415).

We see now that $\langle H \rangle''$ will be less than $\langle H \rangle'$ by virtue of the fact that ϱ'' and P'' are no longer everywhere equal. Comparing the present situation with the case of colouring material in a liquid, we may expect that, as time goes on, ϱ and P will differ more and more and $\langle H \rangle$ will continue to decrease until a distribution has been reached for which either

$$\left. \begin{array}{ll} P = \text{const} \times \delta(\nu - N), & E \leqslant \epsilon \leqslant E + \delta E; \\ P = 0, & \epsilon < E, \quad E + \delta E < \epsilon, \end{array} \right\} \qquad (5.1425)$$

or

$$P = A\,\delta(\nu - N)\,e^{-\beta\epsilon}, \qquad (5.1426)$$

or,

$$P = B\,e^{\alpha\nu - \beta\epsilon}, \qquad (5.1427)$$

depending on the situation we are studying.

Since $\langle H \rangle$ is a minimum when the distributions (5.1425) to (5.1427) are reached, it will no longer decrease. If these distributions are reached we would, on observation, in general, come to the conclusion that the state of the physical system would be best represented by a micro- or a macrocanonical ensemble. We have thus found a justification for using these ensembles to describe systems in thermodynamical equilibrium.

We saw that the decrease of $\langle H \rangle$ resulted from the fact that we came from a state where $\varrho = P$ to a state where ϱ was no longer everywhere equal to P. The decrease of $\langle H \rangle$ thus corresponds to a decrease of the specific information we could obtain about our physical system. Since $\langle H \rangle$ can be regarded to be a "coarse-grained" $\langle \eta \rangle$ and since $\langle \eta \rangle$ was seen to be equal to the negative entropy (apart from a constant factor; see Eq.(5.0309)), we see how *increase of lack of knowledge* corresponds to an increase in entropy. We shall return to this aspect of entropy in §6.4.

Problems

1. Use a canonical ensemble to derive a recursion formula for the following quantity:
$$\frac{\overline{(\mathcal{T} - \overline{\mathcal{T}})^n}}{\overline{\mathcal{T}}^n}, \qquad n = 0, 1, 2, \cdots,$$
where $\mathcal{T}$ is the kinetic energy (see Eq.(5.0102)), and hence evaluate this quantity for the special cases where $n = 2$, 3, or 4.

2. Evaluate the following correlations
$$\langle (A_k - \langle A_k \rangle)(A_l - \langle A_l \rangle) \rangle \qquad \text{and} \qquad \langle (\varepsilon - \langle \varepsilon \rangle)(A_k - \langle A_k \rangle) \rangle.$$

3. If Δ denotes the Jacobian

$$\frac{\partial(\dot{q}_1, \dot{q}_2, \cdots, \dot{q}_{sN})}{\partial(p_1, p_2, \cdots, p_{sN})},$$

and Δ' its reciprocal, we can follow Gibbs and introduce elements of extension in configuration,

$$\sqrt{\Delta'}\, dq_1 \cdots dq_{sN},$$

and elements of extension in velocity,

$$\sqrt{\Delta}\, dp_1 \cdots dp_{sN},$$

which can easily be seen to reduce — apart from a constant factor — to the expressions (5.0506) for the case considered in § 5.5. Show that these two elements of extension are invariants.

4. Evaluate e^{ϕ_p} of Eq.(5.0508).

5. If we consider the band structure of a semiconductor, say, we find that the band widths are functions of the lattice constants. As the lattice constants will be functions of the temperature, we are dealing here with the at first sight paradoxical concept of temperature-dependent energy levels.[40] The reason for this difficulty is, of course, that we are treating the motion of the lattice as being given, before we evaluate the energy levels available to the electrons; that is, instead of considering the complete system, we consider only a subsystem, while we make some assumptions about the influence of the remainder of the system.

Consider a system, for which the energy levels can be written in the form

$$\varepsilon = \varepsilon(p', q') + \varepsilon(p, q),$$

where p, q stands for all coordinates that we want to ignore in our considerations and p', q' for those in which we are interested. For instance, the $\varepsilon(p', q')$ might be the energy of the electrons and $\varepsilon(p, q)$ the energy of the lattice. In many calculations we would be tempted to use as a partition function the expression

$$Z'_\Gamma = \int e^{-\beta\varepsilon(p',q')}\, dp'\, dq'. \tag{A}$$

Prove that a proper averaging over the ignored degrees of freedom (p, q) leads to the conclusion that we should in Eq.(A) replace $\varepsilon(p', q')$ by the Helmholtz free energy $F(p', q')$ which is related to $\varepsilon(p', q')$ through the Gibbs-Helmholtz equation

$$\varepsilon(p'q') = \frac{\partial F}{\partial \beta}.$$

---

[40]See G.S.Rushbrooke, *Trans. Faraday Soc.* **36**, 1055 (1940).

6. Evaluate $\Omega(\varepsilon)$ for a system of N one-dimensional harmonic oscillators of frequency ω, and hence find the average energy of this system as function of the temperature.

7. Use the relation $S = k \ln W$ between the entropy S and the probability W for a situation and also the thermodynamic relation between a change in the entropy and a change in the free energy for a closed system in thermodynamic equilibrium to derive Eq.(5.1005).[41]

8. Use Eq.(5.1005) and the fact that, for small deviations from equilibrium, ΔA can be expanded in a Taylor series (compare Eq.(5.1006)) to prove that the average value of ΔA equals $\frac{1}{2}kT$.

9. Use thermodynamic relations and Eq.(5.1009) to find expressions for the average values of ΔV^2, $\Delta V \Delta T$, and ΔT^2.[42]

10. Do the same for ΔP^2, $\Delta P \Delta S$, and ΔS^2.

11. Do the same for ΔE^2, $\Delta E \Delta V$, and ΔV^2.

12. Do the same for $\Delta S \Delta V$, $\Delta V \Delta P$, and $\Delta S \Delta T$.

13. Express $\langle (\varepsilon - \langle \varepsilon \rangle)^3 \rangle$ in terms of C_v and its temperature derivative. Evaluate this expression for the case of a perfect gas.

14. Express the fluctuations in the magnetisation of a system in terms of the magnetic susceptibility.

15. Consider a system of N spin-$\frac{1}{2}$ particles each with a magnetic moment μ in an external magnetic field $\boldsymbol{B}$. Calculate the fluctuations in the total magnetic moment of the system both directly and by applying the result of the preceding problem.

16. Prove Eq.(5.1028), bearing in mind that the probability for the i-th particle to be at $\boldsymbol{r}_i$ and the j-th particle to be at $\boldsymbol{r}_j$ is obtained by integrating the normalised ensemble density over all variables except $\boldsymbol{r}_i$ and $\boldsymbol{r}_j$.

17. Evaluate the pressure fluctuations, using Eq.(5.1029), taking the derivatives to be meant as being evaluated at constant p, q and a wall potential of the form

$$
\begin{aligned}
U(r) &= 0, & r &< R; \\
&= C(r - R), & r &\geqslant R,
\end{aligned}
$$

where C is a constant, and where we consider a perfect gas in a spherical container.[43]

18. Prove Eq.(5.1112).

19. Use the perfect gas law and Eq.(5.1212) to prove Eq.(5.1214).

20. Prove Statements (1) and (2) from § 5.13.

[41] A. Einstein, *Ann. Physik* **33**, 1275 (1910).

[42] For problems 9 to 12, compare L.D.Landau and E.M.Lifshitz, *Statistical Physics*, Oxford University Press (1958), § 111, and R.Kubo, *Statistical Mechanics*, North-Holland, Amsterdam (1965), pp.381ff.

[43] H.Wergeland, *Proc. Norwegian Acad. Sci.* **28**, 106 (1955).

21. Use arguments similar to the ones leading to Eq. (5.1308) to prove Eq. (5.1117).

22. Prove Statements 1 and 2 from § 5.14.

Bibliographical Notes

This chapter is largely based upon the following references:

1. J.W.Gibbs, *Elementary Principles in Statistical Mechanics* (Vol. II of his Collected Works), New Haven (1948).

2. R.C.Tolman, *The Principles of Statistical Mechanics*, The Clarendon Press, Oxford (1938).

3. P. and T. Ehrenfest, *The Conceptual Foundations of the Statistical Approach in Mechanics*, Cornell University Press, Ithaca, N.Y. (1959).

See also

4. *A commentary on the Scientific Works of J.W.Gibbs* (Ed. A. Haas), New Haven (1936); especially the articles R, S, T, and U by A.Haas and P.S.Epstein.

Independently of Gibbs, Einstein introduced ensemble theory at about the same time:

5. A.Einstein, *Ann. Physik* **9**, 417 (1902).

6. A.Einstein, *Ann. Physik* **11**, 170 (1903).

It should also be mentioned here that Boltzmann had made use of ensembles, especially, of course, in connection with the ergodic theorem. See, for instance,

7. L.Boltzmann, *Wien. Ber.* **63**, 679 (1871).

Section 5.1. Liouville's theorem was announced in 1838:

8. J.Liouville, *J. de Math.* **3**, 348 (1838).

Section 5.2. For a discussion of the term "macrocanonical" see

9. V.F.Lenzen, *Am. J. Phys.* **22**, 638 (1954).

10. D.ter Haar, *Am. J. Phys.* **22**, 638 (1954).

Section 5.9. See References 1 and 2, and also

11. H.A.Kramers, *Proc. Kon. Ned. Akad. Wet.* (*Amsterdam*) **41**, 10, 237 (1938).

12. R.H.Fowler, *Proc. Camb. Phil. Soc.* **34**, 382 (1938).

13. E.A.Guggenheim, *J. Chem. Phys.* **7**, 102 (1939).

14. R.C.Tolman, *Phys. Rev.* **57**, 1160 (1940).

15. R.Becker, *Z. Physik. Chem.* **196**, 181 (1950).

Section 5.10. Fluctuation theory really started with the classical papers of Einstein and Smoluchowski:

16. M.von Smoluchowski, *Ann. Physik* **25**, 205 (1908).

17. A.Einstein, *Ann. Physik* bf 33, 1275 (1910).

For a discussion of the importance of the radial distribution function see:

18. R.Kubo, *Statistical Mechanics*, North-Holland, Amsterdam (1965),

19. R.K.Pathria, *Statistical Mechanics*, Pergamon, Oxford (1977), § 13.2. but especially

20. F.Zernike, *Thesis*, Amsterdam (1915).

21. L.Ornstein and F.Zernike, *Proc. Kon. Ned. Akad. Wet.* **17**, 793 (1914); **18**, 1520 (1916); **19**, 1321 (1917); *Physik. Z.* **19**, 134 (1918); **26**, 761 (1926);

Section 5.12. For a discussion of a simple model related to the Gibbs paradox, see

22. M.J.Klein, *Am. J. Phys.* **26**, 80 (1958); *Ned. Tijds. Natuurk.* **25**, 73 (1959).

It was pointed out by van Kampen that the Gibbs paradox is resolved by replacing the Platonic idea of entropy with an operational definition.

23. N.G.van Kampen, in *Essays in Theoretical Physics* (Ed. W. E. Parry), Pergamon Press, Oxford (1984), p.303.

An extensive bibliography about the Gibbs paradox can be found in

24. С.Д.Хайтин, История парадокса Гиббса [in Russian], Nauka, Moscow (1986).

Sections 5.13 and 5.14. See References 2, 3, and 14, and also

25. D.ter Haar, *Elements of Statistical Mechanics*, 1st ed., Rinehart, New York (1954), Appendix I.

26. D.ter Haar, *Rev. Mod. Phys.* **27**, 289 (1955).

27. G.E.Uhlenbeck and G.W.Ford *Lectures in Statistical Mechanics*, Providence, R.I., Chap.1.

28. R.Jancel, *Foundations of Classical and Quantum Statistical Mechanics*, Pergamon Press, Oxford (1969).

29. I.E.Farquhar, *Ergodic Theory in Statistical Mechanics*, Interscience, New York (1964).

We refer to References 3, 24, 26, and 27 for extensive bibliographies. For a simple, though not completely rigorous, proof of Eq.(5.0704) see

30. H.Wergeland, *Acta Chem. Scand.* **12**, 1117 (1958).

The proof of the impossibility of ergodic systems was given independently by Rosenthal and Plancherel:

31. A.Rosenthal, *Ann. Physik* **42**, 796 (1913).

32. M.Plancherel, *Ann. Physik* **42**, 1061 (1913).

The ergodic theorem in its modern version is due to Birkhoff and von Neumann:

33. G.D.Birkhoff, *Proc. Nat. Acad.* **17**, 650, 656 (1931).

34. G.D.Birkhoff and B.O.Koopman, *Proc. Nat. Acad.* **18**, 279 (1932).

35. J. von Neumann, *Proc. Nat. Acad.* **18**, 70, 263 (1932).

For a general discussion we refer to the following sources, many of which contain further references:[44]

36. E.Hopf, *Ergodentheorie*, Berlin (1937).

37. A.I.Khinchin, *Mathematical Foundations of Statistical Mechanics*, Dover, New York (1949); especially Chaps. II and III.

---

[44]It should be remarked that many of the references quoted are predominantly mathematical in their approach.

38. V.I.Arnold and A.Avez, *Ergodic Problems of Classical Mechanics*, W.A.Benjamin, New York (1968).

39. Ya.G.Sinai, *Introduction to Ergodic Theory*, Princeton University Press (1977).

40. И.П.Корнфельд, Я.Г.Синай, С.В.Фомин, Эргодическая теория [in Russian], Nauka, Moscow (1980).

41. О.В.Кузнецова, История обоснования статистической механики [in Russian], Nauka, Moscow (1988).

42. D.Ruelle, *Chaotic Evolution and Strange Attractors*, Cambridge University Press (1989).

43. R.Balian, *From Microphysics to Macrophysics*, Springer Verlag, Berlin (1991), § 3.4.

CHAPTER 6

THE ENSEMBLES IN QUANTUM STATISTICS

6.1.   The Density Matrix

It is, of course, well known that modern theoretical physics is to a large extent dominated by quantum mechanics. Nevertheless, there are large fields of physics in which one can use classical theory because the quantum-mechanical corrections are far too small to enter into the discussion. Large parts of statistical mechanics belong to classical physics. On the other hand, statistical mechanics is one of the branches of physics in which the successes of quantum mechanics have been most noteworthy. We saw as an instance of the successful intervention of quantum mechanics the solution of the paradox of the specific heats. Other instances are the electron theory of metals and many branches of modern astrophysics. The discussion of quantum statistics was started in Chapters 3 and 4, where we discussed, firstly, the quantum-mechanical analogue of the Maxwell-Boltzmann distribution and, secondly, the implications of the exclusion principle. However, we still must discuss quantum-mechanical ensemble theory, and this will be done in the present chapter.

In discussing quantum-mechanical ensemble theory one must be extremely careful to make a clear distinction between the statistical aspects inherent in quantum mechanics and the statistical aspects introduced by the ensembles. The statistical nature of quantum mechanics is connected with the fact that quantum mechanics gives us a wavefunction to describe a system and that from this wavefunction one then constructs a probability density, a probability current density, and probable (or average, or expectation) values of physical quantities. Only in exceptional cases will it be possible to ascribe to a quantity a completely defined value. If this is the case, it results in a complete lack of definition of the *complementary* quantity. In general, both of the two complementary quantities will be defined only within certain limits, and these limits cannot be chosen independently but must satisfy the so-called Heisenberg relations. We do not wish to discuss this point in any more detail, since it really belongs to a discussion of the principles of quantum mechanics. In the following we shall assume that the reader is familiar with this aspect of quantum mechanics and we shall use it whenever necessary.

216

We now wish to consider the statistical aspects of the ensembles in quantum mechanics. We consider thereto a large number of identical systems. Let $\widehat{\mathcal{H}}$ be the energy (or Hamiltonian) operator and let ψ^k be the normalised wavefunction describing the k-th system. This wavefunction will then satisfy the Schrödinger equation

$$\widehat{\mathcal{H}}\psi^k = i\hbar\,\dot{\psi}^k. \tag{6.0101}$$

We now introduce a complete set of orthonormal functions φ_n into which the wavefunction or any other function defined in the same Hilbert space can be developed.[1]

Expanding ψ^k in terms of the φ_n, we have

$$\psi^k = \sum_n a_n^k \varphi_n, \tag{6.0102}$$

with a_n^k given by

$$a_n^k = \int \varphi_n^* \psi^k \, d\tau, \tag{6.0103}$$

where the asterisk indicates the conjugate complex and where $d\tau$ is a volume element of the relevant Hilbert space.

The coefficient a_n^k can be used to describe the k-th system instead of the ψ^k. The two representations are equivalent and the Schrödinger equation in the a_n^k-representation has the form

$$i\hbar\,\dot{a}_n^k = \sum_m H_{nm} a_m^k, \tag{6.0104}$$

where the H_{nm} are the matrix elements of the Hamiltonian,

$$H_{nm} = \int \varphi_n^* \widehat{\mathcal{H}} \varphi_m \, d\tau. \tag{6.0105}$$

The physical significance of the a_n^k is that they are probability amplitudes, $\left|a_n^k\right|^2$ being the probability that the k-th system is characterised by the function φ_n. From the normalisation of the ψ_k and the fact that the φ_n form an orthonormal system, it follows that

$$\sum_n \left|a_n^k\right|^2 = 1. \tag{6.0106}$$

In the classical ensemble theory we discussed density function $\varrho(p,q)$. Now we introduce a *density operator* which we shall define by its matrix

---

[1] In order not to complicate our equations we shall usually not explicitly introduce degeneracy, spin, or continuous spectra. It is not difficult to extend all formulæ to the most general case, and we leave that to the reader.

elements. This so-called *density matrix* (or *statistical operator*) is defined through its elements by

$$\varrho_{mn} = \frac{1}{\mathcal{N}} \sum_{k=1}^{\mathcal{N}} a_m^k a_n^{k*}, \tag{6.0107}$$

where $\mathcal{N}$ is the number of systems in the ensemble.

From Eq.(6.0104) we can find the equation of motion for the ϱ_{mn}. This equation is

$$i\hbar \dot{\varrho}_{mn} = \sum_{l} (H_{ml}\varrho_{ln} - \varrho_{ml}H_{ln}). \tag{6.0108}$$

In deriving Eq.(6.0108) we used the fact that $\widehat{\mathcal{H}}$ is a Hermitian operator, that is, that

$$H_{kl} = H_{lk}^*. \tag{6.0109}$$

Defining the commutator $[\widehat{A}, \widehat{B}]_-$ of two operators $\widehat{A}$ and $\widehat{B}$ by the equation

$$[\widehat{A}, \widehat{B}]_- = \widehat{A}\widehat{B} - \widehat{B}\widehat{A}, \tag{6.0110}$$

or, in matrix notation,

$$[\widehat{A}, \widehat{B}]_{-,kl} = \sum_{m} [A_{km}B_{ml} - B_{km}A_{ml}], \tag{6.0111}$$

we can write Eq.(6.0108) in matrix form, as follows,

$$i\hbar \, \widehat{\dot{\varrho}} = [\widehat{\mathcal{H}}, \widehat{\varrho}]_-. \tag{6.0112}$$

This equation is the quantum-mecanical counterpart of Liouville's theorem (Eqs.(5.0107) or (5.0115)) of the classical theory. Of course, it is well known that the commutators are the quantum-mechanical counterparts of the Poisson brackets of classical mechanics.

The average value $\overline{G}$ of a physical quantity is defined by the equation

$$\overline{G} = \frac{1}{\mathcal{N}} \sum_{k=1}^{\mathcal{N}} \int \psi^{k*} \, \widehat{G} \, \psi_k \, d\tau, \tag{6.0113}$$

where $\widehat{G}$ is the quantum-mechanical operator corresponding to the classical quantity G. Equation (6.0113) shows that $\overline{G}$ is twice an average. First we take the quantum-mechanical average of G in a system described by the wavefunction ψ^k, and, secondly, we take the average over the ensemble.

Introducing the a_n^k instead of the ψ^k, we can write Eq.(6.0113) in the form

$$\overline{G} = \frac{1}{\mathcal{N}} \sum_{k=1}^{\mathcal{N}} \sum_{m,n} a_m^k a_n^{k*} G_{nm}, \tag{6.0114}$$

or, introducing the density matrix by means of Eq.(6.0107),

$$\overline{G} = \sum_{m,n} \varrho_{mn} G_{nm},$$

or

$$\overline{G} = \sum_{m} (\varrho G)_{mm} = \text{Tr}\,\widehat{\varrho}\,\widehat{G}, \qquad (6.0115)$$

where Tr indicates the trace, that is, the sum of diagonal elements.

A special case of this equation is the normalisation condition which we get by putting $\widehat{G}$ equal to the unit operator ($G_{mn} = \delta_{mn}$, where δ_{mn} is the Kronecker δ-function),

$$\text{Tr}\,\widehat{\varrho} = 1. \qquad (6.0116)$$

This last equation can also be proven directly from the definition of $\widehat{\varrho}$ and Eq.(6.0106),

$$\text{Tr}\,\widehat{\varrho} = \sum_{m} \varrho_{mm} = \frac{1}{\mathcal{N}} \sum_{k} \sum_{m} a_m^k a_m^{k\,*} = \frac{1}{\mathcal{N}} \sum_{k=1}^{N} 1 = 1.$$

We shall restrict our discussion once again mainly to stationary ensembles. In that case $\widehat{\varrho}$ and the energy operator $\widehat{\mathcal{H}}$ must commute. Moreover, we shall assume that $\widehat{\varrho}$ is a function of the energy only,

$$\widehat{\varrho} = f(\widehat{\mathcal{H}}), \qquad (6.0117)$$

and even that we are dealing with a macrocanonical ensemble:

$$\widehat{\varrho} = e^{\beta(\psi - \widehat{\mathcal{H}})}. \qquad (6.0118)$$

This last equation has a meaning only if we understand it as being a short-hand notation for

$$\widehat{\varrho} = e^{\beta\psi} \sum_{n=0}^{\infty} \frac{1}{n!} \left[-\beta\widehat{\mathcal{H}}\right]^n. \qquad (6.0119)$$

Before discussing the statistical implications of our formalism, we shall consider some of the mathematical aspects of the density matrix. First of all, we note that it follows from the definition (6.0107) that $\widehat{\varrho}$ is a Hermitean matrix,

$$\varrho_{mn} = \varrho_{nm}^*, \qquad (6.0120)$$

which is normalised according to Eq.(6.0116). It follows from this (1) that the diagonal elements of $\widehat{\varrho}$ are real, and (2) that they must satisfy the relations

$$\sum_{n} \varrho_{nn} = 1, \quad \text{and thus} \quad 0 \leqslant \varrho_{nn} \leqslant 1. \qquad (6.0121)$$

From the definition (6.0107) it is clear that ϱ_{nn} is the (normalised) probability that φ_n is realised in the ensemble.

Let us now consider a change from the φ-representation to another representation, say, the χ-representation. Let the unitary transformation matrix S_{np} characterise this transformation, so that

$$\chi_p = \sum_n \varphi_n S_{np}, \tag{6.0122}$$

$$S_{np}^* = \left(S^{-1}\right)_{pn}, \tag{6.0123}$$

$$\sum_n S_{np}^* S_{nq} = \delta_{pq}. \tag{6.0124}$$

We now rewrite Eq.(6.0102) as follows:

$$\psi^k = \sum_n a_n^k \varphi_n = \sum_p b_p^k \chi_p, \tag{6.0125}$$

and Eqs.(6.0122) and (6.0125) lead to

$$a_n^k = \sum_q S_{nq} b_q^k, \tag{6.0126}$$

or

$$b_q^k = \sum_n a_n^k S_{nq}. \tag{6.0127}$$

If the transformed density matrix is denoted by $\widehat{\varrho}'$, so that

$$\varrho_{pq}' = \frac{1}{N} \sum_{k=1}^{N} b_p^k b_q^{k*}, \tag{6.0128}$$

we find from Eqs.(6.0127) and (6.0107) that

$$\varrho_{pq}' = \sum_{m,n} \left(S^{-1}\right)_{pm} \varrho_{mn} S_{nq}, \tag{6.0129}$$

or

$$\widehat{\varrho}' = \widehat{S}^{-1} \widehat{\varrho}\, \widehat{S}, \tag{6.0130}$$

which is the normal equation for the transformation of an operator. We can thus treat $\widehat{\varrho}$ in all respects as an ordinary quantum-mechanical operator.

Let us now assume that we have made a transformation to a representation in which $\widehat{\varrho}$ is diagonal. We then have

$$\varrho_{mn} = \varrho_m \delta_{mn}. \tag{6.0131}$$

Consider now the average value of $\hat{\varrho}$. Using Eqs.(6.0121) and (6.0131), we have

$$\overline{\varrho} = \text{Tr}\, \hat{\varrho}^2 = \sum_m \varrho_m^2 \leqslant \left(\sum_m \varrho_m\right)^2 = (\text{Tr}\,\hat{\varrho})^2 = 1, \quad (6.0132)$$

and hence, in any representation,

$$\text{Tr}\, \hat{\varrho}^2 \leqslant 1, \qquad , \qquad \sum_{n,m} \varrho_{nm}\varrho_{mn} = \sum_{n,m} |\varrho_{mn}|^2 \leqslant 1, \quad (6.0133)$$

which imposes a severe limitation upon all elements of the density matrix.

In many cases we are interested in the change with time of the properties of the system. As long as we are working in the Schrödinger representation, that is, describe the systems by means of a time-dependent wavefunction and time-independent operators, the equation of motion for $\hat{\varrho}$ is given by Eq.(6.0112), while we get from Eqs.(6.0115), (6.0112), and the cyclic property of the trace:

$$\begin{aligned}
i\hbar\dot{\overline{G}} &= i\hbar\,\text{Tr}\,\hat{\dot{\varrho}}\hat{G} = \text{Tr}\,[\hat{\mathcal{H}}, \hat{\varrho}]_- \\
&= \text{Tr}\,(\hat{\mathcal{H}}\hat{\varrho}\hat{G} - \hat{\varrho}\hat{\mathcal{H}}\hat{G}) = \text{Tr}\,\hat{\varrho}(\hat{G}\hat{\mathcal{H}} - \hat{\mathcal{H}}\hat{G}) \\
&= \overline{[\hat{G}, \hat{\mathcal{H}}]_-}\,.
\end{aligned} \quad (6.0134)$$

Note that the signs in Eqs.(6.0112) and (6.0134) are different. In fact, in the Heisenberg representation, where we work with time-dependent operators and time-independent state functions, the operators satisfy the equation of motion

$$i\hbar\dot{\hat{G}} = [\hat{G}, \hat{\mathcal{H}}]_-, \quad (6.0135)$$

corresponding to the sign of Eq.(6.0134).

6.2.   Pure Case and Mixed Case

Let us briefly consider once again the reasons for introducing ensembles. As in the classical case, the necessity for introducing representative ensembles arises from the fact that our knowledge about the physical system under consideration is practically always far removed from the maximum attainable knowledge. The situation is now, however, slightly complicated because of the added statistical aspects coming from quantum mechanics. In an ideal (and practically always idealised) case we may know that our system is in a characteristic state of one of the quantum-mechanical operators.[2]   Let us assume for the sake of simplicity that this operator is

---

[2] It is easy to extend the discussion to include the cases where the system is in a characteristic state of two or more commuting operators.

the energy operator, so that we know that the system is in a stationary state. In that case we are dealing with a *pure case*.[3] All the systems in the representative ensemble will possess the same wavefunction, namely, the one corresponding to the characteristic state in question, and of the two averaging processes that are prescribed by Eq.(6.0113) only one remains, namely, that of taking the quantum-mechanical average. In fact, we are in a situation where ensemble theory does not need to be used, at any rate, as long as we are not concerned with the question of how to measure the properties of the system.[4] This situation has no classical analogy because we are still in a position where we must use statistical considerations, even though our knowledge is maximal.[5]

In general, however, the situation will not be so favourable and then we must construct a representative ensemble. Each system in the ensemble will then have a wavefunction which is a superposition, or mixture, of the various characteristic functions of, for instance, the energy operator, and the averages both over ψ^k and over the ensemble must be taken in order to evaluate $\overline{G}$. We are now dealing with a *mixed case*.[6] As in the classical case, we must make certain assumptions about a *priori* probabilities in order to be able to construct the representative ensemble. The assumption that is made is the following one. We assume equal a *priori* probabilities and random a *priori* phases for the various non-degenerate states. Once the repesentative ensemble has been set up, we are in the same position as in the classical case and then can make predictions about the most probable behaviour of a system in the ensemble. An important point is again the smallness of the fluctuations.

At this point we want to discuss briefly two different approaches to the density matrix: the operational approach[7] and the quantum-mechanical approach.[8]

In the operational approach, one recognises, on the one hand, the lack of sufficiently detailed information — which is at the basis of the introduction of representative ensembles, as we discussed in § 5.13 —

---

[3] We assume that the stationary case in question is non-degenerate or, if it is a degenerate state, that it is at the same time a characteristic state of other operators (for example, the operators corresponding to the square and the z-component of the angular momentum) in such a way that there is only one wavefunction that can describe this state of the system.

[4] For a discussion of the theory of measurement and its connection with density matrix theory, see D.ter Haar, *Repts. Progr. Phys.* **24**, 304 (1961), § 10; and especially the treatment by London and Bauer in their monograph *Measurement Theory in Quantum Mechanics*, Princeton University Press.

[5] Professor H.Margenau has pointed out to me that to some extent ordinary mechanics is the classical analogue of a pure case.

[6] The pure and mixed cases are called, respectively, "Reiner Fall" and "Gemisch" by von Neumann.

[7] U. Fano, *Rev. Mod. Phys.* **29**, 74 (1957).

[8] See, for instance, R.McWeeny, *Rev. Mod. Phys.* **32**, 335 (1960).

and, on the other hand, one omits from the start a detailed description of those degrees of freedom in which one is not interested. A typical example is the discussion of polarisation experiments. If one studies the polarisation of an electron beam, the only quantity of interest is the polarisation vector. For a spin-$\frac{1}{2}$ particle the polarisation vector is completely determined by three numbers, and the density matrix to describe the polarisation will therefore be a two-by-two matrix, which by virtue of its Hermiteicity and normalisation has exactly three independent components. Fano (*loc. cit.*) has shown how one can, indeed, define a suitable density matrix to describe polarisation experiments and also how one can find the equation of motion for the polarisation vector from the equation of motion of the density matrix (compare Problem 12 at the end of this chapter).

In the quantum-mechanical approach, one is also led to averaging processes; these lead one to the so-called *reduced density matrices*. To fix our ideas, let us consider an atom in its ground state. Even though we know the Hamiltonian of the atom, we are not able to find the exact ground-state wavefunction — unless it is a hydrogen atom. If we wish to evaluate the ground-sate energy, we must have recourse to approximation methods. As, however, we usually assume that the Hamiltonian will be a sum of operators involving either one electron or two electrons, it is tempting to introduce suitable averages of the many-body wavefunction — or rather the product of this wavefunction and its adjoint as this product occurs in the integrals defining the energy — over all particles but two. Such averages, which determine the energy of the atom, were called by Husimi reduced density matrices. One can show that, in fact, these reduced density matrices are the density matrix, as we have introduced it, for two electrons in the coordinate representation (compare Problem 4 at the end of this chapter).

Let us now derive the necessary and sufficient condition that the density matrix describes a pure case. If we work in a representation in which the density matrix is diagonal, so that Eq.(6.0131) holds, we have from Eq.(6.0107)

$$\varrho_m = \frac{1}{N} \sum_{k=1}^{N} |a_m^k|^2, \tag{6.021}$$

and we see that ϱ_m is the probability of finding an arbitrary system from the ensemble in the state characterised by φ_m. Since all $|a_m^k|^2$ are positive and at most equal to unity, we have

$$0 \leqslant \varrho_m \leqslant 1, \tag{6.022}$$

which also follows from Eqs.(6.0121) and (6.0131). The last equal sign holds only if for all k

$$|a_m^k|^2 = 1, \tag{6.023}$$

which entails that all systems are in the state characterised by φ_m as wavefunction. Hence, $\varrho_m = 1$ is a sufficient condition for a pure case. If $\varrho_m = 1$, we have

$$\varrho_n^2 = \varrho_n \tag{6.024}$$

for any n, or, in matrix notation,

$$\widehat{\varrho}^2 = \widehat{\varrho}. \tag{6.025}$$

Since from Eq.(6.022) it follows that all characteristic values of $\widehat{\varrho}$ are positive and smaller than or at most equal to unity, Eq.(6.025) can be satisfied only if one characteristic value is equal to unity and all the others are equal to zero. Hence, Eq.(6.025) is a different way of expressing a sufficient condition for a pure case.

It is also possible to show that Eq.(6.025) is a necessary condition for a pure case. If we are dealing with a pure case, all a_m^k with a given m are the same, say equal to a_m, and we have

$$\varrho_{mn} = \frac{1}{N} \sum_{k=1}^{N} a_m^k a_n^{k*} = a_m a_n^*. \tag{6.026}$$

Using Eq.(6.0106) we can then show that

$$[\varrho^2]_{mn} = \sum_l \varrho_{ml}\varrho_{ln} = \sum_l a_m a_l^* a_l a_n^* = a_m a_n^* = \varrho_{mn}, \tag{6.027}$$

whence follows Eq.(6.025) as a necessary condition.

6.3.   Macrocanonical Petit Ensembles in Quantum Statistics

In order that we shall be able fully to apply quantum statistics, it is necessary to develop first the formalism of quantum-mechanical ensemble theory to the same extent as we developed that of classical ensemble theory in the preceding chapter. In principle everything is contained in the discussion of § 6.1, but we shall enlarge slightly on that discussion. Let us first consider petit ensembles, and more specifically, macrocanonical petit ensembles with a density matrix

$$\widehat{\varrho} = e^{\beta(\psi - \widehat{\mathcal{H}})}, \tag{6.0301}$$

where $\widehat{\varrho}$ and $\widehat{\mathcal{H}}$ are operators, ψ and β are c-numbers, that is, quantities that commute with any operator, and where the right-hand side of Eq.(6.0301) is understood to stand for the right-hand side of Eq.(6.0119). From the normalising equation (6.0116) we get for ψ the equation

$$e^{-\beta\psi} = \mathrm{Tr}\, e^{-\beta\widehat{\mathcal{H}}}, \tag{6.0302}$$

where we have used the fact that ψ and β are c-numbers. Equation (6.0302) is the quantum-mechanical equivalent of Eq.(5.0302). Once again we shall only indicate the proof that ψ is the (Helmholtz) free energy and that, apart from a multiplying constant, β is equal to the reciprocal of the absolute temperature. We must bear in mind that $\mathcal{H}$ will depend on the external parameters a_k and that the generalised forces are now operators given by the equations

$$\widehat{A}_k = -\frac{\partial \widehat{\mathcal{H}}}{\partial a_k}. \tag{6.0303}$$

We can write Eq.(6.0302) in the form

$$\mathrm{Tr}\,\widehat{\varrho} = e^{\beta\psi}\,\mathrm{Tr}\,e^{-\beta\widehat{\mathcal{H}}} = 1, \tag{6.0304}$$

and, considering two situations with slightly different values of β and the a_k, we get

$$\delta\beta\psi = \delta\beta\,\mathrm{Tr}\,\widehat{\varrho}\widehat{\mathcal{H}} - \beta\sum_k \delta a_k\,\mathrm{Tr}\,\widehat{\varrho}\widehat{A}_k, \tag{6.0305}$$

or, using Eq.(6.0115) for average values, we find

$$\delta\beta\psi = \overline{\mathcal{H}}\,\delta\beta - \beta\sum_k \overline{A}_k\,\delta a_k. \tag{6.0306}$$

If we write

$$\overline{\delta A} = \sum_k \overline{A}_k\,\delta a_k, \tag{6.0307}$$

Eq.(6.0306) can be written in the form

$$\delta\beta\psi = \overline{\mathcal{H}}\,\delta\beta - \beta\overline{\delta A}. \tag{6.0308}$$

This equation is completely analogous to Eq.(5.0305), and it follows therefore that ψ can be considered to be the free energy, that

$$\beta^{-1} = kT, \tag{6.0309}$$

and that, apart from a possible additive constant, the entropy S is given by the equation

$$S = k\beta(\overline{\mathcal{H}} - \psi). \tag{6.0310}$$

Comparing Eqs.(6.0302) and (5.0311), we see that $\mathrm{Tr}\,e^{-\beta\widehat{\mathcal{H}}}$ now takes the place of the partition function of classical theory.

Before we go over to the quantum-mechanical grand ensembles we shall apply the formalism of the quantum-mechanical macrocanonical ensemble to the case of a system of independent particles, for the moment neglecting exclusion effects.

The Hamiltonian of a system consisting of N independent particles is the sum of the Hamiltonians of the individual particles,

$$\widehat{\mathcal{H}} = \sum_{i=1}^{N} \widehat{\mathcal{H}}_i, \tag{6.0311}$$

where $\widehat{\mathcal{H}}_i$ is the energy operator of the i-th particle.

In order to simplify our calculations we shall assume that the φ_n are the characteristic functions of the Hamiltonian. This brings the energy matrix on diagonal form,

$$H_{kl} = E_k \delta_{kl}, \tag{6.0312}$$

and the density matrix is thus also diagonal,

$$\varrho_{kl} = \delta_{kl} \, e^{\beta(\psi - E_k)}, \tag{6.0313}$$

so that we get from Eq.(6.0302)

$$e^{-\beta\psi} = \sum_{k} \gamma_k e^{-\beta E_k}, \tag{6.0314}$$

where γ_k is the multiplicity of the energy level E_k. We see that Eq.(6.0314) resembles closely Eq.(3.401) with one important difference, namely, that in Eq.(3.401) the ε_n were the energy levels of single molecules, whereas now the E_k are the energy levels of the whole system. This corresponds to the transition from μ-space to Γ-space. Since the particles in the system are independent, we have (compare Eq.(6.0311))

$$E_k = \sum_{i=1}^{N} \varepsilon_i. \tag{6.0315}$$

In Eq.(6.0315) the summation is over all particles in the system. We can replace this by a summation over all energy levels, by taking together all particles with the same energy. If there are n_1 particles with energy ε_1, n_2 with energy ε_2, $\cdots$, we can clearly write Eq.(6.0315) in the form

$$E_k = \sum_{\sum n_s = N} n_s \varepsilon_s, \tag{6.0316}$$

where all ε_s are now different.

The degeneracy γ_k arises from two factors. The first one is the fact that the levels ε_s themselves may be degenerate with a degeneracy, say, g_s. The second factor comes from the degeneracy which derives from the fact that the N can be divided in several ways into groups containing, respectively, $n_1, n_2, \cdots, n_s, \cdots$ particles. We obtain

$$\gamma_k = \frac{N!}{\prod_s n_s!} \prod_s g_s^{n_s}, \tag{6.0317}$$

and substituting Eqs.(6.0316) and (6.0317) into Eq.(6.0314), we get

$$e^{-\beta\psi} = \sum_{\sum n_s = N} N! \left(\prod_s \frac{g_s^{n_s}}{n_s!} \right) e^{-\beta \sum n_s \varepsilon_s},$$

or, using Eq.(1.704),

$$e^{-\beta\psi} = \left[\sum_s g_s e^{-\beta\varepsilon_s} \right]^N. \tag{6.0318}$$

The expression inside the square brackets is the partition function for one particle. In §4.6 we calculated this partition function for a system of spinless point particles under the assumption that the energy levels were lying so densely that the sum could be replaced by an integral. There is no point in writing down the result here, since we see immediately from Eq.(6.318) that for a system of N independent particles the free energy is equal to N times the free energy of one particle, as would be expected. We are thus back to the situation encountered in Capter 3 and no new features will be found.

As in the case of the classical macrocanonical ensembles it is possible to calculate the fluctuations in energy and generalised forces. The formulæ, which are derived by the same methods as before, are entirely the same as those found in §5.4, the only difference being in the actual value of ψ. We refer therefore to the discussion in that section.

6.4.   Canonical Grand Ensembles in Quantum Statistics

We come now to one of the most powerful tools of statistical mechanics, the quantum-mechanical canonical grand ensembles. In the preceding chapter we discussed the classical grand ensembles and saw that in many cases these must be preferred to the petit ensembles. It is thus desirable to develop their quantum-mechanical counterparts. Since in quantum mechanics the indistinguishability of identical particles finds its ready expression in the exclusion principle, we shall now have no reason to distinguish between generic and specific densities. This makes the quantum-mechanical ensemble theory simpler than the classical one. Moreover, due to the statistical aspects of quantum mechanics the physical interpretation of the quantum-mechanical ensemble theory will be much more natural.

We shall assume that our system is confined to a finite volume so that the energy levels will form a discrete spectrum. The energy levels will, of course, depend on the ν_i, the number of particles of the various kinds that may be present in the system. We shall denote the energy levels by $E(\nu_i; n)$ where n numbers the levels.

The density matrix of a stationary ensemble must now commute not only with $\widehat{\mathcal{H}}$ but also with the number operators $\widehat{\nu}_i$, which are the Jordan-Klein and Jordan-Wigner matrices from quantum mechanics.[9] Their characteristic values are the non-negative integers ν_i.

We shall discuss only canonical grand ensembles. Their density operator is given by the equation

$$\widehat{\varrho} = e^{-q + \sum_i \alpha_i \widehat{\nu}_i - \beta \widehat{\mathcal{H}}}, \qquad (6.0401)$$

where q, β, and the α_i are c-numbers. The grand potential q is determined by the normalisation condition, and we have

$$e^q = \mathrm{Tr}\, e^{\sum_i \alpha_i \widehat{\nu}_i - \beta \widehat{\mathcal{H}}}. \qquad (6.0402)$$

In Eqs.(6.0401) and (6.0402) the exponentials involving operators are once again defined by the power series expansion. The right-hand side of Eq. (6.0402) is called the (quantum-mechanical) *grand partition function*.

If we choose for the φ_n the characteristic functions of both $\widehat{\mathcal{H}}$ and all the $\widehat{\nu}_i$, eq.(6.0402) becomes

$$e^q = \sum_{\nu_i, n} e^{\sum_i \alpha_i \nu_i - \beta E(\nu_i, n)}, \qquad (6.0403)$$

which is the quantum-mechanical equivalent of Eq.(5.0907). Since $\widehat{\mathcal{H}}$, and thus also the $E(\nu_i; n)$, depend on the external parameters, we see that q is a function of β, the α_i, and the a_k, as in the classical case,

$$q = q(\beta, \alpha_i, a_k). \qquad (6.0404)$$

The average $\langle G \rangle$ of a quantity G over the canonical grand ensemble is given by Eq.(6.0113) which can be put in the form

$$\langle G \rangle = \sum_{\nu_i} \sum_n G(\nu_i; n)\, \varrho(\nu_i; n), \qquad (6.0405)$$

where $\varrho(\nu_i; n)$ is the diagonal element of the density matrix which, due to our choice of the φ_n, is on diagonal form, and where $G(\nu_i; n)$ is the quantum-mechanical expectation value of G for the state, characterised by the ν_i and by n, with wavefunction $\psi(\nu_i; n)$,

$$G(\nu_i; n) = \int \psi^*(\nu_i; n)\, \widehat{G}\, \psi(\nu_i; n)\, d\tau. \qquad (6.0406)$$

By using Eq.(6.0401) for $\widehat{\varrho}$, we can write Eq.(6.0405) in the form

$$\langle G \rangle = e^{-q} \sum_{\nu_i} e^{\sum_i \alpha_i \nu_i} \sum_n G(\nu_i; n)\, e^{-\beta E(\nu_i; n)}. \qquad (6.0407)$$

---

[9] H.A.Kramers, *Quantum Mechanics*, North-Holland, Amsterdam (1957), § 72.

In order to see the physical meaning of β and α_i we again consider two situations with slightly different values of β, the α_i, and the a_k. We shall give only the relevant results without their derivations, since the calculations are exactly the same as those given on numerous previous occasions. We have for the variation of q the equation

$$\delta q \;=\; \sum_i \langle \nu_i \rangle \, \delta\alpha_i - \langle \varepsilon \rangle \, \delta\beta + \beta \sum_k \langle A_k \rangle \, \delta a_k. \qquad (6.0408)$$

The entropy S and (Helmholtz) free energy F are given by the equations

$$\frac{S}{k} \;=\; q - \sum_i \alpha_i \langle \nu_i \rangle + \beta \langle \varepsilon \rangle, \qquad (6.0409)$$

$$F \;=\; \langle \varepsilon \rangle - ST \;=\; -\frac{q}{\beta} + \frac{\sum_i \alpha_i \langle \nu_i \rangle}{\beta}. \qquad (6.0410)$$

Introducing

$$\mu_i \;=\; \frac{\alpha_i}{\beta}, \qquad (6.0411)$$

one can show that the μ_i are the partial free energies or the partial thermal potentials,

$$\mu_i \;=\; \left(\frac{\partial F}{\partial \langle \nu_i \rangle} \right)_{\beta,\,a_k} \;=\; \left(\frac{\partial G}{\partial \langle \nu_i \rangle} \right)_{\beta,\,P,\,a_k (\neq V)}, \qquad (6.0412)$$

where $G = F + PV$ is the thermal potential, or Gibbs free energy.

The temperature T is related to β by the familiar equation

$$\beta^{-1} \;=\; kT, \qquad (6.0413)$$

and in the case of a homogeneous system we have

$$q \;=\; \frac{PV}{kT}. \qquad (6.0414)$$

The grand potential gives us all thermodynamic quantities and we have, for instance, from Eq.(6.0408)

$$\langle \nu_i \rangle \;=\; \frac{\partial q}{\partial \alpha_i}, \qquad (6.0415)$$

$$\langle \varepsilon \rangle \;=\; -\frac{\partial q}{\partial \beta}, \qquad (6.0416)$$

$$\langle A_k \rangle \;=\; \frac{1}{\beta} \frac{\partial q}{\partial a_k}, \qquad (6.0417)$$

$$P \;=\; \frac{1}{\beta} \frac{\partial q}{\partial V}. \qquad (6.0418)$$

One can calculate the fluctuations similarly. The easiest way is to use the normalisation condition and take the derivative with respect to two of the quantities β, α_i, or a_k. In this way we get, for instance,

$$\left\langle (\nu_i - \langle \nu_i \rangle)^2 \right\rangle = \frac{\partial^2 q}{\partial \alpha_i^2}, \qquad \left\langle (\varepsilon - \langle \varepsilon \rangle)^2 \right\rangle = \frac{\partial^2 q}{\partial \beta^2}, \qquad \cdots . \qquad (6.0419)$$

In general the dispersions will be small if the $\langle \nu_i \rangle$ are large, and in such cases we are dealing with ensembles with practically completely determined values of E and the ν_i.

Before applying the formulæ of this section to a perfect gas and investigating the consequences of the exclusion principle, we shall briefly discuss a few aspects of quantum-mechanical ensemble theory. Both points that we shall examine also enter classical ensemble theory, albeit sometimes in a slightly different way, and we have discussed them before in connection with the classical theory.

The first point is the question of fluctuations. In a phenomenological thermodynamical theory all quantities such as temperature and energy, thermal potentials and numbers of particles of a given kind, volume and pressure, and so on, have well defined values. However, if a canonical ensemble is considered, only the first of each of these pairs has a sharply defined value, the second quantity in each pair fluctuating around an average value. In some cases these fluctuations enter because our very inadequate knowledge about the physical system forces us to use a representative ensemble and we cannot expect to obtain completely defined predictions from very incomplete data. Sometimes, however, the fluctuations may not only present the uncertainty of our predictions, but also correspond to a physical phenomenon. If we consider, for instance, our system embedded in a large thermostat which at the same time can exchange paricles with the system, the fluctuations in energy and numbers of particles will occur and under favourable cuircumstances even be measurable.[10] In most cases these fluctuations will be extremely small and actually too small to be observable. This is important, since we can then safely apply the theory to thermodynamical systems.[11]

It might be feared that some fluctuations, for instance, in the pressure or in other forces might be large (compare similar considerations in § 5.4). There are two reasons why these fluctuations are not important. First of all, the large fluctuations often arise due to modifications introduced to simplify calculations, such as, for instance, idealised walls and rigid molecules. Secondly, any measurement of a physical quantity will require a

---

[10]It is possible to consider our system to be one of an ensemble, the other members of the ensemble forming the temperature and particle bath. We do not wish to stress this point of view, since it does not fully acknowledge the importance of representative ensembles.

[11]At the moment we are not considering critical fluctuations which can be measurably large; compare the discussion in § 5.10.

finite time interval so that we are not really concerned with $\langle G \rangle$, but with the time average of $\langle G \rangle$. From quantum mechanics it follows, however, that fluctuations in a time average are extremely small and decrease rapidly with increasing length of the period over which the average is taken.[12]

It is possible to take a point of view complementary to the one taken when one considers canonical grand ensembles. One then considers systems with a constant energy and constant numbers of particles. Such systems are represented by microcanonical petit ensembles which it is possible to consider in quantum mechanics, though we shall not do so. If the temperature or the thermal potentials are measured, they will show fluctuations of the same order of magnitude as the fluctuations in energy and particle numbers in a canonical grand ensemble. As long as the number of particles in the system is very large, all fluctuations will be negligible and the results obtained by using microcanonical petit ensembles are the same as those obtained by using canonical grand ensembles.

The second point that we wish to discuss at this juncture is the entropy. Our considerations will be similar to those of §5.5. Let dZ be the total number of energy levels for which the energy lies between ε and $\varepsilon + d\varepsilon$ and for which the ν_i have values between ν_i and $\nu_i + d\nu_i$. If we write

$$dZ = W(\nu_i, \varepsilon) \, d\varepsilon \prod_i d\nu_i, \qquad (6.0420)$$

the normalisation equation can be written in the form

$$\int e^{-q + \sum_i \alpha_i \nu_i - \beta \varepsilon} \, W(\nu_i; \varepsilon) \, d\varepsilon \prod_i d\nu_i = 1, \qquad (6.0421)$$

where, if necessary, the integral on the left-hand side must be treated as a Stieltjes integral.

Consider now a situation where the $\langle \nu_i \rangle$ are so large that fluctuations are negligible. In that case the values of ε and ν_i for which the integrand in Eq.(6.0421) is a maximum will be practically equal to $\langle \varepsilon \rangle$ and $\langle \nu_i \rangle$ and we can write Eq.(6.0421) in the form

$$\left[e^{-q + \sum_i \alpha_i \nu_i - \beta \varepsilon} \, W(\nu_i; \varepsilon) \right]_{\substack{\nu_i = \langle \nu_i \rangle \\ \varepsilon = \langle \varepsilon \rangle}} \Delta \varepsilon \prod_i \Delta \nu_i = 1, \qquad (6.0422)$$

where $\Delta \varepsilon$ and $\Delta \nu_i$ are the half-widths of the fluctuations, that is,

$$(\Delta \varepsilon)^2 = C \left\langle (\varepsilon - \langle \varepsilon \rangle)^2 \right\rangle, \qquad (\Delta \nu_i)^2 = C_i \left\langle (\nu_i - \langle \nu - i \rangle)^2 \right\rangle, \qquad (6.0423)$$

---

[12]Compare H.A.Kramers, *Quantum Mechanics*, North-Holland, Amsterdam (1957) §52.

with C and the C_i constants of the order of magnitude unity, and where we have approximately

$$\Delta \varepsilon \sim \frac{\langle \varepsilon \rangle}{\sqrt{\sum \langle \nu_i \rangle}}, \qquad \Delta \nu_i \sim \sqrt{\langle \nu_i \rangle}. \qquad (6.0424)$$

Taking the logarithm of both sides of Eq.(6.0422), we have

$$q - \sum_i \langle \nu_i \rangle \alpha_i + \beta \langle \varepsilon \rangle = \ln \left[W(\langle \nu_i \rangle; \langle \varepsilon \rangle) \Delta \varepsilon \prod_i \Delta \nu_i \right]. \qquad (6.0425)$$

The left-hand side of this equation is equal to S/k, as can be seen from Eq.(6.0409), and we have thus

$$S = \ln \left[W(\langle \nu_i \rangle; \langle \varepsilon \rangle) \Delta \varepsilon \prod_i \Delta \nu_i \right]. \qquad (6.0426)$$

From the general relation between entropy and probability it follows that $W(\nu_i; \varepsilon)$ can be considered to be the probability of finding certain values of ε and ν_i, or that $W(\nu_i; \varepsilon) \Delta \varepsilon \prod \Delta \nu_i$ is the probability of finding ε and ν_i within certain limits. This corresponds closely to the definition of W. Indeed, $W(\nu_i; \varepsilon) \Delta \varepsilon \prod \Delta \nu_i$ was defined as the number of energy states corresponding to ε and ν_i within given limits, and this together with the fact that earlier we mentioned that all (non-degenerate) stationary states should have equal a *priori* probabilities shows that, indeed, $W(\nu_i; \varepsilon) \Delta \varepsilon \prod \Delta \nu_i$ can be considered to be the probability for the energy and the numbers of particles to lie inside given intervals. We can express this slightly differently by stating that the entropy measures our *lack of knowledge* or *lack of detailed information*, since $W(\nu_i; \varepsilon) \Delta \varepsilon \prod \Delta \nu_i$ gives us the number of possibilities that are left, once we know that ε and ν_i are confined to certain given intervals. A similar situation is present in the classical case (compare Eq.(5.0524); $(\partial \Omega / \partial \varepsilon) \Delta$ gives us a measure for the volume in Γ-space in which the representative point can be found, once the energy is stated to lie within a given interval).

The relationship between entropy and lack of information has led many authors, notably Shannon, to introduce "entropy" as a measure for the information transmitted by telephone lines, and so on, and in this way entropy has figured largely in discussions in information theory. It must be stressed here that the entropy introduced in information theory is *not* a thermodynamical quantity and that the use of the same term is rather misleading. It was probably introduced because of a rather loose use of the term "information".

In this connection we may briefly discuss *Maxwell's demon*. In 1871 Maxwell introduced his famous demon, "a being whose faculties are so sharpened that he can follow every molecule in its course, and would be able to do what is at present impossible to us. ... Let us suppose that a vessel is divided into two portions A and B by a division in which there is a small hole, and that a being who can see the individual molecules opens and closes this hole, so as to allow only the swifter molecules to pass from A to B, and only the slower ones to pass from B to A. He will thus, without expenditure of work raise the temperature of B and lower that of A, in contradiction to the second law of thermodynamics".

Maxwell's demon has been widely discussed and various authors have set out to show that various attempts to circumvent the second law by using the demon are bound to fail.[13] Although their discussions differ in some respects, they have a few points in common. The first point is the observation that one should take the demon to be part of the total system, and then one must consider the total entropy of the original system and the demon. The second point which was most clearly developed for the first time by Szilard[14] is that the demon, in order to be able to operate the trap-door through which the molecules pass, must receive information. Its own entropy increases therefore; it is now the question whether the increase in the demon's entropy is smaller or larger than the decrease in the entropy of the gas. Both Szilard and Brillouin[15] consider possible arrangements and show that in those cases the net change in the entropy is positive. Szilard analyses the problem very thoroughly and shows that one can describe Maxwell's demon as follows. By some means an operation on a system is determined by the result of a measurement on the system that immediately precedes the operation. In Maxwell's original scheme the operation was the opening of the trap-door and the measurement was the determination of the velocity of an approaching molecule. The result of the operation will be a decrease in entropy, but the preceding measurement will be accompanied by an increase in entropy, and once again one must consider the balance.

Wiener takes a simpler point of view.[16] He considers the situation, where the demon acts, as a metastable state and writes: "In the long run, the Maxwell demon is itself subject to a random motion corresponding to the temperature of its environment and it receives a large number of small impressions until it falls into 'a certain vertigo' and is incapable of clear perceptions. In fact, it ceases to act as a Maxwell demon." This point of view is probably too simplified and we prefer that of Szilard's and refer the reader to his paper for a more extensive discussion.

---

[13]Let us briefly point out that these proofs of the general validity of the second law are different from the proofs of the general validity of the Heisenberg relations in quantum mechanics. In both cases one sets up an idealised experiment which would be suited to get around theoretical limitations, but whereas in quantum mechanics one proves that it is *never* possible to break the limitations, in the thermodynamical case one can prove only that *on the average* it is not possible to violate the second law.

[14]L.Szilard, *Zs Physik* **53**, 840 (1929).

[15]L.Brillouin, *J. Appl. Phys.* **22**, 334 (1951).

[16]N.Wiener, *Cybernetics*, New York (1948) p.72; see also Brillouin, *loc. cit.*, §I.

In §5.5 we saw that there were several ways of defining the entropy, which is also the case here. One can prove that, because of the very steep increase of $W(\nu_i; \varepsilon)$ with both ε and ν_i, the entropy can also be defined by the equation

$$S = k \ln V, \qquad (6.0427)$$

where

$$V = \int_{\varepsilon_{min}}^{(\varepsilon)} \int_{\nu_i=0}^{(\nu_i)} \cdots \int W(\nu_i; \varepsilon) \, d\varepsilon \prod_i d\nu_i. \qquad (6.0428)$$

Equation (6.0427) corresponds to Eq.(5.0535) of the classical theory.

We have included the factor $\Delta\varepsilon \prod \Delta\nu_i$ in Eq.(6.0426) in order that the expression under the logarithm sign would be dimensionless. Actually, Eq.(6.0427) is probably an easier one to use for calculations, since otherwise $\Delta\varepsilon$ and the $\Delta\nu_i$ must be evaluated first. In Chapter 5 we were not concerned with the actual value of the entropy and therefore did not pay any attention to the fact that neither $\partial\Omega/\partial\varepsilon$ nor Ω is dimensionless. We could have corrected this by writing instead of Eqs.(5.0525) and (5.0535)

$$S = k \ln \left[\frac{\partial\Omega}{\partial\varepsilon} \cdot \Delta \cdot h^{-3sN} \right] \qquad (6.0429)$$

and

$$S = k \ln \left(\Omega \cdot h^{-3sN} \right), \qquad (6.0430)$$

where we can see now even more clearly the correspondence between the classical and the quantum-mechanical cases. In Eq.(6.0427), for instance, V is the total number of states with $\varepsilon \leqslant \langle \varepsilon \rangle$ and $\nu_i \leqslant \langle \nu_i \rangle$, whereas in Eq. (6.0430) which relates to a petit ensemble, Ω is the total volume in Γ-space with $\varepsilon \leqslant \bar{\varepsilon}$. The factor h^{-3sN} arises from the fact that each non-degenerate state corresponds to a volume h^{3sN} in Γ-space, and we see that, indeed, the two Eqs.(6.0427) and (6.0430) will give the same result in the limiting case of very high temperatures or very low densities, when all quantum-mechanical effects are negligible.

6.5.   The H-Theorem in Quantum Statistics

In §§ 5.13 and 5.14 we discussed the question of representative ensembles and the approach to equilibrium from the point of view of classical ensemble theory. We shall now similarly consider these problems from the point of view of quantum-mechanical ensemble theory. It is convenient for our discussion to introduce once again the index of probability which is now an operator defined by the equation

$$\widehat{\varrho} = e^{\widehat{\eta}}. \qquad (6.0501)$$

From Eq.(6.0501) we have once again for equilibrium ensembles the relation

$$S = -k\langle\eta\rangle \qquad (6.0502)$$

(compare Eqs.(6.0310) and (6.0409)), which is the quantum-mechanical counterpart of Eq.(5.1107) with the difference that $\widehat{\eta}$ is now an operator. As in the relevant discussion in Chapter 5 we shall consider grand ensembles only, leaving the reader to make the necessary adjustments needed for the discussion of petit ensembles, even though when suitable we shall give results pertaining to petit ensembles.

We remind ourselves that the discussion in Chapter 5, which showed that the microcanonical, the macrocanonical, or the canonical grand ensemble were the equilibrium ensembles, depending on the conditions, fell into three parts. First of all, we showed that $\langle \eta \rangle$ was a minimum for the above-mentioned ensembles. Secondly, we showed that $d\langle \eta \rangle / dt \leqslant 0$, and thirdly, we showed that from the condition that $\langle \eta \rangle$ be a minimum and that the other conditions hold, the canonical ensemble densities follow as the representative ensemble densities, proving that the representative ensembles are, indeed, the equilibrium ones. We leave the last part to the reader as a problem, and shall only discuss here the proofs that $\langle \eta \rangle$ is a minimum for the canonical density matrices and that $d\langle \eta \rangle / dt$ is non-positive. As in Chapter 5, we shall give the proof only for the case where both the average energy and the average number of particles are given, while we also again restrict our discussion to the case where there is only one kind of particles present. The conditions that the average energy, E, and the average number of particles, N, are given are expressed by the equations

$$\mathrm{Tr}\,\widehat{\varrho}\,\widehat{\mathcal{H}} \;=\; E, \tag{6.0503}$$

$$\mathrm{Tr}\,\widehat{\varrho}\,\widehat{\nu} \;=\; N. \tag{6.0504}$$

We must now prove that $\langle \eta \rangle$ is a minimum, if $\widehat{\varrho}$ satisfies the equation

$$\widehat{\varrho} \;=\; e^{-q+\alpha\widehat{\nu}-\beta\widehat{\mathcal{H}}}. \tag{6.0505}$$

To prove this, we consider two ensembles with density matrices $\widehat{\varrho}_1$ and $\widehat{\varrho}_2$, with $\widehat{\varrho}_1$ given by Eq.(6.0505) and $\widehat{\varrho}_2$ by the equation

$$\widehat{\varrho}_2 \;=\; \widehat{\varrho}_1 e^{\widehat{\Delta\eta}}. \tag{6.0506}$$

We now consider the quantity $\langle \eta \rangle_2 - \langle \eta \rangle_1$ which is given by the equation

$$\langle \eta \rangle_2 - \langle \eta \rangle_1 \;=\; \mathrm{Tr}\left(\widehat{\eta}_2 e^{\widehat{\eta}_2} - \widehat{\eta}_1 e^{\widehat{\eta}_1} \right). \tag{6.0507}$$

The right-hand side of Eq.(6.0507) can be transformed as follows:

$$\begin{aligned}
\mathrm{Tr}\left(\widehat{\eta}_2 e^{\widehat{\eta}_2} - \widehat{\eta}_1 e^{\widehat{\eta}_1} \right) &= \mathrm{Tr}\left[\left(\widehat{\eta}_2 - \widehat{\eta}_1 \right) e^{\widehat{\eta}_2} \right] + \mathrm{Tr}\left[\widehat{\eta}_1 \left(\widehat{\varrho}_2 - \widehat{\varrho}_1 \right) \right] \\
&= \mathrm{Tr}\left[\left(\widehat{\eta}_2 - \widehat{\eta}_1 \right) e^{\widehat{\eta}_2} \right] + \mathrm{Tr}\left[\left(-q + \alpha\widehat{\nu} - \beta\widehat{\mathcal{H}} \right) \left(\widehat{\varrho}_2 - \widehat{\varrho}_1 \right) \right] \\
&= \mathrm{Tr}\left[\left(\widehat{\eta}_2 - \widehat{\eta}_1 \right) e^{\widehat{\eta}_2} \right], \tag{6.0508}
\end{aligned}$$

where we have used the fact that both $\widehat{\varrho}_1$ and $\widehat{\varrho}_2$ satisfy Eqs.(6.0503), (6.0504), and the normalisation condition (6.0116). It thus follows that

$$\langle \eta \rangle_2 - \langle \eta \rangle_1 \;=\; \mathrm{Tr}\Big[\big(\widehat{\eta}_2 - \widehat{\eta}_1\big)e^{\widehat{\eta}_2}\Big]. \qquad (6.0509)$$

We now must prove that the right-hand side of Eq.(6.0509) is positive, except when $\widehat{\Delta\eta} = 0$. We first of all note that for any Hermitean operator $\widehat{A}$ we have[17] (see Problem 26 at the end of the chapter)

$$\Big[e^{\widehat{A}}\Big]_{kk} \;\leqslant\; e^{A_{kk}}, \qquad (6.0510)$$

the equal sign holding only when $\widehat{A}$ is on diagonal form.

We now choose a representation in which η_2 is diagonal; then, by using the normalisation condition for $\widehat{\varrho}_1$ and $\widehat{\varrho}_2$, Eq.(6.0510), and Eq. (5.1407), we find

$$
\begin{aligned}
\mathrm{Tr}\Big[\big(\widehat{\eta}_2 - \widehat{\eta}_1\big)e^{\widehat{\eta}_2}\Big] &= \mathrm{Tr}\Big[\big(\widehat{\eta}_2 - \widehat{\eta}_1\big)e^{\widehat{\eta}_2} - e^{\widehat{\eta}_2} + e^{\widehat{\eta}_1}\Big] \\
&= \sum_k \Big[\big(\eta_{2k} - \eta_{1kk}\big)e^{\eta_{2k}} - e^{\eta_{2k}} + \big(e^{\eta_1}\big)_{kk}\Big] \\
&\leqslant \sum_k \Big[\big(\eta_{2k} - \eta_{1kk}\big)e^{\eta_{2k}} - e^{\eta_{2k}} + e^{\eta_{1kk}}\Big] \\
&= \sum_k e^{\eta_{1kk}}\Big[\big(\eta_{2k} - \eta_{1kk} - 1\big)e^{\eta_{2k}-\eta_{1kk}} + 1\Big] \\
&\geqslant 0, \qquad\qquad\qquad\qquad\qquad\qquad (6.0511)
\end{aligned}
$$

which concludes our proof.

As all averages are invariant under unitary transformatons and as we can transform from one instant, t', to another, t'', by a unitary transformation, it follows that, as in the classical case, $\langle \eta \rangle' = \langle \eta \rangle''$. Hence, we cannot prove that $d\langle \eta \rangle/dt \leqslant 0$ and, as in the classical case, we must have recourse to a coarse-grained density.

We introduce this coarse-grained density, or rather the coarse-grained density matrix $\widehat{P}$ (read again "capital rho"), as follows. We choose a representation in which $\widehat{\mathcal{H}}$ and $\widehat{\nu}$ are diagonal. Furthermore, we shall divide the stationary states into groups. Let Z_i be the number of levels in the i-th group.[18] We now define $\widehat{P}$ by its matrix elements in the chosen representation, as follows:

$$P_{kl} \;=\; \frac{\delta_{kl}}{Z_i} \sum_j \varrho_{jj}, \qquad (6.0512)$$

[17]R.Peierls, *Phys. Rev.* **54**, 918 (1938).

[18]We note the difference between the grouping here and that done in Chapter 4. Now we group energy levels of the whole system; then we grouped energy levels of individual particles.

where the energy level E_k belongs to the i-th group and where the summation is over all the states in the i-th group. We see that $Z_i P_{kk}$ is the probability of finding a system of the ensemble in a state belonging to the i-th group. Equation (6.0512) defines the coarse-grained density matrix in a particular representation; its matrix elements in any other representation follow by the usual transformation rules.

From Eq.(6.0512) and the normalisation of $\widehat{\varrho}$ it follows that $\widehat{P}$ is also normalised,

$$\operatorname{Tr} \widehat{P} = 1. \tag{6.0513}$$

We now define a quantity $\widehat{H}$ (read "capital eta") by the equation

$$\widehat{H} = \ln \widehat{P}, \tag{6.0514}$$

and we have

$$\langle H \rangle = \operatorname{Tr} \widehat{\varrho} \ln \widehat{P} = \sum_k \varrho_{kk} \ln P_{kk} = \sum_k P_{kk} \ln P_{kk} = \operatorname{Tr} \widehat{P} \ln \widehat{P}, \tag{6.0515}$$

where we have used Eq.(6.0512) and where we have worked throughout in the representation in which Eq.(6.0512) was written.

If, as in the classical case, we assume the coarse-graining to be in accordance with experimental limitations, which means that we assume that by means of our observational methods we can discriminate between the different groups of energy levels, but not within them, we can at a time t' determine through our observations the coarse-grained density matrix $\widehat{P}$. The fine-grained density matrix $\widehat{\varrho}$ can then only be chosen to be given by the equation (we use here the assumption of equal a *priori* probabilities for non-degenerate states; compare the discussion of Eq.(5.1420)):

$$\widehat{\varrho}' = \widehat{P}', \tag{6.0516}$$

while we have for $\langle H \rangle'$:

$$\langle H \rangle' = \operatorname{Tr} \widehat{P}' \ln \widehat{P}' = \operatorname{Tr} \widehat{\varrho}' \ln \widehat{\varrho}'. \tag{6.0517}$$

If the situation at t' did not correspond to an equilibrium situation so that $\widehat{\varrho}'$ (or $\widehat{P}'$) did not correspond to a canonical density matrix, we cannot expect Eq.(6.0516) to hold at a later time t'', but we would have

$$\widehat{\varrho}'' \neq \widehat{P}'', \tag{6.0518}$$

from which follows

$$\langle H \rangle'' < \langle H \rangle'. \tag{6.0519}$$

Equation (6.0519) can be proved as follows. We first of all rewrite the expression for $\langle H \rangle' - \langle H \rangle''$:

$$\begin{aligned}
\langle H \rangle' - \langle H \rangle'' &= \operatorname{Tr} \widehat{P}' \ln \widehat{P}' - \operatorname{Tr} \widehat{P}'' \ln \widehat{P}'' \\
&= \operatorname{Tr} \widehat{\varrho}' \ln \widehat{\varrho}' - \operatorname{Tr} \widehat{\varrho}'' \ln \widehat{P}'' \\
&= \sum_k \varrho'_{kk} \ln \varrho'_{kk} - \sum_n \varrho''_{nn} \ln P''_{nn}.
\end{aligned} \tag{6.0520}$$

We now add to the right-hand side of Eq.(6.0520) first of all the expression

$$\sum_n \varrho''_{nn} \ln \varrho''_{nn} - \sum_k \varrho'_{kk} \ln \varrho'_{kk}, \qquad (6.0521)$$

which by Klein's lemma[19] is never positive and only vanishes when $\widehat{\varrho}''$ is a diagonal matrix (see Problem 27 at the end of this chapter), and secondly the expression $\operatorname{Tr} \widehat{P}'' - \operatorname{Tr} \widehat{\varrho}''$ which vanishes as both $\widehat{\varrho}''$ and $\widehat{P}''$ are normalised. We thus get

$$\langle H \rangle' - \langle H \rangle'' = \sum_n \left(\varrho''_{nn} \ln \varrho''_{nn} - \varrho''_{nn} \ln P''_{nn} - \varrho''_{nn} + P''_{nn} \right) \geqslant 0, \quad (6.0522)$$

where the last inequality follows from Eq.(5.1407) after the substitution $x = \ln(\varrho''_{nn}/P''_{nn})$ and Eq.(6.0121) from which follows that $P''_{nn} \geqslant 0$.

We want to discuss also a different way of looking at the behaviour of $\langle H \rangle$. First of all, we note that the probability P_i of finding a system in the ensemble in a state of the i-th group is given by

$$P_i = Z_i P_{kk}, \qquad (6.0523)$$

where P_{kk} is one of the Z_i equal diagonal elements belonging to the i-th group.

We can now express $\langle H \rangle$ in terms of the P_i, using Eq.(6.0523), as follows:

$$\langle H \rangle = \sum_i \sum_k \frac{P_i}{Z_i} \ln \frac{P_i}{Z_i} = \sum_i P_i \ln \frac{P_i}{Z_i}, \qquad (6.0524)$$

where the summation over i is over all groups and the summation over k over the Z_i members of the i-th group.

As $\sum P_i = 1$ (see Eq.(6.0513)) so that $\sum dP_i/dt = 0$, we have for the time-derivative of $\langle H \rangle$

$$\frac{d\langle H \rangle}{dt} = \sum_i \left(\ln P_i - \ln Z_i \right) \frac{dP_i}{dt}. \qquad (6.0525)$$

We now need an expression for the average transition probability $N_{i \to j}$ for transitions from the i-th to the j-th group. This is given by the equation[20]

$$N_{i \to j} = A_{i \to j} Z_j P_i. \qquad (6.0526)$$

If we accept the assumptions of equal a priori probabilities and random a priori phases for the various states, we can show[21] that, if we exclude the case where a magnetic field is present, as we shall do,

$$A_{i \to j} = A_{j \to i}. \qquad (6.0527)$$

[19]O.Klein, *Zs. Physik* **72**, 767 (1931).

[20]See R.C.Tolman, *The Principles of Statistical Mechanics*, Oxford University Press (1938), §99, or D.ter Haar, *Rev. Mod. Phys.* **27**, 289 (1955), App.V.

[21]R.C.Tolman, *ibid.*

If we use this equation it follows from Eq.(6.0526) that

$$\frac{dP_i}{dt} = \sum_j A_{i \to j} \left[Z_i P_j - Z_j P_i \right], \qquad (6.0528)$$

and we finally get after a few manipulations (substituting Eq.(6.0528) into Eq.(6.0525) and taking the arithmetic mean of the ensuing expression and the expression obtained by interchanging i and j)

$$\frac{d\langle H \rangle}{dt} = \frac{1}{2} \sum_{i,j} A_{i \to j} \left[P_j Z_i - Z_i P_j \right] \ln \frac{P_i Z_j}{P_j Z_i}. \qquad (6.0529)$$

The right-hand side of Eq.(6.0529) is negative unless $P_i Z_j = P_j Z_i$ (compare the discussion of Eq.(1.504)), as the $A_{i \to j}$ are never negative. We note that $d\langle H \rangle/dt$ vanishes only when $P_i Z_j = P_j Z_i$, that is, when

$$N_{i \to j} = N_{j \to i}, \qquad (6.0530)$$

and we see that at equilibrium there are as many transitions from the i-th to the j-th group as the other way round. This is an example of the so-called *principle of detailed balancing.*

The principle of detailed balancing states that at equilibrium the number of processes destroying situation A and creating situation B will be equal to the number of processes producing A and destroying B. This principle is valid in a great many cases, but not universally so. If it is not valid, one can still prove an H-theorem by considering cycles of transitions. The importance of the principle of detailed balancing lies in its many applications. For instance, it can happen that one can more easily find the rate at which a process goes in one direction by calculating the rate at which the reverse process takes place. Another example comes from nuclear physics where the principle of detailed balancing has been used for determining the spin of the pion.

Equation (6.0530), expressing the principle of detailed balancing, was a consequence of the vanishing of $d\langle H \rangle/dt$ and of Eq.(6.0527) which is an example of *microscopic reversibility.* We emphasise that one should not confuse the two principles.

6.6.   The Perfect Boltzmann Gas

As in Chapters 1, 2, 4, and 5, we shall apply our formalism to the case of a perfect gas. We shall apply here the method of the canonical grand ensembles. We saw in Chapter 4 that all gases behave like a Boltzmann gas at sufficiently high temperatures. We shall therefore first consider a perfect gas without taking into account exclusion effects, even though this is an apparent contradiction, since the exclusion principle is an integral part of quantum mechanics. This can be justified, however, by observing that the formulæ obtained in the present section are, indeed, the limiting cases of the formulæ obtained when exclusion effects are duly considered. We shall not discuss the limiting process, but refer to Chapter 4, where a similar situation occurred and was fully discussed.

In the case of a perfect gas of only one constituent we can drop the indices on the α_i and the ν_i. The total energy and number of particles are given by the equations

$$\varepsilon = \sum_s n_s \varepsilon_s, \tag{6.0601}$$

$$\nu = \sum_s n_s, \tag{6.0602}$$

where n_s is the number of particles in the energy level ε_s.

The degeneracy γ' of a state with energy ε is given by the expression (compare Eq.(6.0317))

$$\gamma' = \nu! \prod_s \frac{g_s^{n_s}}{n_s!}, \tag{6.0603}$$

where g_s is the degeneracy of the level ε_s. We mentioned that we are not including the exclusion principle in our discussion. However, since we are interested in obtaining formulæ that will turn out to be the limiting cases of the formulæ obtained when the exclusion principle is taken into account, we must realise that, because of symmetry effects, there is only one specific phase for each generic phase so that we must introduce an extra factor $1/\nu!$ into the equation for the degeneracy, as we did in Chapter 4. Instead of Eq.(6.0603) we thus have

$$\gamma = \prod_s \frac{g_s^{n_s}}{n_s!}, \tag{6.0604}$$

and we get from Eq.(6.0403) for the grand potential the equation

$$e^q = \sum_{\nu=0}^{\infty} e^{\alpha\nu} \sum_{\Sigma n_s = \nu} \gamma e^{-\beta \Sigma n_s \varepsilon_s}, \tag{6.0605}$$

or

$$\begin{aligned}
e^q &= \sum_{n_s=0}^{\infty} \cdots \sum \left(\prod_s \frac{g_s^{n_s}}{n_s!} \right) e^{\Sigma n_s(\alpha - \beta\varepsilon_s)} \\
&= \prod_s \left[\sum_{n_s=0}^{\infty} \frac{g_s^{n_s}}{n_s!} e^{n_s(\alpha - \beta\varepsilon_s)} \right] \\
&= \exp\left[\sum_s g_s e^{\alpha - \beta\varepsilon_s} \right],
\end{aligned}$$

or

$$q = \sum_s g_s e^{\alpha - \beta\varepsilon_s}. \tag{6.0606}$$

Equation (6.0606) looks very much the same as Eq.(4.607). There is, however, one important difference. While the energies in Eq.(4.607) were representative values for a group of levels, the energies in Eq.(6.0606) are the separate levels. Moreover, for the derivation of Eq.(6.0606) it is not necessary to assume that the levels are lying densely. Equation (6.0606) is thus much more general and has a much wider field of application.

The average number of particles $\langle \nu \rangle$ is given by Eq.(6.0415) and we have

$$\langle \nu \rangle = \frac{\partial q}{\partial \alpha} = q. \tag{6.0607}$$

We can use Eq.(6.0419) for the fluctuations in the number of particles and thus find

$$\left\langle (\nu - \langle \nu \rangle)^2 \right\rangle = \frac{\partial^2 q}{\partial \alpha^2} = q = \langle \nu \rangle, \tag{6.0608}$$

or

$$\frac{\left\langle (\nu - \langle \nu \rangle)^2 \right\rangle}{\langle \nu \rangle^2} = \frac{1}{\langle \nu \rangle}. \tag{6.0609}$$

We can also calculate the fluctuations in the numbers of particles occupying the various energy levels. The energy values ε_s are then considered as external parameters, since we have from Eq.(6.0601)

$$n_s = \frac{\partial \varepsilon}{\partial \varepsilon_s}, \tag{6.0610}$$

and we see that $-\varepsilon_s$ plays the rôle of an external parameter. From Eqs. (6.0417) and (6.0606) we get thus for $\langle n_s \rangle$ the formula

$$\langle n_s \rangle = \frac{1}{\beta} \frac{\partial q}{\partial(-\varepsilon_s)} = g_s \, e^{\alpha - \beta \varepsilon_s}. \tag{6.0611}$$

We see that we obtain the formula for the Maxwell-Boltzmann distribution and we have found here one of the most sophisticated and most general derivations of that formula. Not only that, but we are now also in a position to compute the fluctuations around the Maxwell-Boltzmann distribution. We have for the fluctuations in n_s the equation (compare Eq.(5.1002))

$$\left\langle (n_s - \langle n_s \rangle)^2 \right\rangle = \frac{1}{\beta^2} \frac{\partial^2 q}{\partial \varepsilon_s^2} + \frac{1}{\beta} \left\langle \frac{\partial^2 \varepsilon}{\partial \varepsilon_s^2} \right\rangle. \tag{6.0612}$$

Since

$$\frac{\partial^2 \varepsilon}{\partial \varepsilon_s^2} = 0,$$

Eq.(6.0612) gives us

$$\left\langle (n_s - \langle n_s \rangle)^2 \right\rangle = \frac{1}{\beta^2} \frac{\partial^2 q}{\partial \varepsilon_s^2} = g_s \, e^{\alpha - \beta \varepsilon_s} = \langle n_s \rangle, \tag{6.0613}$$

and we see that, as soon as $\langle n_s \rangle$ is large compared to one, the fluctuations in n_s will be small.

The results for the n_s and for ν were independent of the kind of particles that were present in the gas. This will no longer be true when we wish to consider the average energy and its fluctuations. Let us therefore assume now that we are dealing with a perfect monatomic gas. We can then use the fact that for all temperatures the energy levels will be lying so densely that the sum in Eq.(6.0606) can be replaced by an integral; then, of course, we get the same equation for q as in Chapter 4,

$$ q = \left(\frac{2\pi m}{\beta h^2} \right)^{3/2} e^\alpha V. \tag{6.0614} $$

From Eq.(6.0614) we find that the perfect gas law follows in the usual way, and for the average energy we have

$$ \langle \varepsilon \rangle = \frac{3\langle \nu \rangle}{2\beta}. \tag{6.0615} $$

The fluctuations in the energy follow again from Eq.(6.0419), and we have

$$ \left\langle (\varepsilon - \langle \varepsilon \rangle)^2 \right\rangle = \frac{\partial^2 q}{\partial \beta^2} = \frac{15q}{4\beta^2}, \tag{6.0616} $$

or, using Eqs.(6.0607) and (6.0615),

$$ \left\langle (\varepsilon - \langle \varepsilon \rangle)^2 \right\rangle = \frac{5}{3\langle \nu \rangle}. \tag{6.0617} $$

That the fluctuations in the energy are now larger than those given by Eq.(5.0804) is due to the fact that there is an additional source of fluctuations, namely, the number of particles.

6.7.   The Perfect Bose-Einstein Gas

Apart from the calculations of the fluctuations the preceding section brought no new results. This could have been expected, since exclusion effects had been neglected and finally even the sum in Eq.(6.0606) was replaced by an integral, thus carefully removing any chance we may have had of obtaining new results. In the present and the following section we shall see how the exclusion principle will change the considerations of the preceding section. The results will be practically the same as those obtained in Chapter 4, but our results will be less restricted, just as Eq.(6.0606) was more general than Eq.(4.607). We shall assume for the sake of simplicity that all energy levels ε_s are non-degenerate, that is, that all g_s are equal to one.

The case of a perfect Bose-Einstein gas is rather simple. We can use Eq.(6.0605) for the grand potential, but γ is now not given by Eq.(6.0604) but is equal to one for all values of ε, or

$$\gamma = 1. \tag{6.0701}$$

We then have for q the equation

$$e^q = \sum_{\nu=0}^{\infty} e^{\alpha\nu} \sum_{\Sigma n_s = \nu} e^{-\beta \Sigma n_s \varepsilon_s}, \tag{6.0702}$$

or

$$e^q = \sum_{n_s=0} \cdots \sum e^{\Sigma n_s(\alpha - \beta\varepsilon_s)}$$

$$= \prod_s \left(1 - e^{\alpha - \beta\varepsilon_s}\right)^{-1},$$

or

$$q = -\sum_s \ln\left[1 - e^{\alpha - \beta\varepsilon_s}\right]. \tag{6.0703}$$

Again, this equation is practically the same as Eq.(4.707), but it was not necessary in deriving Eq.(6.0703) to assume the energy levels to lie densely, and the ε_s are the individual energy levels and not a representative average of a group of energy levels as were the E_j in Eq.(4.707).

We find for $\langle \nu \rangle$ and the fluctuations in ν

$$\langle \nu \rangle = \frac{\partial q}{\partial \alpha} = \sum_s \left(e^{-\alpha + \beta\varepsilon_s} - 1\right)^{-1}, \tag{6.0704}$$

and

$$\left\langle (\nu - \langle \nu \rangle)^2 \right\rangle = \frac{\partial^2 q}{\partial \alpha^2} = \langle \nu \rangle + \sum_s \left(e^{-\alpha + \beta\varepsilon_s} - 1\right)^{-2}. \tag{6.0705}$$

We see here that the fluctuations are larger than in the case of a Boltzmann gas.

Of special interest again are the $\langle n_s \rangle$ and the fluctuations in the n_s, and we have

$$\langle n_s \rangle = -\frac{1}{\beta} \frac{\partial q}{\partial \varepsilon_s} = \frac{1}{e^{-\alpha + \beta\varepsilon_s} - 1}, \tag{6.0706}$$

and

$$\left\langle (n_s - \langle n_s \rangle)^2 \right\rangle = \frac{1}{\beta^2} \frac{\partial^2 q}{\partial \varepsilon_s^2} = \langle n_s \rangle + \langle n_s \rangle^2. \tag{6.0707}$$

We note here that the fluctuations in n_s may become quite large. Let us consider for a moment the case of a perfect boson gas below its condensation temperature T_0 (see §4.7). If we are well below T_0, practically all bosons will be in the ground state so that $\langle n_0 \rangle \equiv N$, and we get then from Eq.(6.0707)

$$\left\langle (n_0 - \langle n_0 \rangle)^2 \right\rangle = \langle n_0 \rangle + \langle n_0 \rangle^2 \approx N^2. \qquad (6.0708)$$

This means that at very low temperatures we can expect large fluctuations in density. The fact that they have never been observed in liquid helium is a strong argument against identifying liquid helium with a perfect boson gas.

Let us now consider the situation well above T_0. In that case, α will be large and negative so that $\langle n_s \rangle \ll 1$. There is thus little point in discussing the fluctuations in n_s. However, we can group together a number of levels in such a way that the quantity

$$N_i = \sum{}' n_s, \qquad (6.0709)$$

with the summation over the i-th group of levels, is large compared to unity. We can now consider the fluctuations in N_i. We have, first of all,

$$N_i = \sum{}' \frac{\partial \varepsilon}{\partial \varepsilon_s}, \qquad (6.0710)$$

and thus

$$\langle N_i \rangle = -\frac{1}{\beta} \sum{}' \frac{\partial q}{\partial \varepsilon_s} = \sum{}' \frac{1}{e^{-\alpha + \beta \varepsilon_s} - 1}. \qquad (6.0711)$$

As the fluctuations in the n_s are independent, we have for the fluctuations in the N_i

$$\left\langle (N_i - \langle N_i \rangle)^2 \right\rangle = \frac{1}{\beta^2} \sum{}' \frac{\partial^2 q}{\partial \varepsilon_s^2}, \qquad (6.0712)$$

or

$$\begin{aligned} \left\langle (N_i - \langle N_i \rangle)^2 \right\rangle &= \langle N_i \rangle + \sum{}' \left(e^{-\alpha + \beta \varepsilon_s} - 1 \right)^{-2} \\ &= \langle N_i \rangle + \sum{}' \langle n_s \rangle^2. \end{aligned} \qquad (6.0713)$$

As the $\langle n_s \rangle$ are small compared to unity, the second term on the right-hand side of Eq.(6.0713) will be small compared to the first one, and we see that, at sufficiently high temperatures, the relative fluctuations in the N_i are small.

We note that the first term on the right-hand sides of Eqs.(6.0707) and (6.0713) is the term that also occurs in the case of Boltzmann statistics. In his paper on the perfect boson gas Einstein pointed out that this term arises from the particle nature of the bosons. The second term, on the other hand, arises from the wave nature of the bosons — the diffraction aspect of wave mechanics. To illustrate this, let us consider the case of black-body radiation so that we can identify n_s and N_i, respectively, with the number of oscillators of frequency ω (compare § 3.3) and the number of oscillators with frequencies between ω and $\omega + d\omega$. Multiplying Eq.(6.0713) by $(\hbar\omega)^2$, we get from it the relation

$$\left\langle (\varrho(\omega)\, d\omega - \langle \varrho(\omega)\rangle\, d\omega^2 \right\rangle = \hbar\omega\, \langle \varrho(\omega)\rangle\, d\omega + \frac{V\langle\varrho(\omega)\rangle^2\, d\omega}{D(\omega)}$$

$$= \hbar\omega\, \langle \varrho(\omega)\rangle\, d\omega + \frac{\pi^2 c^3 \langle\varrho(\omega)\rangle^2\, d\omega}{\omega^2}, \quad (6.0714)$$

where $\varrho(\omega)$ is the radiation energy density while $D(\omega)$ is given by Eq. (3.307). The first term is again the "particle term". On the other hand, in the limit as $\hbar \to 0$, when we should regain the classical wave picture of electromagnetic radiation, only the second term remains.[22]

The above results are of importance for photon fluctuation experiments which can be performed much more easily as a result of the high-intensity laser beams that have now become available for such experiments. Purcell[23] has given the following argument showing why one may expect large fluctuations: Consider the interference of two coherent wave trains, with each train containing one photon. If the two trains overlap, a packet may emerge with anything between 0 and 4 photons, since the amplitudes can be added and the number of photons is proportional to the intensity.

Let us now restrict ourselves to the case of a perfect monatomic gas. We can then use the results of Chapter 4 for q, since the sum in Eq.(6.0703) can be replaced by an integral, and we have

$$q = \left(\frac{2\pi m}{\beta h^2}\right)^{3/2} e^\alpha\, V\, g(\alpha), \quad (6.0715)$$

where

$$g(\alpha) = \sum_{n=1}^{\infty} \frac{e^{n\alpha}}{n^{5/2}}. \quad (6.0716)$$

Using this formula for q we get easily from Eqs.(6.0419)

$$\frac{\left\langle (\nu - \langle\nu\rangle)^2 \right\rangle}{\langle\nu\rangle^2} = \frac{g''(\alpha)}{g'(\alpha)} \frac{1}{\langle\nu\rangle}, \quad \left(g'(\alpha = \frac{dg}{d\alpha}, \quad g''(\alpha) = \frac{d^2 g}{d\alpha^2} \right)$$

$$\quad (6.0717)$$

and

$$\left\langle (\varepsilon - \langle\varepsilon\rangle)^2 \right\rangle = \frac{5}{3\langle\nu\rangle} \frac{g'(\alpha)}{g(\alpha)}, \quad (6.0718)$$

---

[22] Heisenberg (*Physical Principles of Quantum Theory*, University of Chicago Press (1930), § 5.7) has given a quantum-mechanical derivation of Eq.(6.0714). This equation can also be derived directly from Eq.(5.1013).

[23] E.M.Purcell, *Nature* **178**, 1449 (1956).

where we used the fact that

$$\langle \nu \rangle = \frac{\partial q}{\partial \alpha} = q \frac{g'(\alpha)}{g(\alpha)}. \tag{6.0719}$$

Since $g''(\alpha)$ diverges as $\alpha \to 0$, the fluctuations in ν can become arbitrarily large as soon as $\alpha \to 0$.

Of course, these large fluctuations occur in a grand ensemble, that is, they can be expected in open systems. Because the grand ensemble formalism is so convenient it is widely used and, since in most cases there are no differences between the results obtained from different kinds of ensemble, this does usually not lead to any difficulties. However, strictly speaking, experimentally one usually deals with systems with a macroscopically conserved amount of matter.

6.8.   The Perfect Fermi-Dirac Gas

The calculations are similar to those of the preceding section. We now have for γ

$$\left.\begin{array}{ll} \gamma = 1, & \sum n_s^2 = \nu; \\[2mm] \gamma = 0, & \sum n_s^2 > \nu, \end{array}\right\} \tag{6.0801}$$

and we thus have

$$e^q = \sum_{\nu=0}^{\infty} e^{\alpha\nu} \sum_{\substack{\Sigma n_s = \nu \\ \Sigma n_s^2 = \nu}} e^{-\beta\Sigma n_s \epsilon_s}, \tag{6.0802}$$

or

$$e^q = \sum_{n_s=0}^{1} \cdots \sum e^{\Sigma n_s(\alpha-\beta\epsilon_s)}$$

$$= \prod_s \left(1 + e^{\alpha-\beta\epsilon_s} \right),$$

or

$$q = \sum_s \ln\left[1 + e^{\alpha-\beta\epsilon_s} \right]. \tag{6.0803}$$

This equation corresponds to Eq.(4.802), and again Eq.(6.0803) is less restricted than its counterpart of Chapter 4.

Since the calculations are completely analogous to those of § 6.7, we shall give only the more important results. We have for the fluctuations in ν

$$\left\langle (\nu - \langle \nu \rangle)^2 \right\rangle = \langle \nu \rangle - \sum_s \left(e^{-\alpha+\beta\epsilon_s} + 1 \right)^{-2} < \langle \nu \rangle, \tag{6.0804}$$

which shows that the fluctuations are now smaller than in the Boltzmann case.

We have for $\langle n_s \rangle$ and the fluctuations in the n_s

$$\langle n_s \rangle = \frac{1}{e^{-\alpha+\beta\varepsilon_s}+1}, \tag{6.0805}$$

$$\left\langle (n_s - \langle n_s \rangle)^2 \right\rangle = \langle n_s \rangle - \langle n_s \rangle^2. \tag{6.0806}$$

Since the n_s can only be zero or one, n_s is certainly not a large number, and although the fluctuations in the n_s are smaller than in the Boltzmann case, the relative fluctuations $\langle (n_s - \langle n_s \rangle)^2 \rangle / \langle n_s \rangle^2$ can never be very small — unless $\langle n_s \rangle$ is practically equal to one, that is, unless we are dealing with the lowest energy levels which are practically completely occupied as long as the temperature is not too high. However, it is possible to consider again a group of levels such that the average number of particles in this group of levels is large as compared to unity. For the number of fermions in the i-th group of levels we again have Eqs.(6.0709) and (6.0710). Instead of Eqs.(6.0711) and (6.0713) we have now

$$\langle N_i \rangle = -\frac{1}{\beta} {\sum}' \frac{\partial q}{\partial \varepsilon_s} = {\sum}' \frac{1}{e^{-\alpha+\beta\varepsilon_s}+1}. \tag{6.0807}$$

$$\begin{aligned}
\left\langle (N_i - \langle N_i \rangle)^2 \right\rangle &= \frac{1}{\beta^2} {\sum}' \frac{\partial^2 q}{\partial \varepsilon_s^2} \\
&= \langle N_i \rangle - {\sum}' \left(e^{-\alpha+\beta\varepsilon_s} + 1 \right)^{-2} \\
&= \langle N_i \rangle - {\sum}' \langle n_s \rangle^2, \tag{6.0808}
\end{aligned}$$

and we see that the relative fluctuations in the N_i are, indeed, small as compared to unity, as soon as $\langle N_i \rangle$ is large.

Considering once more the case of a perfect monatomic gas, we can replace the sum in Eq.(6.0803) by an integral and we get for q the equation

$$q = \left(\frac{2\pi m}{\beta h^2} \right)^{3/2} e^\alpha V h(\alpha), \tag{6.0809}$$

where

$$h(\alpha) = -\sum_{n=1}^{\infty} \frac{(-e^\alpha)^n}{n^{5/2}}. \tag{6.0810}$$

After an easy calculation we get for the fluctuations in the energy

$$\left\langle (\varepsilon - \langle \varepsilon \rangle)^2 \right\rangle = \frac{5}{3\langle \nu \rangle} \frac{h'(\alpha)}{h(\alpha)}, \tag{6.0811}$$

where we have used for $\langle \nu \rangle$ the equation

$$\langle \nu \rangle = q \frac{h'(\alpha)}{h(\alpha)}. \tag{6.0812}$$

6.9.   The Saha Equilibrium

In the preceding three sections we considered a system consisting of one kind of particle only. In the present section, we shall consider systems consisting of three kinds of particles, A, B, and C, between which the following reactions can take place:

$$A + B \leftrightarrows C. \qquad (6.0901)$$

In reaction (6.0901), A and B may be atoms and C the diatomic molecule AB, or C may be an atom, A an electron, and B the C^+ ion, and so on.

We shall use Eq.(6.0403) for the grand potential; this equation can now be written in the form

$$e^q = \sum_{\nu_A=0}^{\infty} \sum_{\nu_B=0}^{\infty} \sum_{\nu_C=0}^{\infty} e^{\alpha_A \nu_A + \alpha_B \nu_B + \alpha_C \nu_C} \sum_k \gamma_k \, e^{-\beta E_k(\nu_A, \nu_B, \nu_A)}, \qquad (6.0902)$$

where the last summation extends over all energy levels $E_k(\nu_A, \nu_B, \nu_A)$ of a system consisting of ν_A particles A, ν_B particles B, and ν_C particles C, and where γ_k is the multiplicity of the level $E_k(\nu_A, \nu_B, \nu_A)$. It must be borne in mind that α_A, α_B, and α_C are not independent, but satisfy the equilibrium condition (compare Eq.(5.1117))

$$\alpha_C = \alpha_A + \alpha_B. \qquad (6.0903)$$

We now make the first of a nmber of assumptions. This first assumption is that we may treat all particles in the system to be independent. In that case we can write

$$E_k(\nu_A, \nu_B, \nu_A) = E_{kA}(\nu_A) + E_{kB}(\nu_B) + E_{kC}(\nu_C), \qquad (6.0904)$$

and

$$\gamma_k = \gamma_{kA} \cdot \gamma_{kB} \cdot \gamma_{kC}, \qquad (6.0905)$$

so that we can write Eq.(6.0902) in the form

$$e^q = \{A\} \cdot \{B\} \cdot \{C\}, \qquad (6.0906)$$

with

$$\{A\} = \sum_{\nu_A=0}^{\infty} e^{\alpha_A \nu_A} \sum_{kA} \gamma_{kA} \, e^{-\beta E_{kA}(\nu_A)}, \qquad (6.0907)$$

and similar expressions for $\{B\}$ and $\{C\}$.

Since all particles are assumed to be non-interacting, we have further (compare Eq.(6.0601))

$$E_{kA}(\nu_A) = \sum_{sA} n_{sA} \varepsilon_{sA}, \qquad \sum_{sA} n_{sA} = \nu_A. \qquad (6.0908)$$

We now make our second assumption: that we may neglect all symmetry effects. In that case we get for γ_{kA} the expression (compare Eq.(6.0604))

$$\gamma_{kA} = \prod_{sA} \frac{g_{sA}^{n_{sA}}}{n_{sA}!}.$$ (6.0909)

Substituting Eqs.(6.0908) and (6.0909) into Eq.(6.0906), we find

$$q = e^{\alpha_A} \cdot Z_A + e^{\alpha_B} \cdot Z_B + e^{\alpha_C} \cdot Z_C,$$ (6.0910)

where $Z_A, \cdots$ are the partition functions

$$Z_A = \sum_{sA} g_{sA} e^{-\beta \varepsilon_{sA}}, \quad \cdots .$$ (6.0911)

We note that it follows from Eqs.(6.0910) and (6.0414) that the total pressure is the sum of the partial pressures due to A, B, and C, as we should have expected for non-interacting particles.

From Eqs.(6.0910), (6.0415), and (6.0903) we get

$$\frac{\langle \nu_A \rangle \langle \nu_B \rangle}{\langle \nu_C \rangle} = \frac{Z_A Z_A}{Z_C}.$$ (6.0912)

Let us now remind ourselves that the energies ε_{sA}, ε_{sB}, and ε_{sC} can be split into at least two independent parts: the kinetic energy of the centre of mass and the internal energy. As far as the kinetic energy is concerned, we can perform the summation involved, and the result is (see Eq.(4.609)) a factor

$$\left(\frac{2\pi m}{\beta h^2} \right)^{3/2} V,$$ (6.0913)

where V is the volume occupied by the system and m the mass of the particle concerned. If we now reckon the internal energies of the particles from their ground states as zero, we have (ε_{kin} denotes the kinetic energies)

$$\varepsilon'_{sA} = \varepsilon_{sA} - \varepsilon_{kinA}, \qquad \varepsilon'_{sB} = \varepsilon_{sB} - \varepsilon_{kinB}, \qquad \varepsilon'_{sC} = \varepsilon_{sC} + \chi - \varepsilon_{kinC},$$ (6.0914)

where we have primed the internal energies reckoned from their ground state as zero and where χ is the dissociation energy of the reaction (6.0901). If $c_A = \langle \nu_A \rangle / V, \cdots$ denote the concentrations, we find (compare Problem 3 at the end of Chapter 4)

$$\frac{c_A c_B}{c_C} = \left(\frac{2\pi \mu}{\beta h^2} \right)^{3/2} e^{-\beta \chi} \frac{Z'_A Z'_B}{Z'_C},$$ (6.0915)

where $Z'_A, \cdots$ are the internal partition functions,

$$Z'_A = \sum \gamma'_{sA} e^{-\beta \varepsilon'_{sA}}, \quad \cdots$$ (6.0916)

with $\gamma'_{sA}, \cdots$ referring to the internal energies only, while μ is the reduced mass,

$$\mu = \frac{m_A\, m_B}{m_C}. \tag{6.0917}$$

Equation (6.0915) is essentially Saha's formula. Special cases of this formula are considered in Problems 31 and 32 at the end of this chapter.

6.10.   The Relativistic Electron Gas

Although so far we have used only non-relativistic formulæ, it can easily be shown that most formulæ can immediately be adapted to the case of relativistic statistics. In the present section we shall discuss how this adaptation can be made and we shall apply it to the case of a gas of free electrons, since this is the case that has been of most interest in the application of relativistic statistics. These applications have been especially in the field of astrophysics, first and foremost to the theory of white dwarfs by such authors as Fowler and Chandrasekhar.[24] This means that we are confining ourselves here to the case of independent Fermi-Dirac particles. We can use most of the results of the earlier sections of the present chapter; for instance, since in the derivation of Eq.(6.0803) for the grand potential q no use was made of the fact that so far we were interested only in the non-relativistic case, but only of the fact that we were dealing with a system of independent particles, we can apply it to the present discussion. We have thus for q the equation

$$q = \sum_s \ln\left(1 + e^{\alpha - \beta\varepsilon_s}\right), \tag{6.1001}$$

where the summation is over all single-particle energy levels.

If the energy levels are lying sufficiently densely, the sum in Eq.(6.1001) may be replaced by an integral and we have

$$q = \int \ln\left(1 + e^{\alpha - \beta\varepsilon}\right) dZ, \tag{6.1002}$$

where dZ is once again the number of energy levels between ε and $\varepsilon + d\varepsilon$.

It is at this juncture that the relativistic case differs from the non-relativistic one. We can no longer use Eq.(4.604) for dZ but we must derive its relativistic counterpart. We shall again assume the gas to be enclosed in a cube of edge length L and volume $V = L^3$. The wavefunctions

---

[24]We refer to an account of the application of relativistic statistics to the theory of white dwarfs to S.Chandrasekhar, *An Introduction to the Study of Stellar Structure*, University of Chicago Press (1939), Chaps. X and XI, where also an extensive bibliography can be found. See also S.Chandrasekhar, *Selected Papers*, University of Chicago Press (1989), Vol.1.

of a free particle which satisfy the Dirac equations with periodic boundary conditions can be written as[25]

$$\psi_\lambda = a_\lambda \, e^{i[(p \cdot r) - \varepsilon t]},$$

(6.1003)

where λ runs from 1 to 4, where the ψ_λ are the four components of the Dirac wavefunction, and where the momentum p and the energy ε are related to each other as follows:

$$\frac{\varepsilon^2}{c^2} = p^2 + m^2 c^2,$$

(6.1004)

where m is the electron (rest) mass. Moreover, for a given p there are two linearly independent solutions of the Dirac equation, which means that each energy level is twofold degenerate. This degeneracy is the usual spin degeneracy $(g_s = 2s + 1, s = \frac{1}{2})$.

The fact that the particle is enclosed in a cube of edge length L entails that not all values of p_x, p_y, and p_z are permissible, but only those which satisfy the equations

$$p_x = \frac{k_1 h}{L}, \qquad p_y = \frac{k_2 h}{L}, \qquad p_z = \frac{k_3 h}{L},$$

(6.1005)

where the k_i can take on the values $\pm 1, \pm 2, \cdots$.[26]

Substituting expressions (6.1005) into Eq.(6.1004), we have

$$\frac{\varepsilon^2}{c^2} = \frac{h^2 k^2}{L^2} + m^2 c^2,$$

(6.1006)

where

$$k^2 = k_1^2 + k_2^2 + k_3^2.$$

(6.1007)

In terms of k we have for dZ

$$dZ = 8\pi k^2 \, dk;$$

(6.1008)

the difference of a factor 16 with Eq.(4.603) stems (a) from the spin degeneracy and (b) from the fact that now the k_i can take on both positive and negative integral values. From Eqs.(6.1006) and (6.1008) we now get

$$dZ = \frac{8\pi}{h^3} V \left(\frac{\varepsilon^2}{c^2} - m^2 c^2 \right)^{1/2} \frac{\varepsilon \, d\varepsilon}{c^2}.$$

(6.1009)

---

[25] See, for instance, A.S.Davydov, *Quantum Mechanics*, Pergamon Press, Oxford (1991), §60.

[26] The situation is slightly different here from the case discussed in §4.6. In the non-relativistic case the wavefunctions with negative values of the k_i are not linearly independent of those with only positive values of the k_i, but in the relativistic case they are.

If the energy is only just larger than the rest energy mc^2, we can write $\varepsilon = mc^2 + \varepsilon'$, where ε' is now the usual non-relativistic energy. Substituting this expression into Eq.(6.1009) and neglecting all terms but the leading one, we get

$$dZ = 4\pi \left(\frac{2m}{h^2}\right)^{3/2} V \sqrt{\varepsilon'}\, d\varepsilon', \qquad (6.1010)$$

which is Eq.(4.604), apart from a factor 2 due to the spin.

If we introduce the absolute magnitude of the momentum, p, instead of ε and use Eq.(6.1004), we can write Eq.(6.1009) in the form[27]

$$dZ = 2 \cdot \frac{V \cdot 4\pi p^2\, dp}{h^3}. \qquad (6.1011)$$

Since $4\pi p^2\, dp \cdot V$ is the volume in phase space corresponding to momenta of absolute magnitude between p and $p + dp$, Eq.(6.1011) expresses the fact that to each volume h^3 in μ-space there correspond two stationary states, the factor 2 arising from the spin degeneracy.

From Eqs.(6.1001) and (6.1009) we now have for the grand potential

$$q = \frac{8\pi}{h^3 c^3} V \int \varepsilon \left(\varepsilon^2 - m^2 c^4\right)^{1/2} \ln\left(1 + e^{\alpha - \beta\varepsilon}\right) d\varepsilon, \qquad (6.1012)$$

where the integration extends from mc^2 to ∞.

We note that the energy ε contains the rest energy. As a consequence, the α in the equations in the present section differs from the α encountered in the preceding sections by a term βmc^2. If we denote the earlier α by α', we have $\alpha - \beta\varepsilon = \alpha' - \beta\varepsilon'$, where ε' is again the energy reckoned from the rest energy as zero, or

$$\alpha = \alpha' + \beta mc^2. \qquad (6.1013)$$

Our equations will simplify if we use p, instead of ε, as integration variable. Equation (6.1012) then becomes

$$q = \frac{8\pi}{h^3} V \int_0^\infty \ln\left(1 + e^{\alpha - \beta\varepsilon}\right) p^2\, dp, \qquad (6.1014)$$

where ε as function of p is given by Eq.(6.1004).

Using Eqs.(6.0415), (6.0416), and (6.0414) we now have for the total number of particles, N, the total energy (including the rest energy Nmc^2),

---

[27]It can easily be verified by using the non-relativistic relation $\varepsilon' = p^2/2m$ that Eq.(6.1011) is valid both in the non-relativistic and in the relativistic case. That such should be the case follows also from the dicussion of Eq.(6.1011).

E, and the pressure, P, the following equations

$$N = \frac{8\pi}{h^3} V \int_0^\infty \frac{p^2\, dp}{e^{-\alpha+\beta\varepsilon} + 1}, \tag{6.1015}$$

$$E = \frac{8\pi}{h^3} V \int_0^\infty \frac{\varepsilon\, p^2\, dp}{e^{-\alpha+\beta\varepsilon} + 1}, \tag{6.1016}$$

$$P = \frac{kT}{V} q = \frac{8\pi}{h^3} kT \int_0^\infty \ln\left(1 + e^{\alpha-\beta\varepsilon}\right) p^2\, dp. \tag{6.1017}$$

It is more convenient for the further discussions to follow Jüttner[28] and introduce a variable θ defined by the equation

$$\theta = \text{arsinh} \frac{p}{mc}. \tag{6.1018}$$

From Eqs.(6.1004) and (6.1018) we get

$$\varepsilon = mc^2 \cosh\theta, \tag{6.1019}$$

and Eqs.(6.1015) and (6.1016) can be written in the form

$$N = 8\pi \frac{m^3 c^3}{h^3} V \int_0^\infty \frac{\sinh^2\theta \cosh\theta\, d\theta}{e^{-\alpha+\beta mc^2 \cosh\theta} + 1}, \tag{6.1020}$$

$$E = 8\pi \frac{m^4 c^5}{h^3} V \int_0^\infty \frac{\sinh^2\theta \cosh^2\theta\, d\theta}{e^{-\alpha+\beta mc^2 \cosh\theta} + 1}. \tag{6.1021}$$

After integrating by parts, we can write Eq.(6.1017) in the form

$$P = \frac{8\pi}{3} \frac{m^4 c^5}{h^3} \int_0^\infty \frac{\sinh^4\theta\, d\theta}{e^{-\alpha+\beta mc^2 \cosh\theta} + 1}. \tag{6.1022}$$

We shall now discuss a few limiting cases. First of all, we shall distinguish the degenerate and the non-degenerate cases for which, respectively, we have $e^\alpha \gg 1$ and $e^\alpha \ll 1$.

Let us consider the case of a completely degenerate relativistic fermion gas, that is, the limit as $e^\alpha \to \infty$. In that case we can split the integrals in Eqs.(6.1020) to (6.1022) in two parts. We have in that way, for instance, for N:

$$N = 8\pi \frac{m^3 c^3}{h^3} V \left\{ \int_0^{\theta_0} \frac{\sinh^2\theta \cosh\theta\, d\theta}{e^{-\alpha[1-\kappa mc^2 \cosh\theta]} + 1} \right.$$
$$\left. + \int_{\theta_0}^\infty \frac{\sinh^2\theta \cosh\theta\, d\theta}{e^{\alpha[\kappa mc^2 \cosh\theta-1]} + 1} \right\}, \tag{6.1023}$$

---

[28]F.Jüttner, Zs. Physik **47**, 542 (1928).

where $\kappa = \beta/\alpha$ and where θ_0 is given by the equation

$$\alpha = \beta mc^2 \cosh \theta_0. \qquad (6.1024)$$

In the limit as $\alpha \to \infty$ the second integral will be equal to zero, and in the first integral the denominator reduces to 1 (compare the discussion of Eq.(4.820)). To a first approximation we get thus for large α the following equations for N, E, and P:

$$N = 8\pi \frac{m^3 c^3}{h^3} V \int_0^{\theta_0} \sinh^2 \theta \cosh \theta \, d\theta, \qquad (6.1025)$$

$$E = 8\pi \frac{m^4 c^5}{h^3} V \int_0^{\theta_0} \sinh^2 \theta \cosh^2 \theta \, d\theta, \qquad (6.1026)$$

$$P = \frac{8\pi}{3} \frac{m^4 c^5}{h^3} \int_0^{\theta_0} \sinh^4 \theta \, d\theta. \qquad (6.1027)$$

These equations are exact for a completely degenerate relativistic Fermi-Dirac gas. The integrals can be expressed in terms of a quantity x given the equation

$$x = \sinh \theta_0, \qquad (6.1028)$$

and we have

$$N = \frac{8\pi}{3} \frac{m^3 c^3}{h^3} V x^3, \qquad (6.1029)$$

$$E = \frac{\pi}{3} \frac{m^4 c^5}{h^3} V g(x), \qquad (6.1030)$$

$$P = \frac{\pi}{3} \frac{m^4 c^5}{h^3} f(x), \qquad (6.1031)$$

where $f(x)$ and $g(x)$ are given by the equations[29]

$$f(x) = (2x^3 - 3x)(x^2 + 1)^{1/2} + 3 \operatorname{arsinh} x, \qquad (6.1032)$$

$$g(x) = 8x^3 (x^2 + 1)^{1/2} - f(x). \qquad (6.1033)$$

The physical meaning of the variable x can be seen as follows. From Eqs.(6.1029), (6.1028), (6.1018), and (6.1011), it follows that

$$x = \frac{p_0}{mc}, \qquad (6.1034)$$

and

$$N = \frac{8\pi}{3} \frac{V}{h^3} p_0^3 = \int_0^{p_0} dZ. \qquad (6.1035)$$

---

[29]It should be noted that our $g(x)$ differs from the function $g(x)$ introduced by Chandrasekhar: $g_{\text{here}} = g_{\text{Chand}} + 8x^3$. This difference is due to the fact that Chandrasekhar's U does not include the rest mass energy, while E does.

This equation, like Eqs.(6.1025) to (6.1027) and (6.1029) to (6.1031), is exact in the case of complete degeneracy. It expresses the fact that all the lowest enery levels up to a limiting energy are fully occupied, as should be expected at the absolute zero, which is the temperature at which absolute degeneracy exists (compare the considerations in §4.8 and also Eq.(6.1040)). The limiting energy corresponds to a threshold momentum of absolute magnitude p_0. From Eqs.(6.1029) and (6.1034) it follows that p_0 is determined by the mean density n $(= N/V)$ of the electron gas:

$$x = \frac{h}{mc} \left(\frac{3n}{8\pi} \right)^{1/3}, \tag{6.1036}$$

or

$$p_0 = h \left(\frac{3n}{8\pi} \right)^{1/3}. \tag{6.1037}$$

Equations (6.1029), (6.1031), and (6.1032) give us together the equation of state of a completely degenerate (relativistic) electron gas in parameter form. It can be shown (see Problem 40 at the end of this chapter) that Eqs.(6.1025) to (6.1027) are a fair approximation as long as

$$4\pi^2 \frac{x(x^2+1)^{1/2}}{f(x)} \ll (\beta mc^2)^2. \tag{6.1038}$$

From Eq.(6.1036) it follows that the left-hand side of this inequality depends only on the mean density of the electron gas. Since the Compton wavelength of the electron, h/mc, is 2×10^{-12} m, we see that, as long as n is less than of the order of 10^{36} m^{-3} — corresponding to mass densities of at least 10^9 kg/m^3 — x will be small compared to unity. For small x the function $f(x)$ behaves as $\frac{8}{5}x^5$ so that for not very large densities inequality (6.1038) can be written in the form

$$c \frac{h^2}{mk} \frac{n^{2/3}}{T} \gg 1, \tag{6.1039}$$

where c is a constant of the order of 10. Expressing T in °K and n in m^{-3}, we have from inequality (6.1039)

$$4 \times 10^{-13} \frac{n^{2/3}}{T} \gg 1 \tag{6.1040}$$

as the condition for the validity of Eqs.(6.1029) to (6.1031).[30]

---

[30]It can easily be verified that this condition is equivalent to the condition $e^\alpha \gg 1$ (compare Eq.(4.703)) or to the condition $\eta \ll 1$ which we met in §4.8 as the condition for the applicability of the formulæ for the case of complete degeneracy in the non-relativistic case.

As long as x is small as compared to unity we can simplify Eqs.(6.1030) and (6.1031) by using the asymptotic expressions for $f(x)$ and $g(x)$,

$$f(x) \approx \tfrac{8}{5} x^5, \qquad g(x) \approx 8x^3 + \tfrac{12}{5} x^5. \tag{6.1041}$$

In that way we get for E and P the expressions

$$E = Nmc^2 + \frac{4\pi}{5} \frac{m^4 c^5}{h^3} V x^5, \qquad x \ll 1, \tag{6.1042}$$

$$P = \frac{8\pi}{15} \frac{m^4 c^5}{h^3} x^5, \qquad x \ll 1. \tag{6.1043}$$

Denoting the kinetic energy $E - Nmc^2$ by $\mathcal{T}$ and using Eq.(6.1036) for x, we get the equations

$$\mathcal{T} = \frac{3}{40} \left(\frac{3}{\pi}\right)^{2/3} \frac{h^2}{m} V n^{5/3}, \qquad x \ll 1, \tag{6.1044}$$

$$PV = \tfrac{2}{3} \mathcal{T}, \qquad x \ll 1. \tag{6.1045}$$

Equation (6.1045) is the same as Eq.(4.715).

In some astrophysical applications we may encounter cases where $x \gg 1$, for instance in stellar interiors where n may exceed 10^{36} m^{-3}. In that case we have for $f(x)$ and $g(x)$ the asymptotic expressions

$$f(x) \approx 2x^4, \qquad g(x) \approx 6x^4. \tag{6.1046}$$

Inequality (6.1038) reduces also in this case to an inequality of the form (6.1039) or (6.1040), the only difference being a factor $\tfrac{1}{2}\sqrt{5}$ in the numerical constant.

We easily get for E, which is now to a first approximation equal to $\mathcal{T}$, and for P the expressions

$$\mathcal{T} \cong E = 2\pi \frac{m^4 c^5}{h^3} V x^4, \qquad x \gg 1, \tag{6.1047}$$

$$P = \frac{2\pi}{3} \frac{m^4 c^5}{h^3} x^4, \qquad x \gg 1, \tag{6.1048}$$

or

$$\mathcal{T} = \frac{3}{8} \left(\frac{3}{\pi}\right)^{1/3} hcV n^{4/3}, \qquad x \gg 1, \tag{6.1049}$$

$$PV = \tfrac{1}{3} \mathcal{T}, \qquad x \gg 1. \tag{6.1050}$$

and we see that now Eq.(4.715) is no longer valid. The reason is that while in the $x \ll 1$ case the kinetic energy is small as compared to the rest energy so that the non-relativistic case is a good approximation, in the

$x \gg 1$ case the rest energy is small as compared to the kinetic energy and the non-relativistic approximation breaks down.

We want to point that Eqs.(6.1045) and (6.1050) can be derived in an elementary way, completely similar to the derivation of Eq.(4.715). The energy levels ε_k now satisfy Eq.(6.1006). If the kinetic energy is small as compared to the rest mass energy, we have approximately $\varepsilon_k = mc^2 + \varepsilon'_k$ where ε'_k is proportioal to $V^{-2/3}$ and hence P will be equal to $2\mathcal{I}/3V$. If, on the other hand, the rest mass energy is small as compared to the kinetic energy, ε_k is proportional to $V^{-1/3}$ and P will be equal to $E/3V$, or $\mathcal{I}/3V$. The question whether or not the gas is degenerate does not enter into these considerations and we can thus expect to find a similar situation in the case of a non-degenerate electron gas (see Eqs.(6.1063) and (6.1067))

We shall now consider a non-degenerate electron gas, that is, the case where the temperature is so high or the density so low that $e^\alpha \ll 1$. In that case the 1 in the denominator of the integrands in Eqs.(6.1020) and (6.1021) can be neglected and the logarithm in the integrand in Eq.(6.1017) expanded, after the substitution (6.1018) has been made. The result is

$$N = 8\pi \frac{m^3 c^3}{h^3} V e^\alpha \int_0^\infty e^{-\beta mc^2 \cosh \theta} \sinh^2 \theta \cosh \theta \, d\theta, \quad (6.1051)$$

$$E = 8\pi \frac{m^4 c^5}{h^3} V e^\alpha \int_0^\infty e^{-\beta mc^2 \cosh \theta} \sinh^2 \theta \cosh^2 \theta \, d\theta, \quad (6.1052)$$

$$P = 8\pi \frac{m^3 c^3}{h^3} kT e^\alpha \int_0^\infty e^{-\beta mc^2 \cosh \theta} \sinh^2 \theta \cosh \theta \, d\theta. \quad (6.1053)$$

After integrating by parts, we can write Eq.(6.1017) in the form

$$P = \frac{8\pi}{3} \frac{m^4 c^5}{h^3} \int_0^\infty \frac{\sinh^4 \theta \, d\theta}{e^{-\alpha + \beta mc^2 \cosh \theta} + 1}. \quad (6.1022)$$

First of all, we note that the perfect gas law is identically true, since it follows from Eqs.(6.1051) and (6.1053) that

$$PV = NkT. \quad (6.1054)$$

We see that, irrespective of the inclusion or exclusion of relativistic effects, the Boyle-Gay-Lussac formula applies in the non-degenerate case.

We shall now discuss briefly the two limiting cases $\beta mc^2 \ll 1$ and $\beta mc^2 \gg 1$. Putting

$$\beta mc^2 = a \quad (6.1055)$$

and introducing as a new integration variable the quantity u defined by the equation

$$u = a \cosh \theta, \quad (6.1056)$$

we have from Eqs.(6.1051) to (6.1053):

$$N = \frac{PV}{kT} = A \int_0^\infty e^{-u} \sqrt{u^2 - a^2}\, u\, du, \qquad (6.1057)$$

$$E = AkT \int_0^\infty e^{-u} \sqrt{u^2 - a^2}\, u^2\, du, \qquad (6.1058)$$

where

$$A = 8\pi \frac{m^3 c^3}{h^3 a^3} V e^\alpha. \qquad (6.1059)$$

We now consider the two limiting cases $a \to 0$ and $a \to \infty$. In the limit as $a \to 0$ we have from Eqs.(6.1057) and (6.1058)

$$N = \frac{PV}{kT} \cong A \int_0^\infty e^{-u} u^2\, du = 2A, \qquad (6.1060)$$

$$E \cong AkT \int_0^\infty e^{-u} u^3\, du = 6AkT. \qquad (6.1061)$$

Since it follows from Eqs.(6.1054) and (6.1055) that

$$\frac{Nmc^2}{PV} = a, \qquad (6.1062)$$

we see that $Nmc^2 \ll PV = 2AkT = \frac{1}{3}E$ for $a \ll 1$. This means that the rest mass energy is small as compared to E. We get thus for the relation between the kinetic energy $\mathcal{J}$ and PV

$$PV = \tfrac{1}{3}E \cong \tfrac{1}{3}\mathcal{J}. \qquad (6.1063)$$

We could have expected a deviation from the non-relativistic relation, since we are now in the region of the very high temperatures and energies where relativistic effects make Eq.(6.1045) inapplicable.

In the $a \to \infty$ limit it is more advantageous to introduce instead of u a new variable v through the equation

$$v^2 = u - a, \qquad (6.1064)$$

and to expand in powers of v^2/a. Doing this, we get from Eqs.(6.1057) and (6.1058)

$$N = \frac{PV}{kT} = Ba\sqrt{2a} \int_0^\infty e^{-v^2} v^2\, dv = \tfrac{1}{2}Ba\sqrt{2\pi a}, \qquad (6.1065)$$

$$E = BkT \left[a^2\sqrt{2a} \int_0^\infty e^{-v^2} v^2\, dv + a\sqrt{2a} \int_0^\infty e^{-v^2} v^4\, dv \right]$$
$$= Nmc^2 + \tfrac{3}{4}BkTa\sqrt{2\pi a}, \qquad (6.1066)$$

and we get for the relation between $\mathcal{T}$ and PV

$$E - Nmc^2 \;=\; \mathcal{T} \;=\; \tfrac{3}{2}\,PV, \tag{6.1067}$$

as we should expect, since now $\mathcal{T} \ll Nmc^2$.

Problems

1. Prove that Eq.(6.0115) is invariant under a unitary transformation.

In the Schrödinger representation the wavefunction ψ_{S} satisfies the wave equation

$$i\hbar\dot\psi_{\mathrm{S}} \;=\; \widehat{\mathcal{H}}\psi_{\mathrm{S}},$$

and, if we exclude explicit time dependence, the operators $\widehat{G}_{\mathrm{S}}$ are time-independent operators.

In the Heisenberg representation the system is described by a wavefunction ψ_{H} which is related to ψ_{S} by the relation

$$\psi_{\mathrm{H}} \;=\; e^{i\widehat{\mathcal{H}}t/\hbar}\,\psi_{\mathrm{S}},$$

from which it follows that ψ_{H} is time-independent, while the operators $\widehat{G}_{\mathrm{H}}$ in the Heisenberg representation are related to those in the Schrödinger representation through the equation

$$\widehat{G}_{\mathrm{H}} \;=\; e^{i\widehat{\mathcal{H}}t/\hbar}\,\widehat{G}_{\mathrm{S}}\,e^{-i\widehat{\mathcal{H}}t/\hbar}.$$

In the so-called interaction representation one is dealing with a Hamiltonian $\widehat{\mathcal{H}}$ which can be split into two parts:

$$\widehat{\mathcal{H}} \;=\; \widehat{\mathcal{H}}_0 + \widehat{\mathcal{H}}'.$$

Often $\widehat{\mathcal{H}}_0$ corresponds to a system of non-interacting particles and $\widehat{\mathcal{H}}'$ to the interaction between the particles — hence the name "interaction representation" — but other situations may also occur (compare Problems 8 and 9). The wavefunction ψ_{int} is now related to ψ_{S} by the equation

$$\psi_{\mathrm{int}} \;=\; e^{i\widehat{\mathcal{H}}_0 t/\hbar}\,\psi_{\mathrm{S}},$$

so that ψ_{int} satisfies the equation of motion

$$i\dot\psi_{\mathrm{int}} \;=\; \widehat{\mathcal{H}}'_{\mathrm{int}}\,\psi_{\mathrm{int}},$$

where $\widehat{\mathcal{H}}'_{\mathrm{int}}$ is $\widehat{\mathcal{H}}'$ in the interaction representation in which the operators $\widehat{G}_{\mathrm{int}}$ are related to the $\widehat{G}_{\mathrm{S}}$ by the equation

$$\widehat{G}_{\mathrm{int}} \;=\; e^{i\widehat{\mathcal{H}}_0 t/\hbar}\,\widehat{G}_{\mathrm{S}}\,e^{-i\widehat{\mathcal{H}}_0 t/\hbar}.$$

2. Prove that Eq.(6.0134) holds in the Heisenberg representation.

3. Prove that Eq.(6.0134) holds in the interaction representation.

4. Consider a system of many degrees of freedom. Denote those degrees of freedom in which we are interested collectively by x and those that we wish to ignore collectively by q. The total normalised wavefunction of our system will then be a function of both q and x: $\psi(q, x)$. Let $\widehat{G}$ be an operator that is a function of x only. Introduce the density matrix $\widehat{\varrho}$ in the Dirac notation[31] by the equation

$$\langle x|\widehat{\varrho}|x'\rangle = \int \psi^*(q, x')\,\psi(q, x)\,dq;$$

then prove that $\widehat{\varrho}$ is normalised and find an expression for the average value of $\widehat{G}$. Find also the equation of motion for $\widehat{\varrho}$ assuming that the Hamiltonian can be split into two parts, one pertaining to the q and one pertaining to the x.

5. Show that, if for the system considered in the previous problem $\psi(q, x)$ can be written as the product $\varphi(q)\chi(x)$, we are dealing with a pure case.

6. Introduce the density matrix in the coordinate representation by using the eigenfunctions $\varphi_n(x)$ (which in the Dirac notation can be written $\langle n|x\rangle$) as transformation matrix.[32]

7. Write down the general expression in terms of the φ_n for the density matrix in the coordinate representation for the case where the ensemble is a macrocanonical ensemble.

8. Consider a Hamiltonian given by

$$\widehat{\mathcal{H}} = \widehat{\mathcal{H}}_0 + \widehat{\mathcal{H}}', \qquad \widehat{\mathcal{H}}' = \widehat{V}\,\mathrm{e}^{-i\omega t}.$$

Find and solve the equation of motion for $\Delta\widehat{\varrho}$ ($= \widehat{\varrho} - \widehat{\varrho}(-\infty)$) in the interaction representation for the case where at $t = -\infty$ the density matrix $\widehat{\varrho}(-\infty)$ is given by expression (6.0301) with $\widehat{\mathcal{H}} = \widehat{\mathcal{H}}_0$, neglecting the second-order term involving the product of $\Delta\widehat{\varrho}$ and $\widehat{V}$. Hence prove the following expression for the average value of the operator $\widehat{G}$ at time t (Kubo formula[33]):

$$\langle G(t)\rangle = \langle G\rangle_0 - \frac{i}{\hbar} \int_{-\infty}^{t} \langle [\widehat{G}_{\mathrm{int}}(t), \widehat{V}_{\mathrm{int}}(\tau)]_-\rangle_0\,\mathrm{e}^{-i\omega\tau}\,d\tau,$$

where $\langle\cdots\rangle_0$ indicates an average over $\widehat{\varrho}(-\infty)$.

[31] See, for instance, A.S.Davydov, *Quantum Mechanics*, Pergamon Press, Oxford (1991), § 27.

[32] Compare the discussion by Davydov (*Quantum Mechanics*, Pergamon Press, Oxford (1991), §§ 28 and 30) of changes in representation.

[33] R.Kubo, *J. Phys. Soc. Japan* **12**, 570 (1957).

9. Use the result of the preceding problem to find an expression for the electrical conductivity tensor, putting

$$\widehat{G} = \widehat{j}_k, \quad k = x, y, z, \quad \text{and} \quad \widehat{V} = -e \sum_l (\boldsymbol{E} \cdot \boldsymbol{r}_l),$$

where $\boldsymbol{E}$ is an external electric field, the $\widehat{j}_k$ are the components of the electric current operator, and the summation is over all charge carriers (with charge e) in the system.

10. Find the density matrix of the macrocanonical ensemble in the coordinate representation for the case of a single free particle in a volume V.

11. Find the density matrix of the macrocanonical ensemble in the coordinate representation for the case of N identical free particles, distinguishing carefully between bosons, fermions, and Boltzmann particles.[34]

12. If $\widehat{\sigma}_x$, $\widehat{\sigma}_y$, and $\widehat{\sigma}_z$ are the Pauli matrices,

$$\widehat{\sigma}_x = \begin{pmatrix} 0 & 1 \\ 1 & 0 \end{pmatrix}, \quad \widehat{\sigma}_y = \begin{pmatrix} 0 & -i \\ i & 0 \end{pmatrix}, \quad \widehat{\sigma}_z = \begin{pmatrix} 1 & 0 \\ -1 & 0 \end{pmatrix},$$

one can for the case of spin-$\frac{1}{2}$ particles define the polarisation vector $\boldsymbol{P}$ by the equation

$$\boldsymbol{P} = \langle \boldsymbol{\sigma} \rangle.$$

Use this definition of $\boldsymbol{P}$ to find an expression for the spin density matrix, which in this case is a 2×2 matrix, in terms of the three components of $\boldsymbol{\sigma}$ and the unit matrix, and the value of $\boldsymbol{P}$
Using the equation of motion (6.0134) for the average value of an operator and the expression

$$\widehat{\mathcal{H}}_\mathrm{s} = -\tfrac{1}{2}\gamma\hbar (\widehat{\boldsymbol{\sigma}} \cdot \boldsymbol{B})$$

for the spin Hamiltonian $\widehat{\mathcal{H}}_\mathrm{s}$, where γ is the magnetogyric ratio and $\boldsymbol{B}$ an external magnetic field, find the equation of motion of $\boldsymbol{P}$.

13. Evaluate the fluctuations in the energy for a system of independent localised harmonic oscillators. Consider the asymptotic expressions for $\beta\hbar\omega \ll 1$ and $\beta\hbar\omega \gg 1$.

14. Evaluate, up to terms of the order β, the high-temperature expression for the fluctuations in the energy of a system of independent localised three-dimensional rotators. (Use the series expansion given in problem 16 of Chapter 3.)

15. By comparing the average value of the number of particles in the s-th single-particle state in a canonical ensemble for a system of N independent particles with the same quantity in a canonical ensemble for a system

---

[34]See, for example, D.ter Haar, *Elements of Statistical Mechanics*, Rinehart, New York (1954), § VIII.4.

of $N+1$ independent particles, find an expression for that average number for the case of Boltzmann particles, bosons, or fermions. Use the fact that the two averages should be the same in the limit of large N.[35]

16. A microcanonical quantum-mechanical ensemble describes a system with a fixed number of particles, N, and with a fixed total energy, E. Let Ω be the total number of quantum states compatible with these conditions. According to the principle of equal a *priori* probabilities for all quantum states, the probability P_i for finding a system of the ensemble in the i-th state will be equal to $1/\Omega$. This corresponds to a density matrix $\hat{\varrho}$ of the following form:

$$\varrho_{kl} = \frac{\delta_{kl}}{\Omega}, \quad \text{if the } k\text{-th state is one of the } \Omega \text{ states compatible}$$

$$\text{with } N \text{ and } E;$$

$$= 0, \quad \text{otherwise,}$$

where we use the energy representation.
Prove that the entropy S is given by the equation

$$S = k \ln \Omega.$$

17. Consider a system of N distinguishable one-dimensional harmonic oscillators of frequency ω.[36] Neglect the zero-point energy and write $E = N_0 \hbar \omega$; then find an expression for Ω (see previous problem), and hence for the entropy S, using Stirling's formula for the factorial. Use the thermodynamic relation (compare Eq.(4.411)):

$$\beta = \left(\frac{\partial (S/k)}{\partial E} \right)_{V,N},$$

to find a relation between β and E, and thus derive the Planck expression for the energy of a system of N oscillators.

18. Discuss why α vanishes in the case of a gas of photons or phonons.

19. Use the method of § 5.5 to derive the following relation between the energy level density $\varrho(\langle \varepsilon \rangle, \langle \nu_i \rangle)$ and the entropy

$$\varrho(\langle \varepsilon \rangle, \langle \nu_i \rangle) \approx e^{S/k} \left[-\frac{\pi}{2} \frac{\partial \langle \varepsilon \rangle}{\partial \beta} \prod_i \left(\frac{\pi}{2} \frac{\partial \langle \nu_i \rangle}{\partial \alpha_i} \right) \right]^{-1/2},$$

where $\varrho(\varepsilon, \nu_i) d\varepsilon \, \Pi \, d\nu_i$ is the number of levels with energies between ε and $\varepsilon + d\varepsilon$ for a system with the number of particles of the i-th kind lying between ν_i and $\nu_i + d\nu_i$, and where we have used Eq.(6.0409).

---

[35] F.Ansbacher and W.Ehrenberg, *Phil. Mag.* **40**, 626 (1949); H.Schmidt, *Zs. Physik* **134**, 430 (1953).

[36] A.V.Tobolsky, *Am. J. Phys.* **32**, 799 (1964).

20. Consider a nucleus. which for the sake of simplicity is considered to be a system of Z $(= \frac{1}{2}A)$ protons and N $(= \frac{1}{2}A)$ neutrons which move without interactions in a volume V equal to Ar_0^3, where r_0 is a length of nuclear dimensions of the order of a fermi $(= 10^{-15}$ m$)$. First, verify that as long as the nuclear temperature which can be defined by the equation $(kT)^{-1} = \beta = \partial S/\partial E$, expressed in energy units, does not exceed 100 MeV, the proton and neutron gases are degenerate.

Secondly, bearing in mind the fact that microcanonical ensembles lead to the same results as canonical grand ensembles and using the results of §4.8, evaluate the energy level density from the result of the preceding problem.

21. Repeat the calculations of the preceding problem for the case where the nucleons, while being independent, have a single particle energy spectrum which is not that of free point particles in a volume V, but is such that the number of levels dZ with energies between ε and $\varepsilon + d\varepsilon$ is given by the equation

$$dZ = f(\varepsilon)\, d\varepsilon.$$

22. Prove that if all systems consisting of particles of one kind in an ensemble have the same number of particles, N, and an energy lying within a given interval $E, E + \delta E$, the quantity $\langle \eta \rangle$ is a minimum for the density matrix $\hat{\varrho}$ given by the equation (we use the Dirac notation):

$$\langle k, \nu | \hat{\varrho} | l, \nu' \rangle = c\, \delta_{kl}\, \delta_{\nu N}\, \delta_{\nu' N}, \qquad \text{if} \qquad E \leqslant E_k(\nu) \leqslant E + \delta E;$$
$$\langle k, \nu | \hat{\varrho} | l, \nu' \rangle = 0, \qquad\qquad \text{otherwise.}$$

23. Prove that if all systems consisting of one kind of particles in an ensemble have the same number of particles, N, while only the average energy is given, so that Eq.(6.0503) holds, $\langle \eta \rangle$ is a minimum for the density matrix $\hat{\varrho}$ given by the equation

$$\langle k, \nu | \hat{\varrho} | l, \nu' \rangle = \delta_{kl}\, \delta_{\nu N}\, \delta_{\nu' N}\, e^{\beta[\psi - E_k(\nu)]}.$$

24. Prove that $d\langle H \rangle / dt < 0$ for the cases considered in the preceding two problems.

25. Prove that the canonical ensembles are the proper representative ensembles, following the line of argument of §5.13.

26. Prove Eq.(6.0510) by considering the unitary transformation that brings $\hat{A}$ onto diagonal form.

27. Prove Klein's lemma by considering the unitary transformation that transforms the probability amplitudes at t' to those at t''.

28. Prove that, if w_{kl} is the probability for a transition per unit time from a state k to a state l with the same energy, and if $w_{kl} = w_{lk}$, the P_i given in problem 16 are attained exponentially (quantum-mechanical ergodic theorem in its simplest form).

29. Use the Darwin-Fowler method to prove Eqs.(6.0612), (6.0707), and (6.0806).

30. If we consider a system of N bosons, assumed to be independent point particles, below the transition temperature T_0, we can treat the part of the system consisting of particles in excited states by using a canonical grand ensemble, the particles in the ground state acting as a "particle bank". Prove that the fluctuations in the total number of particles in excited states, N_1, satisfy the equation

$$\left\langle \left(N_1 - \langle N_1 \rangle \right)^2 \right\rangle = C \langle N_1 \rangle^{4/3},$$

where C is a constant.

31. Apply formula (6.0912) to the case of an ionisation equilibrium, considering especially the case where the temperature is relatively low.

32. Apply Eq.(6.0912) to the case of the dissociation equilibrium of a diatomic molecule AB into its constituent atoms A and B. Assume that the electronic, vibrational, and rotational energies of the molecule are completely independent.

33. At a given temperature and volume a gas of AB molecules is 50 per cent dissociated into A and B atoms. If at constant temperature the volume is doubled, evaluate the relative change in pressure.

34. Discuss the changes to be made in Eq.(6.0912) if symmetry effects are taken into account.

35. Consider the reaction

$$A + 2B \; \leftrightarrows \; AB_2,$$

where A, B, and AB_2 are all bosons. We introduce N_A atoms A and N_B atoms B into a volume V at temperature T. Find the fraction of A atoms in molecular form in the high-temperature limit.

36. Discuss the changes to be made in the formalism of § 6.10, if the particles obeyed Boltzmann or Bose-Einstein rather than Fermi-Dirac statistics.

37. Prove Eqs.(6.1030) and (6.1031).

38. Prove Eq.(6.1046).

39. Use the method of § 4.8 to find the first correction to Eqs.(6.1029) to (6.1031).

40. Use the results of the preceding problem to prove inequality (6.1038).

41. Find expressions for the Fermi level, up to the first correction term, and for the specific heat of a nearly completely degenerate relativistic fermion gas.

42. Find an expression for the paramagnetic susceptibility of a nearly completely degenerate strongly relativistic electron gas.

43. Prove that the perfect gas law $PV = NkT$ holds for a perfect Boltzmann gas, independently of whether it is relativistic or not.

44. For a non-relativistic Boltzmann gas the average kinetic energy per particle equals $\frac{3}{2}kT$. Find the corresponding expression that equals $\frac{3}{2}kT$ for a relativistic Boltzmann gas.

45. Prove that for a Boltzmann gas of non-interacting particles the ultrarelativistic specific heat is twice that of the non-relativistic one.

46. Prove that a two-dimensional ultrarelativistic perfect boson gas shows the Bose-Einstein condensation phenomenon, while a one-dimensional ultrarelativistic perfect boson gas does not condense.[37]

47. Prove that the specific heats of two-dimensional ultrarelativistic perfect boson and fermion gases are identically the same.[37]

48. Show that if the temperature is sufficiently high so that pair production can take place, in the limit as $T \to \infty$ the particle density will be proportional to T^3, independent of statistics.

49. Consider a system with an electron number N_0 at $T = 0$. Find the ratio N^+/N^- of the number of positrons to the number of electrons, as a function of temperature.

Bibliographical Notes

The density matrix was introduced independently by Landau, who used the quantummechanical approach, and by von Neumann:

1. L.D.Landau, *Zs. Physik* 45, 430 (1927); *Collected Papers*, Pergamon Press, Oxford (1965), p.8.
2. J. von Neumann, *Göttinger Nachr.* 1927, 1, 24, 273.

Dirac discussed both the quantum-mechanical approach:

3. P.A.M.Dirac, *Proc. Camb. Phil. Soc.* **26**, 376 (1930); **27**, 240 (1931),

and the statistical approach:

4. P.A.M.Dirac, *Proc. Camb. Phil. Soc.* **25**, 62 (1929); **26**, 361 (1930),
without, however, mentioning the connection between the two kinds of density matrices, and normalising the density matrix in different ways.

For general surveys of the density matrix and its applications, see:

5. K.Husimi, *Proc. Phys.-Math. Soc. Japan* **22**, 264 (1940).
6. D.ter Haar, *Rept. Progr. Phys.* **24**, 304 (1961).

Section 6.2. See Reference 2 and also

6. J. von Neumann, *Mathematical Foundations of Quantum Mechanics*, Princeton University Press (1955).

Section 6.4. Pauli was the first to introduce canonical grand ensembles in quantum statistics:

7. W.Pauli, *Zs Physik* **41**, 81 (1927).

Our treatment is largely based upon Kramers's work:

---

[37]See R.M.May, *Phys. Rev.* A135, 1515 (1964).

8. H.A.Kramers, *Proc. Kon. Ned . Akad. Wet. (Amsterdam)* **41**, 10 (1938); *Collected Scientific Papers* North-Holland, Amsterdam (1956), p.738.

We refer to Balian for a discussion of the relation between statistical mechanics and information theory:

9. R.Balian, *From Microphysics to Macrophysics*, Springer, Heidelberg-Berlin (1991), Chapter 3.

Section 6.5. We refer to the bibliographical notes at the end of the previous chapter.

Section 6.7. For a discussion of fluctuations in the radiation field see:

10. A.Einstein, *Physik. Z.* **10**, 185 (1909).
11. L.Rosenfeld, *Osiris* **2**, 149 (1936).
12. M.J.Klein, *Proc. Kon. Ned. Akad. Wet.* **B62**, 41 (1959).
13. D.ter Haar, *The Old Quantum Theory*, Pergamon, Oxford (1966), Chap.2.

Early experiments showing large fluctuations in the radiation field were carried out by Hanbury Brown and Twiss:

14. R.Hanbury Brown and R.Q.Twiss, *Nature* **177**, 27 (1956).
15. R.Q.Twiss, A.D.Little, and R.Hanbury Brown, *Nature* **180**, 324 (1957).

For a discussion of fluctuations in a perfect boson gas when a closed system of N bosons is considered rather than an open system see:

16. I.Fujiwara, D. ter Haar, and H.Wergeland, *J. Stat. Phys.* **2**, 329 (1970).
17. D.ter Haar, *Lectures on Selected Topics in Statistical Mechanics*, Pergamon, Oxford (1977), § 4.1.

Section 6.9. Saha's formula was first published in 1920:

18. M.N.Saha, *Phil. Mag.* **40**, 472 (1920).

See also

19. M.N.Saha, *Zs. Physik* **6**, 40 (1921).

For a bibliography and a discussion of the various assumptions involved, see:

20. D.ter Haar, *Am. J. Phys.* **23**, 326 (1955).

Section 6.10. The treatment in this section is mainly based upon the following reference to which we also refer for a more extensive bibliography:

21. S.Chandrasekhar, *An Introduction to the Study of Stellar Structure*, University of Chicago Press (1939), Chap. X.

Planck and his pupil Jüttner developed relativistic statistics for the classical case:

22. M.Planck, *Ann. Physik* **26**, 1 (1908).
23. F.Jüttner, *Ann. Physik* **34**, 856 (1911); **35**, 145 (1911).

Reference to applications of relativistic statistics can be found in

24. R.C.Tolman, *Phil. Mag.* **28**, 583 (1914),

while there are some general remarks on relativistic statistics in

25. W.Pauli, *Enzyklopädie der mathematischen Wissenschaften*, Leipzig-Berlin (1921), Vol.V, Part 19, §§ 48 and 49.

Landsberg has studied the relativistic Einstein condensation:

26. P.T.Landsberg and J.Dunning-Davies, *Phys. Rev.* **138A**, 1049 (1965).

27. P.T.Landsberg, in *Statistical Mechanics of Quarks and Hadrons* (Ed. H.Satz), North-Holland, Amsterdam (1981), p.355.

Frankel, Hines, Kowalenko, and coworkers have made extensive studies of relativistic fermion and boson plasmas, both without and in a magnetic field:

28. A.E.Delsante and N.E.Frankel, *Phys. Rev.* **D8**, 1795 (1979).

29. V.Kowalenko, N.E.Frankel, and K.C.Hines, *Phys. Rept.* **126**, 109 (1985).

30. N.E.Frankel, K.C.Hines, and V.Kowalenko, *Laser Part. Beams* **3**, 251 (1985).

31. N.S.Witte, V.Kowalenko, and K.C.Hines, *Phys. Rev.* **D38**, 3667 (1988); **D40**, 1370 (1989).

32. J.Daicic and N.E.Frankel, *Prog. Theor. Phys.* **88**, 1 (1992).

33. J.Daicic and N.E.Frankel, *J. Phys.* **A26**, 1 (1993).

34. J.Daicic, N.E.Frankel, and V.Kowalenko, *Phys. Rev. Lett.* **71**, 1779 (1993).

35. J.Daicic, N.E.Frankel, R.M.Gailis, and V.Kowalenko, *Phys. Rept.* **237** (1994).

Further references can be found in Refs. 27, 31, and 35.

CHAPTER 7

THE EQUATION OF STATE OF AN IMPERFECT GAS

7.1. The Equation of State

We have seen in Chapters 5 and 6 that one can derive all thermodynamic properties of a system in equilibrium once one knows the grand potential q. For classical systems this quantity is given by Eq.(5.0907),

$$e^q = \sum_\nu \frac{e^{\alpha\nu}}{\nu!} \int_\Gamma e^{-\beta\varepsilon} \, d\Omega, \tag{7.101}$$

and for quantum-mechanical systems by Eq.(6.0402),

$$e^q = \mathrm{Tr}\, e^{\alpha\widehat{\nu} - \beta\widehat{\mathcal{H}}}. \tag{7.102}$$

We have assumed here that our system consists of only one kind of particles.

We shall restrict ourselves in this section to classical monatomic gases, and we shall assume that the forces between the atoms in the gas are binary, central forces which can be derived from a potential energy function $\phi(r)$ where r is the distance apart of the two interacting atoms. Under those assumptions the intermolecular forces are additive, and the Hamiltonian of the system is given by the equation

$$\mathcal{H} = \mathcal{H}_0 + \mathcal{H}_{\text{int}}, \tag{7.103}$$

with $\mathcal{H}$ the kinetic energy and $\mathcal{H}_{\text{int}}$ the potential energy which is given by

$$\mathcal{H}_{\text{int}} = \sum_{i=1}^\nu F(r_i) + \sum_{i<j} \phi(r_{ij}), \tag{7.104}$$

where r_{ij} is the distance betwen the i-th and the j-th atoms, where i and j run from 1 to ν, where the last summation extends over all pairs of atoms in the gas, and where we have introduced again a function $F(r)$ to take account of the fact that the gas is confined to a finite volume V:

$$\left.\begin{array}{ll} F(r) = 0, & \text{if } r \text{ lies inside } V, \\ F(r) = \infty, & \text{if } r \text{ lies outside } V. \end{array}\right\} \tag{7.105}$$

268

The energy of the system is thus given by the equation (compare Eq. (5.1201))

$$\varepsilon = \sum_{i=1}^{\nu} \left(\frac{p_i^2}{2m} + F(r_i) \right) + \sum_{i<j} \phi(r_{ij}). \tag{7.106}$$

In order that the right-hand side of Eq.(7.101) be dimensionless we shall use for the element of extension in phase, $d\Omega$, instead of Eq.(5.0106) the formula

$$d\Omega = \prod_{i=1}^{\nu} \frac{d^3p_i \, d^3r_i}{h^3}. \tag{7.107}$$

Substituting (7.107) into Eq.(7.101) and integrating over the momenta we find

$$e^q = \sum_{\nu=0}^{\infty} \frac{1}{\nu!} \frac{e^{\alpha\nu}}{v_0^{\nu}} \int \cdots \int \exp\left[-\beta \sum_{i<j} \phi(r_{ij}) \right] d^3r_1 \cdots d^3r_{\nu}, \tag{7.108}$$

where due to the occurrence of the $F(r_i)$ all r_i are confined to the volume V and where v_0 is given by Eq.(4.736),

$$v_0 = \left(\frac{\beta h^2}{2\pi m} \right)^{3/2}. \tag{7.109}$$

We saw in §4.7 that v_0 is essentially the cube of the thermal de Broglie length. We also saw there that as long as v_0 is small as compared to the average volume occupied by an atom in the gas we can use the classical theory.

For actual gases the potential energy $\phi(r)$ is practically equal to zero as soon as r is larger than a few ångstrom. For the sake of simplicity we shall assume that we have rigorously

$$\phi(r) = 0, \quad \text{if } r > D, \tag{7.110}$$

where D is of the order of 10 Å.

The integral over the position coordinates which occurs in Eq.(7.108) is called the *configurational integral* or *configurational partition function*. We shall denote it by $Z_{\nu}^{(\text{conf})}$; it is a function of the temperature, or β, the volume V, and the number of particles ν:[1]

$$Z_{\nu}^{(\text{conf})}(\beta, V, \nu) = \int \cdots \int e^{-\beta \mathcal{H}_{\text{int}}} d^3r_1 \cdots d^3r_{\nu}$$

$$= \int \cdots \int \prod_{i<j} e^{-\beta \phi(r_{ij})} d^3r_1 \cdots d^3r_{\nu}. \tag{7.111}$$

---

[1] One should note that different authors use slightly different definitions, the difference consisting in a factor V^{ν}.

As long as the gas is not too compressed, the average volume available per atom will be large as compared to D^3 and practically all configurations will be such that most atoms are lying at a distance apart from all other atoms which is larger than D, a few will be lying in pairs, even fewer will form groups (or clusters) of three, and so on. We define a *cluster* (a term first introduced by Mayer) as follows: we say that a cluster of n atoms is formed if from any of the n atoms one can reach all the other $n-1$ atoms by travelling over a chain of atoms such that two consecutive atoms in the chain are always lying at a distance apart less than D while all the other atoms in the gas are at distances greater than D from all the n atoms in the cluster.

We shall now use the fact that the atoms in the gas are grouped in clusters to expand the right-hand side of Eq.(7.108) in powers of $\langle \nu \rangle v_0 / V$. To do this we first of all introduce a set of functions $W_N(\mathbf{r}_1, \cdots, \mathbf{r}_N)$ by the equation

$$W_N(\mathbf{r}_1, \cdots, \mathbf{r}_N) = \frac{1}{v_0^N} \exp\left[-\beta \sum_{i<j} \phi(r_{ij})\right], \qquad (7.112)$$

where i and j run from 1 to N.

The important property of the $W_N(\mathbf{r}_1, \cdots, \mathbf{r}_N)$ is that if the N atoms fall into two clusters consisting, respectively, of N_1 and N_2 atoms, we have

$$W_N = W_{N_1} \cdot W_{N_2}, \qquad (7.113)$$

where W_{N_1} depends only on the coordinates of the atoms in the first cluster and W_{N_2} only on the coordinates of the atoms in the second cluster. This follows immediately since there will be no terms in the sum in the exponential in Eq.(7.112) for which i and j belong to different clusters.

We now introduce a new set of functions $U_N(\mathbf{r}_1, \cdots, \mathbf{r}_N)$ which have the property that they vanish, unless the N atoms to which they refer all belong to the same cluster. Their introduction is based upon the way one can split up the N atoms on which the W_N depend into clusters. The defining equations are:

$$\begin{aligned}
W_1(\mathbf{r}_1) &= U_1(\mathbf{r}_1), \\
W_2(\mathbf{r}_1, \mathbf{r}_2) &= U_1(\mathbf{r}_1) U_1(\mathbf{r}_2) + U_2(\mathbf{r}_1, \mathbf{r}_2), \\
W_3(\mathbf{r}_1, \mathbf{r}_2, \mathbf{r}_3) &= U_1(\mathbf{r}_1)U_1(\mathbf{r}_2)U_1(\mathbf{r}_3) + U_2(\mathbf{r}_1, \mathbf{r}_2)U_1(\mathbf{r}_3) \\
&\quad + U_2(\mathbf{r}_2, \mathbf{r}_3)U_1(\mathbf{r}_1) + U_2(\mathbf{r}_3, \mathbf{r}_1)U_1(\mathbf{r}_2) \\
&\quad + U_3(\mathbf{r}_1, \mathbf{r}_2, \mathbf{r}_3), \\
&\cdots\cdots\cdots\cdots\cdots\cdots\cdots\cdots\cdots\cdots\cdots\cdots \\
W_N(\mathbf{r}_1, \cdots, \mathbf{r}_N) &= \sum U_{n_1} U_{n_2} \cdots U_{n_r}, \qquad \left(\sum_k n_k = N\right)
\end{aligned}$$

$$(7.114)$$

where the summation on the right-hand side of the last equation extends, first, over all possible partitiones of N as a sum of positive integers and, second, for a given partitio over all possible ways in which N atoms can be divided into groups consisting, respectively, of $n_1, n_2, \ldots$ atoms.

The number of terms corresponding to a particular partitio which is such that there are λ_1 1's, λ_2 2's, $\cdots$, λ_k k's, $\cdots$, will be equal to $N!\,C(\lambda_1, \lambda_2, \cdots)$ with

$$C(\lambda_1, \lambda_2, \cdots) = \left[\prod_k \lambda_k!\,(k!)^{\lambda_k} \right]^{-1}. \tag{7.115}$$

The third equation (7.114) illustrates how the equations are constructed. One can split the atoms 1, 2, and 3 up into either three groups of 1 atom each, or two groups, one containing 1 atom and the other 2 atoms, where the single atom can either be atom 1, or atom 2, or atom 3, or, finally, into one group of three atoms.

If we solve Eqs.(7.114), we get the following equations, giving us the U's in terms of the W's:

$$\left.\begin{aligned}
U_1(\boldsymbol{r}_1) &= W_1(\boldsymbol{r}_1),\\
U_2(\boldsymbol{r}_1, \boldsymbol{r}_2) &= W_2(\boldsymbol{r}_1, \boldsymbol{r}_2) - W_1(\boldsymbol{r}_1)\,W_1(\boldsymbol{r}_2),\\
U_3(\boldsymbol{r}_1, \boldsymbol{r}_2, \boldsymbol{r}_3) &= W_3(\boldsymbol{r}_1, \boldsymbol{r}_2, \boldsymbol{r}_3) - W_2(\boldsymbol{r}_1, \boldsymbol{r}_2)W_1(\boldsymbol{r}_3)\\
&\quad - W_2(\boldsymbol{r}_2, \boldsymbol{r}_3)W_1(\boldsymbol{r}_1) - W_2(\boldsymbol{r}_3, \boldsymbol{r}_1)W_1(\boldsymbol{r}_2)\\
&\quad + 2W_1(\boldsymbol{r}_1)W_1(\boldsymbol{r}_2)W_1(\boldsymbol{r}_3),\\
\end{aligned}\right\} \tag{7.116}$$

$$\cdots\cdots\cdots\cdots\cdots\cdots\cdots\cdots\cdots\cdots\cdots\cdots\cdots\cdots\cdots\cdots\cdots$$

We now introduce a set of integrals I_n by the equation

$$I_n = \int U_n(\boldsymbol{r}_1, \cdots, \boldsymbol{r}_n)\,d^3\boldsymbol{r}_1 \cdots d^3\boldsymbol{r}_n. \tag{7.117}$$

Since U_n vanishes unless all the atoms, the coordinates of which are its arguments, belong to the same cluster[2] these I_n also vanish, unless all n atoms belong to a single cluster. Therefore, if we integrate over the $\boldsymbol{r}_i$ one by one, only the last integration will give us a factor V, since only one of the atoms will have the full volume at its disposal, the others being restricted by belonging to a single cluster. We can therefore introduce dimensionless quantities b_n, which will depend on the temperature, but not on the volume, through the relations

$$I_n = \frac{V}{v_0}\,n!\,b_n, \tag{7.118}$$

where v_0 is given by Eq.(7.109).

---

[2] We leave the proof of this property of the U_n to the reader; see Problem 1 at the end of the present chapter.

Combining Eqs.(7.108), (7.112), (7.114), (7.115), (7.117), and (7.118) we get after some calculations

$$
\begin{aligned}
e^q &= \sum_\nu \frac{e^{\nu\alpha}}{\nu!} \int W_\nu(\mathbf{r}_1, \cdots, \mathbf{r}_\nu)\, d^3\mathbf{r}_1 \cdots d^3\mathbf{r}_\nu \\
&= \sum_\nu \frac{e^{\nu\alpha}}{\nu!} \sum_{\lambda_i} \nu!\, C(\lambda_1, \lambda_2, \cdots)\, I_1^{\lambda_1} I_2^{\lambda_2} \cdots \\
&= \prod_k \sum_{\lambda_k=0}^\infty \frac{(b_k e^{k\alpha} V/v_0)^{\lambda_k}}{\lambda_k!} \\
&= \exp\left[\frac{V}{v_0} \sum_{n=1}^\infty b_n e^{n\alpha} \right],
\end{aligned}
\tag{7.119}
$$

or

$$
q = \frac{V}{v_0} \sum_{n=1}^\infty b_n e^{n\alpha}.
\tag{7.120}
$$

We note that Eqs.(6.0614), (6.0714), and (6.0809) are special cases of Eq.(7.120) with, respectively,[3]

$b_1 = 1, \quad b_n = 0 \ (n > 1)$ for a perfect Boltzmann gas;
$b_n = n^{-5/2}$ for a perfect Bose-Einstein gas;
$b_n = (-1)^{n+1} n^{-5/2}$ for a perfect Fermi-Dirac gas.

The pressure follows from Eq.(5.0926),

$$
P = \frac{kTq}{V},
\tag{7.121}
$$

or

$$
P = \frac{kT}{v_0} \sum_{n=1}^\infty b_n e^{n\alpha}.
\tag{7.122}
$$

This equation gives P as a power series in the activity (e^α), but it is more usual and more convenient for comparing the theoretical and experimental equations of state to express P as a power series in the density, or $1/V$ (compare Eq.(1.307)). If we denote the average number of atoms, $\langle \nu \rangle$ by N, we have from Eq.(5.0928)

$$
N = \frac{\partial q}{\partial \alpha} = \frac{V}{v_0} \sum_{n=1}^\infty n b_n e^{n\alpha},
\tag{7.123}
$$

or

$$
\frac{N v_0}{V} = \sum_{n=1}^\infty n b_n e^{n\alpha}.
\tag{7.124}
$$

---

[3] Of course, the perfect Bose-Einstein and Fermi-Dirac gases are not really covered by our present, classical discussion.

Combining Eqs.(7.122) and (7.124) we find

$$
\begin{aligned}
PV &= NkT \, \frac{\sum b_n e^{n\alpha}}{\sum n b_n e^{n\alpha}} \\
&= NkT \left(1 - b_2 e^\alpha - (2b_3 - 4b_2^2)e^{2\alpha} + \cdots \right),
\end{aligned}
\tag{7.125}
$$

where we have used the fact that

$$
b_1 = 1,
\tag{7.126}
$$

as follows from Eqs.(7.118), (7.117), (7.116), and (7.112).

Using Eq.(7.124) we can write the equation of state in the form

$$
PV = NkT \sum_{n=0}^{\infty} a_n \left(\frac{Nv_0}{V}\right)^n,
\tag{7.127}
$$

with

$$
\left.
\begin{aligned}
a_0 &= 1, \\
a_1 &= -b_2, \\
a_2 &= -2b_3 + 4b_2^2, \\
a_3 &= -3b_4 + 18b_2 b_3 - 20b_2^3, \\
&\quad \cdots\cdots\cdots\cdots\cdots\cdots\cdots
\end{aligned}
\right\}
\tag{7.128}
$$

One can, of course, derive Eqs.(7.128) straightforwardly, but there is an alternative, much more elegant method, due to Kramers.

If we introduce the notation

$$
e^\alpha = z, \qquad \frac{Nv_0}{V} = x, \qquad F(z) = \sum_{n=1}^{\infty} b_n z^n, \qquad F' = \frac{dF}{dz},
\tag{7.129}
$$

Eqs.(7.124) and (7.127) take the form

$$
x = zF'(z),
\tag{7.130}
$$

and

$$
\frac{F}{zF'(z)} = \frac{F}{x} = \sum a_n x^n.
\tag{7.131}
$$

Consider now the integral

$$
\frac{1}{2\pi i} \oint F(z) \, d\ln(zF'(z) - x),
\tag{7.132}
$$

where the integration contour is in the complex z-plane enclosing the origin and the point z_0 which is the value of z satisfying Eq.(7.130). The integral (7.132) is equal to $F(z_0)$ and, if we write

$$
zF'(z) - x = zF'(z) \left[1 - \frac{x}{zF'(z)}\right],
\tag{7.133}
$$

and use the power series expansion for $\ln(1-y)$ we have

$$
F(z_0) = \frac{1}{2\pi i} \oint F(z)\,d\left\{\ln zF'(z) - \sum_{n=0}^{\infty} \frac{1}{n+1}\left[\frac{x}{zF'(z)}\right]^{n+1}\right\}
$$

$$
= \frac{1}{2\pi i} \sum_{n=0}^{\infty} \frac{x^{n+1}}{n+1} \oint \frac{[F'(z)]^{-n}\,dz}{z^{n+1}}, \tag{7.134}
$$

where we have integrated by parts. We now see that a_n is the coefficient of z^n in $[F'(z)]^{-n}$ divided by $n+1$. Equations (7.128) then follow straightaway.

7.2. The van der Waals Equation of State

In § 1.3 we very briefly discussed the van der Waals equation of state. As an application of the theory of the earlier sections of the present chapter we shall show, following van Kampen, how one can derive Eq.(1.305) *together with the Maxwell rule*.

In the present section we shall use macrocanonical ensemble theory to consider a classical gas of hard sphere molecules which attract one another weakly. The Hamiltonian of such a gas is given by Eq.(7.103) with the potential energy satisfying the equations

$$
\left.\begin{array}{ll}
\phi(r) = \infty, & \text{if } r \leqslant a; \\
\phi(r) < 0, & \text{if } r > a.
\end{array}\right\} \tag{7.201}
$$

Before giving a more rigorous treatment, we shall first follow Ornstein's treatment in his Leiden thesis to show how one can use the Hamiltonian (7.103) to derive the van der Waals law. We separate the interatomic potential into two parts, a repulsive part, ϕ_r, amd an attractive part, ϕ_a:

$$
\phi(r) = \phi_r(r) + \phi_a(r). \tag{7.202}
$$

The configurational partition punction (7.111) can now be written as

$$
Z_N^{(\text{conf})} = \int \cdots \int \prod_i d^3r_i \, \exp\left[-\beta\sum_{j<k}\phi_r(r_{jk}) - \beta\sum_{j<k}\phi_a(r_{jk})\right], \tag{7.203}
$$

where the integration over the r_i extends over the volume V and where N is the number of particles in the system..

If $S(x)$ is the function defined by

$$
S(x) = \theta(x-a), \quad x \neq a; \qquad S(a) = 0, \tag{7.204}
$$

where $\theta(y)$ is the Heaviside function,

$$
\left.\begin{array}{ll}
\theta(y) = 0, & y < 0 \\
= 1, & y > 0,
\end{array}\right\} \tag{7.205}
$$

we can in the case of the potential (7.201) write

$$\exp\left[-\beta \sum_{j<k} \phi_r(r_{jk})\right] = \prod_{j<k} S(r_{jk}). \qquad (7.206)$$

To deal with the attractive part we write

$$\sum_{j<k} \phi_a(r_{jk}) \approx \frac{1}{2} \sum_j \left(\frac{N}{V} d^3r\right) \phi_a(|r - r_j|) = -\frac{N^2 C}{2V}, \qquad (7.207)$$

where we have used the fact that $\sum_{j<k} \equiv \frac{1}{2} \sum_{j\neq k}$ and where

$$C = -\int \phi_a(r) \, d^3r. \qquad (7.208)$$

Using Eqs.(7.207) and (7.206) we find for the configurational partition function

$$Z_N^{(\text{conf})} = D \, e^{\beta C N^2/2V}, \qquad (7.209)$$

with

$$D = \int \cdots \int \left(\prod_i d^3r_i\right) \prod_{j<k} S(r_{jk}). \qquad (7.210)$$

If $V \gg Nv_1$, where v_1 is the volume of one of the hard spheres,

$$v_1 = \frac{4\pi}{3} a^3, \qquad (7.211)$$

D is approximately equal to V^N, while if V approaches Nv_1, we would expect D to behave more or less like $V - Nv_1$. We interpolate this behaviour by writing

$$D = (V - b)^N, \qquad (7.212)$$

where b will be of the order of Nv_1. We thus get

$$Z_N^{(\text{conf})} = (V - b)^N \, e^{\beta C N^2/2V}. \qquad (7.213)$$

Since the configurational partition function is the only factor of Z_Γ which depends on the volume, we find by using Eq.(5.0317) for the pressure that

$$\beta P = \frac{\partial \ln Z_N^{(\text{conf})}}{\partial V} = -\frac{\beta C N^2}{2V^2} + \frac{N}{V - b}, \qquad (7.214)$$

or

$$\beta \left(P + \frac{a}{V^2}\right)(V - b) = N, \qquad (7.215)$$

where

$$a = \frac{1}{2} C N^2. \qquad (7.216)$$

Equation (7.215) is the van der Waals equation.

We know that if $\beta < \beta_{cr} = 27Nb/8a$ the isotherms given by Eq.(7.215) are well behaved and everywhere satisfy the thermodynamic stability condition $\partial P/\partial V \leqslant 0$. However, if $\beta > \beta_{cr}$ the isotherm has the shape shown in Fig.7.1, and one must supplement Eq.(7.215) with the Maxwell rule which leads to an isotherm with a horizontal part drawn under the equal area rule.

Let us briefly consider two aspects of Ornstein's derivation. Firstly, we would expect that a correct evaluation of the configurational partition function should always lead to a stable isotherm without having to invoke Maxwell's rule. However, we have (tacitly) assumed that we are dealing with a single-phase system, and this has led to an isotherm with an unstable part.

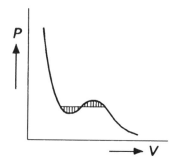

Fig.7.1. The isotherm of a van der Waals gas below the critical temperature. The horizontal part has been drawn using the equal area rule.

Secondly, we must draw attention to the fact that Eq.(7.212) is only a first approximation. Indeed, if we put $\phi_a \equiv 0$ so that we are dealing with a hard-sphere gas, we can calculate the various virial coefficients exactly. It turns out that the second virial coefficient agrees with the one following from Eq.(7.215), provided we put $b = 4Nv_1$, but higher virial coefficients do not agree with their van der Waals values.

We shall now consider a model due to van Kampen which produces the correct kind of isotherms, including the horizontal parts. It involves very-long-range attractive interactions between the particles and therefore still leaves open the question why actual gases, for which the attractive forces have a short range, show phase transitions. On the other hand, the essential feature of the condensation of a gas which is the competition between low-energy liquid configurations and gas configurations corresponding to large volumes in phase space is clearly shown in this model.

As the equation of state follows, once we know the configurational partition function $Z_N^{(\text{conf})}$, we shall concentrate on it. In order to do evaluate it we divide the volume V into cells of size Δ where Δ is so large that each cell contains a large number of particles, but at the same time so small that ϕ_a is practically constant within Δ. This means that, on the one hand, the range of ϕ_a must be large as compared to the hard-sphere radius a and, on the other hand, the density is sufficiently high so that many particles interact simultaneously.

A given configuration $\{N_i\}$ will now be characterised by a set of numbers N_i which give the number of particles in the i-th cell, which we take to be situated at the point r_i. The N_i must satisfy the condition

$$\sum_i N_i = N. \tag{7.217}$$

We can write for $Z_N^{(\text{conf})}$ the equation (compare Eq.(7.209))

$$Z_N^{(\text{conf})} = \frac{1}{N!} \sum_{\{N_i\}} A(N_i) \, B(N_i) \, C(N_i). \tag{7.218}$$

The factors $A(N_i)$, $B(N_i)$, and $C(N_i)$ have the following meaning.
The first factor is a combinatorial expression:

$$A(N_i) = \frac{N!}{\prod_i N_i!}. \tag{7.219}$$

The second factor takes into account the repulsive part of the potential so that, by analogy with Eqs.(7.210) and (7.212) we can to a first approximation write

$$B(N_i) = \int \cdots \int \left(\prod_i d^3 r_i \right) \prod_{j<k} S(r_{jk})$$
$$\cong \prod_i (\Delta - N_i \delta)^{N_i}, \tag{7.220}$$

where we shall assume δ to be a constant of the order of v_1.

Finally, using the fact that ϕ_a is assumed to be constant over the dimensions of a cell, we can write the third factor in the form

$$C(N_i) = \exp\left[-\tfrac{1}{2}\beta \sum_{i,j} N_i N_j \phi_{ij}^{(a)} \right], \tag{7.221}$$

where $\phi_{ij}^{(a)} \equiv \phi_a(r_{ij})$.

Combining these expressions we finally get for the configurational partition function the equation

$$Z_N^{(\text{conf})} = \sum_{\{N_i\}} e^{\Phi(N_i)}, \tag{7.222}$$

where

$$\Phi(N_i) = \sum_i \left[N_i \ln(\Delta - N_i \delta) - N_i \ln N_i + N_i \right] - \tfrac{1}{2}\beta \sum_{i,j} N_i N_j \phi_{ij}^{(a)};$$
(7.223)

when writing down this expression for $\Phi(N_i)$ we have used the fact that the N_i are large numbers so that we can use Stirling's formula (1.708) for the factorials.

We now proceed in much the same way as we did in § 4.3. To find the equilibrium configuration, which will make the free energy, and thus $\ln Z_N^{(\text{conf})}$, a maximum we look for the *absolute* maximum of $\Phi(N_i)$ satisfying the condition (7.217). This condition can, as ususal, be taken into account by means of a Lagrangian multiplier. The equilibrium configuration $\{N_i\}$ is thus the one for which the N_i satisfy the equation

$$\ln \frac{\Delta - N_i \delta}{N_i} - \frac{N_i \delta}{\Delta - N_i \delta} - \beta \sum_j \phi_{ij}^{(a)} N_j = \alpha,$$
(7.224)

where α is the Lagrangian multiplier.

Let us first of all look for a solution of Eq.(7.224) corresponding to a uniform density, that is, a solution of the form

$$N_i = \frac{N\Delta}{V} = n\Delta,$$
(7.225)

where we have introduced the number density n by the equation

$$n = \frac{N}{V}.$$
(7.226)

If we introduce a quantity ϕ_0 by the equation[4]

$$\phi_0 = \int \phi_a(r) \, d^3r,$$
(7.227)

which for the equilibrium solution (7.225) takes the form

$$\phi_0 = \sum_i \phi_{ij}^{(a)} \Delta,$$
(7.228)

the extremum value, Φ_{extr}, of Φ can be written as

$$\Phi_{\text{extr}} = V \left[n \ln \frac{1 - \delta n}{n} + n - \tfrac{1}{2}\beta n^2 \phi_0 \right],$$
(7.229)

where we have used Eqs.(7.223), (7.225), and (7.226).

---

[4] Note that since ϕ_a is everywhere negative, ϕ_0, like the quantity ϕ_2 introduced by Eq.(7.246), is a negative quantity.

The pressure is given by the equation

$$\beta P = -\frac{\partial \Phi_{\text{extr}}}{\partial V} = \frac{n}{1 - n\delta} + \tfrac{1}{2}\beta n^2 \phi_0, \tag{7.230}$$

or

$$\beta \left(P - \frac{\phi_0 N^2}{2V^2} \right)(V - N\delta) = N, \tag{7.231}$$

which is the same as Eq.(7.215). To some extent we can say that so far all we have done is somewhat better to derive the van der Waals equation.

To simplify matters we now introduce units in which δ is unity. In those units n must lie between 0 and 1 and from Eqs.(7.224) and (7.225) it follows that it must satisfy the equation

$$\ln \frac{1 - n}{n} - \frac{1}{1 - n} - \beta n \phi_0 = \alpha, \tag{7.232}$$

while we can write Φ_{extr} in the form

$$\Phi_{\text{extr}} = V f(n), \tag{7.233}$$

with

$$f(n) = n \ln \frac{1 - n}{n} + n - \tfrac{1}{2}\beta n^2 \phi_0. \tag{7.234}$$

In terms of $f(n)$ Eq.(7.232) reads

$$f'(n) = \alpha, \tag{7.235}$$

where the prime indicates differentiation with respect to n.

One can prove (see Problem 7 at the end of this chapter) that the solution (7.225) corresponds to an absolute maximum of Φ, provided the second derivative of f with respect to n is negative, $f''(n) < 0$, that is, provided $f'(n)$ is a monotonically decreasing function of n. This means that we must satisfy the condition

$$\frac{1}{n(1 - n)^2} + \beta \phi_0 > 0. \tag{7.236}$$

Since the maximum value of $1/n(1 - n)^2$ is $27/4$, it follows that provided the temperature, or β, satisfies the condition

$$-\beta \phi_0 < \tfrac{27}{4}, \tag{7.237}$$

Eq.(7.236) will be satisfied for all n. Moreover, one can prove (see again Problem 7 at the end of the chapter) that if Eq.(7.237) is satisfied Φ has only a single maximum. We have thus shown that the stable state of the system at temperatures above the critical temperature T_{cr}, given by the equation

$$kT_{\text{cr}} = \frac{-4\phi_0}{27}, \tag{7.238}$$

corresponds to the equation of state (7.231).

Consider now temperatures such that

$$-\beta\phi_0 > \tfrac{27}{4}, \tag{7.239}$$

so that $f'(n)$ is no longer a monotonically decreasing function of n (see Fig.7.2). There is now a range of α-values for which Eq.(7.232) has three solutions: n_I, n_{II}, and n_{III}. The first two correspond to values for which $f''(n) < 0$ so that the corresponding value of Φ is at least a relative maximum, but for n_{III} the value of $f''(n)$ is positive which means that — as one can readily prove by using arguments similar to the ones used in Problem 7 at the end of the chapter — the corresponding value of Φ is not a maximum and thus does not correspond to a physical state of the system.

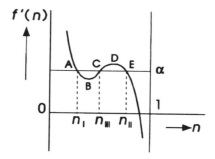

Fig.7.2. The function $f'(n)$ for temperatures below the critical temperature.

We now interpret the densities n_I and n_{II} as the densities of possibly coexisting phases. However, if there are two coexisting phases, we are no longer dealing with a uniform system so that Eq.(7.225) no longer applies, and we must go back to Eq.(7.224). Instead of Eq.(7.225) we now put

$$N_i = n(\mathbf{r}_i)\Delta, \tag{7.240}$$

and we get from Eq.(7.223)

$$\Phi = \int n(\mathbf{r}) \left[\ln \frac{1 - n(\mathbf{r})}{n(\mathbf{r}} + 1\right] d^3r$$
$$- \tfrac{1}{2}\beta \int\int \phi_a(|\mathbf{r} - \mathbf{r}'|)\, n(\mathbf{r})n(\mathbf{r}')\, d^3r\, d^3r', \tag{7.241}$$

or,

$$\Phi = \int f[(n(\mathbf{r})]\, d^3r + \tfrac{1}{4}\beta \int\int \phi_a(\mathbf{r} - \mathbf{r}')\, [n(\mathbf{r}) - n(\mathbf{r}')]^2\, d^3r\, d^3r', \tag{7.242}$$

where we have used Eq.(7.234) for $f(n)$. Instead of Eq.(7.235) we now get the equation

$$\frac{\delta \Phi}{\delta n(r)} \equiv f'[n(r)] + \beta \int \phi_a(|r - r'|)[n(r) - n(r')] d^3r' = \alpha. \quad (7.243)$$

If we assume that $n(r)$ changes so slowly that we can write

$$n(r') - n(r) \approx ((r - r') \cdot \nabla n(r)) + \tfrac{1}{6}|r - r'|^2 \nabla^2 n(r), \quad (7.244)$$

we find from Eq.(7.243)

$$f'[n(r)] - \tfrac{1}{2}\beta \phi_2 \nabla^2 n(r) = \alpha, \quad (7.245)$$

where

$$\phi_2 = \tfrac{1}{3} \int r^2 \phi_a \, d^3r. \quad (7.246)$$

Let us next assume that $n(r)$ depends on x only so that Eq.(7.245) becomes a one-dimensional equation,

$$-\tfrac{1}{2}\beta \phi_2 \frac{d^2 n}{dx^2} = -f'(n) + \alpha. \quad (7.247)$$

We can now obtain all we need from Eq.(7.247) by noting that it can be interpreted as the classical equation of motion of a point particle of "mass" $-\tfrac{1}{2}\beta \phi_2$, with "position" n at "time" x, moving in a "potential" $\psi(n)$ given by (see Fig.7.3)

$$\psi(n) = f(n) - \alpha n = n \ln \frac{1 - n}{n} + (1 - \alpha)n - \tfrac{1}{2}\beta \phi_0 n^2. \quad (7.248)$$

Possible solutions are clearly once again $n = n_I$, n_{II}, or n_{III} but these are uniform density solutions in which we are no longer interested. Instead we shall look for a solution such that

$$\left.\begin{array}{lll} n \to n_I & \text{as } x \to -\infty; \\ n \to n_{II} & \text{as } x \to +\infty. \end{array}\right\} \quad (7.249)$$

Looking at Fig.7.3 and bearing in mind the analogy with the classical motion, this means that we look for the case where the "particle" starts on one hill top and ends up on the other hill top. This can clearly be realised only for that value of α, say α_0, for which

$$\psi(n_I) = \psi(n_{II}), \quad (7.250)$$

or

$$\int_{n_I}^{n_{II}} f'(n) \, dn = \alpha_0(n_{II} - n_I), \quad (7.251)$$

which means that the area ABC in Fig.7.2 is equal to the area CDE: a Maxwell construction, though not as yet in the P-V diagram. We can see as follows that it corresponds to the equal area construction in the P-V diagram. For the van der Waals curve P satisfies the equation

$$\beta P = \frac{\partial \Phi}{\partial V} = f(n) - nf'(n). \qquad (7.252)$$

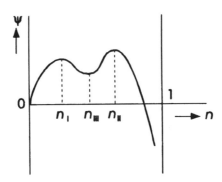

Fig.7.3. The "potential" $\psi(n)$.

From Eqs.(7.252), (7.235), (7.248), and (7.250) it the follows that

$$\beta P = f(n) - \alpha_0 n \qquad (7.253)$$

for $n = n_\mathrm{I}$ or n_II, and hence that

$$P_\mathrm{I} = P_\mathrm{II}. \qquad (7.254)$$

Using Eqs.(7.252) to (7.254) we then find

$$\int_\mathrm{II}^\mathrm{I} P\,dV = \frac{N}{\beta} \int_\mathrm{I}^\mathrm{II} \frac{f - nf'}{n^2}\,dn = \frac{N}{\beta}\left.\frac{f}{n}\right|_{n_\mathrm{I}}^{n_\mathrm{II}} = N\left.\frac{P}{n}\right|_{n_\mathrm{I}}^{n_\mathrm{II}} = P(V_\mathrm{I} - V_\mathrm{II}), \qquad (7.255)$$

which concludes the proof.

Let us now identify the phase with n_I as the gas phase and the phase with n_II as the liquid phase, and replace the indices I and II by "gas" and "liq". We next assume that apart from a negligible part of the volume, where there is a transition from the gas to the liquid phase, the density is either n_gas or n_liq. Let V_gas and V_liq be the volumes filled with densities n_gas and n_liq. We then have

$$V_\mathrm{gas} + V_\mathrm{liq} = V. \qquad (7.256)$$

We have thus a set of two-phase states, all with $\alpha = \alpha_0$. Since we have

$$N = V_{\text{gas}} n_{\text{gas}} + V_{\text{liq}} n_{\text{liq}}, \qquad (7.257)$$

we see that these states occur for all average densities $n \ (= N/V)$ in the range

$$n_{\text{gas}} \leqslant n \leqslant n_{\text{liq}}. \qquad (7.258)$$

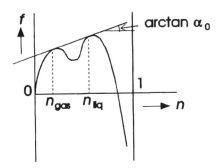

Fig.7.4. The double-tangent construction to find n_{gas} and n_{liq}.

The densities n_{gas} and n_{liq} must both satisfy the equation

$$f'(n) = \alpha_0, \qquad (7.259)$$

whence follows that they must satisfy the equation

$$f'(n_{\text{gas}}) = f'(n_{\text{liq}}). \qquad (7.260)$$

They must also satisfy Eq.(7.250), or

$$f(n_{\text{gas}}) - n_{\text{gas}} f'(n_{\text{gas}}) = f(n_{\text{liq}}) - n_{\text{liq}} f'(n_{\text{liq}}), \qquad (7.261)$$

which means that the gas and liquid densities can be found by constructing in the f-n plot (see Fig.7.4) the double tangent. The value of α_0 is found at the same time, since the slope of the double tangent is equal to α_0.

The derivation of the van der Waals equation given here is instructive but it is, unfortunately, not completely rigorous as it included the approximation (7.220). There exists a very ingenious model of an imperfect gas, due to Kac, Uhlenbeck, and Hemmer, which can be treated exactly and which also produces the van der Waals isotherms together with the Maxwell construction. However, it is one-dimensional and also involves an unphysical very-long-range attractive interaction potential. We refer to the papers quoted in the Bibliographical Notes for details about this model.

Problems

1. Prove (by induction) that the $U_N(r_1, \cdots, r_N)$ defined by Eqs.(7.114) vanish, unless the N atoms belong all to a single cluster.

2. Use the method described in the small-type section at the end of §7.1 to obtain the power series in Eq.(4.719).

3. If we write the equation of state as

$$PV = NkT \left[1 + \frac{B}{V} + \frac{C}{V^2} + \cdots \right], \tag{A}$$

the coefficients $B, C, \ldots$ are called, the second, third, $\ldots$ virial coefficients (see §1.3).

Prove that the second virial coefficient is given by the equation

$$B = 2\pi N \int_0^\infty \left[1 - e^{-\beta\phi(r)} \right] r^2 \, dr.$$

This equation was first derived from ensemble theory in Ornstein's doctoral thesis.

4. Prove that the third virial coefficient is given by the equation

$$C = \frac{N^2}{3} \int (1 - e^{-\beta\phi_{12}})(1 - e^{-\beta\phi_{23}})(1 - e^{-\beta\phi_{31}}) \, d^3r_{12} \, d^3r_{13},$$

where $\phi_{ik} \equiv \phi(r_{ik})$ and $r_{ik} \equiv r_i - r_k$.

5. Introduce the quantities

$$\psi_{ij} = e^{-\beta\phi_{ij}}, \tag{A}$$

so that the configurational partition function (7.111) can be written in the form

$$Z_N^{(\text{conf})} = V^N \, \overline{\psi_{12}\psi_{13} \cdots \psi_{N-1,N}}, \tag{B}$$

where the bar indicates the average over all possible particle positions in V.

Prove that if one assumes that to a first approximation — which should hold at low densities — we have

$$\overline{\psi_{12}\psi_{13} \cdots \psi_{N-1,N}} = \overline{\psi_{12}} \, \overline{\psi_{13}} \, \overline{\psi_{14}} \cdots \overline{\psi_{N-2,N}} \, \overline{\psi_{N-1,N}}, \tag{C}$$

we find that in the thermodynamic limit (that is, as $N \to \infty$, $V \to \infty$, $N/V = n = $ constant) we have

$$Z_N^{(\text{conf})} = V^N \, e^{\frac{1}{2}Nn\beta_1}, \tag{D}$$

with

$$\beta_1 = \int \left\{ e^{-\beta\phi(r)} - 1 \right\} d^3r.$$

Find the next term in the expansion of the exponent on the right-hand side of Eq.(D) in powers of N/V.[5]

6. In classical thermodynamics van der Waals introduced the *law of corresponding states* which stated that if p_{cr}, V_{cr}, and T_{cr} are the values of the pressure, volume, and temperature at the critical point, one can write the equation of state in the form

$$\tilde{P} = f(\tilde{V}, \tilde{T}),$$

where $f(\tilde{V}, \tilde{T})$ is a universal function and where

$$\tilde{P} = \frac{P}{P_{cr}}, \qquad \tilde{V} = \frac{V}{V_{cr}}, \qquad \tilde{T} = \frac{T}{T_{cr}}.$$

A slightly different law of corresponding states holds for classical gases for which the interatomic interaction potential $\phi(r)$ can be written in the form

$$\phi(r) = \sigma\psi(\tau r),$$

where ψ is a universal function and where σ and τ will differ from one substance to another. An example are the gases H_2, He, Ne, N_2, and A for which $\phi(r)$ is given by the Lennard-Jones potential (1.302) with $m = 6$ and $n = 12$.

Let us assume that σ has the dimensions of an energy so that ψ is dimensionless and that τ has the dimensions of an inverse length so that τr is dimensionless. Introduce reduced variables

$$r^* = \tau r, \qquad \beta^* = \beta\sigma, \qquad v_0^* = v_0\tau^3, \qquad W_n^* = v_0^{*n} W_n,$$

and

$$P^* = \frac{P}{\tau^3\sigma}, \qquad V^* = \tau^3 V, \qquad T^* = \frac{1}{\beta^*},$$

to prove that the equation of state can be written in the form

$$P^*V^* = NT^*\chi(\beta^*, V^*), \tag{B}$$

where χ is a universal function for a given potential energy function ψ. Equation (B) is the law of corresponding states in the form first derived by de Boer and Michels.

7. Prove that the real symmetric matrix

$$A_{ij} = a_i\delta_{ij} - b_i j, \qquad b_{ij} = b_{ji} \geqslant 0,$$

---

[5] This problem is based on the following paper: N.G.van Kampen, *Physica* **27**, 783 (1961).

will be positive definite, if for all i we have

$$a_i > \sum_j b_{ij}.$$

Hence, by considering the matrix of the second derivatives of the function $\Phi(N_i)$ given by Eq.(7.223), prove that Φ_{extr} is an absolute maximum for the solution (7.225), provided we have

$$f''(n) < 0. \qquad\qquad\qquad (C)$$

Since the matrix of the second derivatives of Φ is negative definite when condition (C) is satisfied, it follows that Φ has only a single maximum and that is the one we have already found.

8. Prove that the free energy corresponding to the horizontal part of the isotherm is lower than the free energy on the van der Waals curve. Hint: use Fig.7.4.

Bibliographical Notes

Section 7.1. A classical reference for the older literature about the equation of state is

1. H.Kamerlingh Onnes and W.H.Keesom, *Enzyklopädie der mathematischen Wissenschaften*, Leipzig-Berlin, Vol.V, Pt 10.
   See also
2. J.de Boer, *Repts. Progr. Phys.* **12**, 305 (1949).

The first author to use ensemble theory to obtain the equation of state was Ornstein:

3. L.S.Ornstein, *Toepassing der Statistische Mechanica van Gibbs op Molekulair-theoretische Vraagstukken*, Thesis, Leiden (1908).

Ursell developed the principles on which the main developments in the theory of the equation of state followed:

4. H.D.Ursell, *Proc. Camb. Phil. Soc.* **23**, 685 (1927).

Later developments followed in the work of Mayer, Uhlenbeck, Born, de Boer and their collaborators:

5. J.E.Mayer, *J. Chem. Phys.* **5**, 67 (1937).
6. G.E.Uhlenbeck and E.Beth, *Physica* **3**, 729 (1936).
7. E.Beth and G.E.Uhlenbeck, *Physica* **4**, 915 (1937).
8. B.Kahn, *On the Theory of the Equation of State*, Thesis, Utrecht (1938).
9. M.Born and K.Fuchs, *Proc. Roy. Soc. (London)* **A166**, 391 (1938).
10. J.de Boer and A.Michels, *Physica* **5**, 945 (1938).
11. J.de Boer and A.Michels, *Physica* **6**, 409 (1939).
12. J.de Boer, *Contribution to the Theory of Compressed Gases*, Thesis, Amsterdam (1940).

For a discussion of the quantum mechanical equation of state see

13. D.ter Haar, *Elements of Statistical Mechanics*, Rinehart, New York (1954), Chapter 8, especially §§ 4 to 7, and the references given at the end of that chapter.

Section 7.2. The discussion in this section is based on

14. N.G.van Kampen, *Phys. Rev.* **135**, A362 (1964).

See also Ref.3 and

15. J.Lebowitz and O.Penrose, *J. Math. Phys.* **7**, 98 (1966).
16. E.Lieb, *J. Math. Phys.* **7**, 1016 (1966).
17. H.S.Leff, *Phys. Rev.* **148**, 92 (1966).
18. D.ter Haar, *Lectures on Selected Topics in Statistical Mechanics*, Pergamon Press, Oxford (1977), § 5.1.

A one-dimensional model, which can be treated exactly and which leads to the van der Waals equation together with the Maxwell construction was introduced by Kac, Uhlenbeck, and Hemmer:

19. M.Kac, G.E.Uhlenbeck, and P.C.Hemmer, *J. Math. Phys.* **4**, 216 (1963).

See also§ 5.2 of Ref.18 and

20. P.C.Hemmer, M.Kac, and G.E.Uhlenbeck, *J. Math. Phys.* **4**, 229 (1963).
21. G.E.Uhlenbeck, P.C.Hemmer, and M.Kac, *J. Math. Phys.* **5**, 60 (1964).

CHAPTER 8

THE OCCUPATION NUMBER REPRESENTATION

8.1.   Quasi-Particles and Elementary Excitations

In Chapter 1 we stated that in statistical mechanics we attempt to derive the equilibrium or thermal properties of matter in bulk from the properties of the constituent particles but that in the first four chapters of the present book we would be mainly concerned with systems of independent particles, that is, perfect gases. In fact, in those chapters we went slightly beyond the restrictions to perfect gases as we discussed the effect of collisions — which are, after all, interactions between particles — albeit only in as far as they lead to the establishment of the equilibrium distribution function for a perfect gas; we also discussed a very simple approach to the equation of state of an imperfect gas, which led us to the van der Waals equation (§ 1.3). In Chapters 5 and 6 we introduced the possibility that the particles in our systems could interact and we developed a general theory for such systems of interacting particles. However, as far as applications of this general theory are concerned we have so far restricted the discussion to that of a perfect gas. We now wish to go beyond this and look at actual physical systems with interactions. If the interactions can be considered to be in some sense "small", we might hope that we could expand the right-hand side of Eq.(5.0907) in a power series in some small parameter, for instance, the interaction potential.

It has to be borne in mind that a perfect gas in some cases is a fair approximation to an actual physical system, but in many cases this perfect gas is, as we shall see presently, not a perfect gas of the original constituent particles, but a kind of "quasi-perfect gas". In the case where a perfect gas or a "quasi-perfect" gas is a good approximation one would hope that one could use some kind of perturbation theory, starting from the perfect gas as a zeroth approximation.

Before considering the statistical mechanics of systems of interacting particles, we shall briefly discuss some relevant aspects of the quantum mechanics of such systems. Let us remind ourselves that the problem we are concerned with is how to deal with a many-body system. It is well known that both in classical and in quantum mechanics we can solve the so-called two-body system, that is, we can find the behaviour of a system consisting of just two particles which interact with one another. However, a three-body problem is already too complicated to allow an exact solution. As long as the interactions are weak we may hope that some kind of perturbation theory treatment might suffice, but the question then arises: What happens when the interactions are too strong for this? Let us consider for a moment a Debye solid. We saw in Problem 8 of Chapter 3 that at low temperatures it behaves like a black-body. In fact, the density of modes given by Eq.(A) of that problem is the same as the density of modes given by Eq.(3.307) — at low-temperatures only the low-lying modes come into play so that the restriction $\omega < \omega_m$ which distinguishes (A) from (3.307) can be ignored. We know that black-body radiation can be described as a perfect gas of photons and we can thus say that to a good approximation a Debye solid — at low temperatures — can also be considered as a perfect gas, namely, a perfect gas of the lattice-vibration quanta, the so-called *phonons*.

Before returning to the case of a Debye solid, let us consider another example of a system which behaves, at least to a good approximation, as a perfect gas. We mentioned in Chapter 4 that one can apply Fermi-Dirac statistics to the "gas" of conduction electrons in metals — or in intrinsic semiconductors. These conduction electrons behave to a very good approximation like a perfect gas of fermions with an *effective mass* which may differ significantly from the free electron mass. This difference in mass is a result of the interactions of the electrons with the lattice and as a result the electrons behave like *quasi-particles*, that is, they have all the properties of particles, even including their symmetry properties, but at the same time their behaviour has been affected by the interactions with their surroundings. In this case again to zeroth approximation we are dealing with a perfect gas.

If we look at the examples we have so far encountered in this section, we have, on the one hand, the "quasi-perfect gas" of the conduction electrons, where the constituents are quasi-particles which in this case are very similar to electrons in their behaviour, but part of their interaction with the lattice has led to their having an effective mass rather than the *bare* mass of the electrons. This effective mass may even be negative: this is the case for the holes in semiconductors. On the other hand, we have the case of black-body radiation where the quasi-particles are the photons or light quanta which show the particle aspect of light, or we have the quasi-perfect gas of the phonons, where the quasi-particles are the lattice-wave quanta. In the latter case, we start from a system of particles which interact strongly. In the classical description of this system the lowest energy state will correspond to all particles in the lattice being in their equilibrium positions. The excitations of the system correspond to sound waves — collective states in which all particles are involved. Just as the electromagnetic waves correspond to photons, that is, light quanta, the sound waves correspond to phonons, that is, sound quanta. We see that considering the Debye solid as a (quasi-)perfect gas of phonons has involved changing the description from that of a system of interacting particles to that of a system of non-interacting sound waves.

The Debye solid is one example of a broad class of systems where to a good approximation the *elementary excitations* form — at least at not too high temperatures — a perfect gas of quasi-particles. Another example we may mention is liquid helium where the elementary excitations are the phonons and rotons.

Some authors distinguish between quasi-particles and elementary excitations which are collective states, involving a large number of particles. When the constituents of the perfect gas are *dressed* particles like the conduction electrons in a metal or the holes in a semiconductor, they are called quasi-particles, but when the constituents involve obviously many particles as, for instance, the phonons in a Debye solid, they are called elementary (or collective) excitations. This ignores the fact that in the "dressing" of a quasi-particle many other particles are involved.

Let us briefly consider the case of the "dressed" particles. The original particles while moving through the system interact with the other constituents and their behaviour will consequently be altered. One says that they carry with them a "cloud" of interactions, the "dressing", which distinguishes them from their own original "bare" self. As another example we may mention the Debye-Hückel gas where each electron will attract a "cloud" of positive charges around it and as a result its Coulomb interaction with the other electrons is screened. One can see that there is, indeed, a difference between this kind of system where the quasi-particles still remind us of their origin and systems such as the Debye solid where the elementary excitations, the sound quanta, are called quasi-particles because they are the constituents of a perfect — or quasi-perfect — gas. However, having stressed the difference between the two cases, we shall see later on in this chapter that the Hamiltonians describing them are very similar, thus justifying calling the constituents of the system by the same name, as we shall continue to do.

Let us briefly list a few examples, both of collective excitations and of "dressed" particles. As to collective excitations, apart from the phonons which we have already met with, we can mention the *plasmons*, that is, the various waves occurring in a plasma,[1] the *magnons* or *spin waves* which occur in ferromagnets, and the vibrational and rotational excitations of nuclei. Among the "dressed" particles we can mention, apart from the conduction electrons and holes in metals and semiconductors, the *polaron*, an electron moving through an insulating polar crystal accompanied by a cloud of phonons,[2] and the nucleons in the shell model of the nucleus (the cloud here consists of other nucleons, just as the cloud of an electron in a Debye-Hückel gas consists of other charged particles).

---

[1] The term plasmons is often reserved for quanta of one particular kind of plasma wave, namely, the electrostatic Langmuir waves.

[2] See, for instance, H.Fröhlich, *Adv. Phys.* **3**, 325 (1954).

Let us now consider how we can reduce our system to a system resembling a perfect gas. The behaviour of a many-body system is governed by its Hamiltonian $\mathcal{H}$ which will be a function of the coordinates and momenta of all the constituents. In §4.2 we considered a Hamiltonian given by Eq.(4.201),

$$\mathcal{H} = \sum_j \mathcal{H}_j, \tag{8.101}$$

where the sum was over the N particles in the system which were assumed to be identical so that the $\mathcal{H}_j$ were identical functions of their arguments, differing merely in the index numbering the particles. Equation (8.101) assumes that there are no interactions between the particles, as is the case for a perfect gas. In the general case when there are interactions the Hamiltonian will not have the simple form (8.101), but will be of the form

$$\mathcal{H} = \sum_j \mathcal{H}_j + \mathcal{H}_{\text{int}}, \tag{8.102}$$

where $\mathcal{H}_{\text{int}}$ is the interaction Hamiltonian which describes the interactions in the system. To simplify our considerations we shall often restrict our discussion to the case where the only interactions between the constituents are two-body forces. In that case we have a Hamiltonian of the form

$$\mathcal{H} = \sum_{j=1}^{N} \mathcal{H}_i + \tfrac{1}{2} \sum_{j \neq k=1}^{N} V_{j,k}, \tag{8.103}$$

where the $V_{j,k}$ term corresponds to the interaction between the j-th and the k-th particles. In the §7.1 we discussed the case of an imperfect gas considered as a perfect gas with interactions added onto it. In that case the second sum on the right-hand side of Eq.(8.103) will be considered to some extent as a perturbation. At the moment, however, we shall not make any assumptions as to the smallness of that term.

We have stated earlier that many systems can — at least to a good approximation — be considered to be perfect gases. This means that it must be possible to transform the Hamiltonian which originally looks like (8.102) to look like (8.101), or at least to be of the form (8.102), but now with an $\mathcal{H}_{\text{int}}$ which, in some sense, can be considered to be small. Let us assume that this can be done by finding a suitable transformation.

Let us, to simplify the discussion, assume for a moment that the original system consisted of N point particles, corresponding to $3N$ coordinates r_j ($j = 1, \ldots, N$). We now want to find a set of coordinates ξ_n such that, when we carry out the $r_j \to \xi_n$ transformation the Hamiltonian has the form (8.101). If the momenta which are the conjugates of the ξ_n are π_n, it might happen that the resultant Hamiltonian would have the form

$$\mathcal{H} = \sum_n \tfrac{1}{2} \alpha_n \pi_n^2 + \mathcal{H}'_{\text{int}}, \tag{8.104}$$

where the second term on the right-hand side is "small"; we have quasi-particles which behave like the original particles with effective masses α_n^{-1}. However, another possibility would be a Hamiltonian of the form

$$\mathcal{H} = \sum_n \tfrac{1}{2} \alpha_n \pi_n^2 + \sum_n V_n(\xi_n) + \mathcal{H}''_{\text{int}}, \tag{8.105}$$

and the ξ_n would now correspond to the collective excitation kind of quasi-particles. An example of the second kind is given by a system which bears some relationship to the Debye solid, namely, a linear chain of atoms interacting harmonically (Fig.8.1). Its Hamiltonian is

$$\mathcal{H} = \sum_{j=1}^{N} \left[\frac{p_j^2}{2m} + \tfrac{1}{2} \beta \left(x_j - x_{j-1} \right)^2 \right], \tag{8.106}$$

where m is the mass of each of the atoms and where we have neglected end effects.

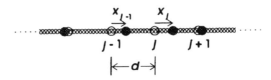

Fig.8.1. Linear chain: x_j is the displacement of the j-th atom from its equilibrium position; d is the equilibrium distance between atoms.

The transformation[3]

$$\left. \begin{aligned} x_j &= \frac{1}{\sqrt{N}} \sum_n \left(\xi_n \cos njd - \frac{1}{m\omega_n} \pi_n \sin njd \right), \\ p_j &= \frac{1}{\sqrt{N}} \sum_n \left(m\omega_n \xi_n \sin njd + \pi_n \cos njd \right), \end{aligned} \right\} \tag{8.107}$$

where d is the equilibrium distance between the atoms in the chain and where the ω_n are given by the relations

$$\omega_n = \sqrt{\left[\frac{2\beta}{m} (1 - \cos nd) \right]}, \tag{8.108}$$

---

[3] See, for instance, R.D.Mattuck, *A Guide to Feynman Diagrams in the Many-Body Problem*, McGraw-Hill, London (1967), p.10.

transforms the Hamiltonian (8.106) into

$$\mathcal{H} = \sum_n \left[\frac{\pi_n^2}{2m} + \tfrac{1}{2} m \omega_n^2 \xi_n^2 \right], \tag{8.109}$$

that is, a Hamiltonian of the form (8.105).

The reduction of the Hamiltonian (8.106) to the form (8.109) was possible, because we had a very good idea of the form of the elementary excitations. In other words, if we know most of the answer already, it is not that difficult to find the complete answer. This case is also much simpler than the general case in that the final Hamiltonian no longer contains any interactions. In the general case, the situation is not that easy. We may have some idea of the nature of some of the elementary excitations — in most cases they will be some kind of waves, like the phonons, the various kinds of plasmons, or the magnons. In that case the $\mathcal{H}_j$ terms are of the form $\tfrac{1}{2}(\alpha_j \pi_j^2 + \beta_j \xi_j^2)$, but it will not always be possible to find from first principles the form of the $\mathcal{H}_{\text{int}}$ term.

Moreover, the collective excitations may not be the only constituents of the system. For instance, in a ferromagnet we should expect apart from the magnons, also phonons, and in a metal, apart from the phonons also the conduction electrons. This means that, in general, the final Hamiltonian will, after suitable transformations which will lead to the explicit appearance of the various quasi-particles, look something like

$$\mathcal{H} = \mathcal{H}^{(1)} + \mathcal{H}^{(2)} + \cdots + \mathcal{H}_{\text{int}}^{(1)} + \cdots + \mathcal{H}_{\text{int}}^{(1,2)} + \cdots, \tag{8.110}$$

where the $\mathcal{H}^{(\alpha)}$ correspond to the sums of single-quasi-particle Hamiltonians of quasi-particles of the kind α, which may look like (8.109), or simply be of the form $(p_k^\alpha)^2 / 2m_\alpha$, where the $\mathcal{H}_{\text{int}}^\alpha$ correspond to residual interactions between quasi-particles of the kind α, while the $\mathcal{H}_{\text{int}}^{\alpha,\beta}$ correspond to (residual) interactions between quasi-particles of the kind α and quasi-particles of the kind β.

It is hardly ever possible to reduce the original Hamiltonian, which we may know from first principles — if we are lucky — to the form (8.110). The usual procedure is to guess what the final Hamiltonian will look like, that is, to guess what kind of collective modes will occur in our system and on that basis write down the Hamiltonian, also making an educated guess as to the form of the various interaction terms. One may hope that in some favourable circumstances one may determine at least some of the various coefficients from experimental data. One of the fundamental difficulties which usually gets scant attention is the fact that it is very easy to end up with a Hamiltonian where the number of variables greatly exceeds the number of degrees of freedom. We shall neglect this problem completely.[4]

---

[4] See D.ter Haar, *Introduction to the Physics of Many-Body Systems*, Interscience, New York (1958), § 6.4 for a discussion of the problem of the so-called redundant variables.

8.2.   The Occupation Number Representation for Bosons

Even if we have been able to reduce the Hamiltonian to the form (8.110) we must still discuss in how far the perfect gas description is a good one and we must discuss the effect of the various interaction terms, that is, we must consider in how far the perfect gas is a good first approximation for our system and how, if this is the case, one can proceed to the next approximation. A formalism which is often used and which is well suited for this is the so-called *occupation number representation*, sometimes called the *second-quantisation representation* which is, however, a misnomer since there is nowhere a further introduction of the quantum of action; it derives from the fact that in this formalism the wavefunctions become operators. Although in this book we only, at the end of the present section, discuss one minor application of this formalism, in view of its general importance we shall give the basic ideas of the formalism. It would take us too far to discuss the important aspects of diagram expansions using this formalism, or to consider other applications, and we refer for this to the general references given at the end of this chapter, especially the books by Abrikosov, Gorkov, and Dzyaloshinskii and by Parry.

We shall start our discussion of systems with possibly strong interactions by introducing this representation. It turns out that the representation differs somewhat in the case of a system of bosons from that for the case of a system of fermions, and we shall discuss these two cases separately. In the present section we shall discuss the boson case, postponing the discussion of the fermion case to the next section.

Let us assume that our system consists of N identical particles and let there be a complete orthonormal set of single-particle kets[5] $|i\rangle \equiv \varphi_i(q_m)$ where q_m stands for the coordinates of the m-th particle.

Let us assume that the particles making up our systems are either fermions or bosons.[6]   In that case we can use as a complete orthonormal set of basic functions for our system of N identical fermions or bosons the set

$$|i_i, i_2, \ldots, i_N\rangle = \frac{1}{N!} \sum_{P} \varepsilon_P \, \varphi_{i_1}(P_1)\varphi_{i_2}(P_2)\cdots\varphi_{i_N}(P_N), \qquad (8.201)$$

where $P_1, P_2, \ldots, P_N$ is a permutation of the N numbers $1, \ldots, N$, and where ε_P is equal to $+1$ for bosons, while for fermions $\varepsilon_P = +1$ or -1 according as to whether P is an even or an odd permutation. The complete orthonormal set $|i_1, \ldots, i_N\rangle$ is a set with the proper symmetry properties for the system under consideration. The bra set corresponding to the ket set (8.201) is

$$\langle i_i, i_2, \ldots, i_N| = \frac{1}{N!} \sum_{P} \varepsilon_P \, \varphi_{i_1}^*(P_1)\varphi_{i_2}^*(P_2)\cdots\varphi_{i_N}^*(P_N), \qquad (8.202)$$

---

[5] We shall use Dirac"s bra and ket notation as it is particularly suitable for our discussion.

[6] This is quite a strong assumption, but for the moment we shall neglect other possibilities; compare the discussion in § 4.9.

where φ_i^* is the complex conjugate of φ_i.[7]

The sets (8.201) and (8.202) satisfy the orthonormality relation

$$\langle i_i', i_2', \ldots, i_N' | i_i, i_2, \ldots, i_N \rangle = \frac{1}{N!} \sum_P \varepsilon_P \, \delta(i_1 - i_{P_1}') \ldots \delta(i_N - i_{P_N}'),$$

(8.203)

as well as a completeness or closure relation.[8]

If $|\Psi\rangle$ is a wavefunction of the N-particle system with the correct symmetry properties, we can expand it in terms of the set (8.201):

$$|\Psi\rangle = \int \cdots \int di_1 \cdots di_N \, |i_i, i_2, \ldots, i_N \rangle \langle i_i, i_2, \ldots, i_N | \Psi\rangle. \quad (8.204)$$

Similarly, an operator $\widehat{G}$, operating on N-particle functions in the Hilbert space spanned by the set (8.201) will have the form

$$\widehat{G} = \int \cdots \int di_1 \cdots di_N \, di_1' \cdots di_N' \, |i_i, \ldots, i_N \rangle$$

$$\langle i_i, \ldots, i_N | \widehat{G} | i_i', \ldots, i_N' \rangle \langle i_i', \ldots, i_N' |. \quad (8.205)$$

So far we have only considered systems with N particles, that is, all wavefunctions were given in a Hilbert space $\mathfrak{H}_N$. If we want to work with grand ensembles, or if we want to consider systems in which the number of particles is not fixed but can vary due to creation or annihilation processes we must work in a Hilbert space $\mathfrak{H}$ which is the direct product of the Hilbert spaces corresponding to $0, 1, \ldots, N, \ldots$ particles:

$$\mathfrak{H} = \mathfrak{H}_0 \otimes \mathfrak{H}_1 \otimes \mathfrak{H}_2 \otimes \cdots \mathfrak{H}_N \otimes \cdots. \quad (8.206)$$

The Hilbert space $\mathfrak{H}$ is spanned by the sets (8.201) with N ranging over all non-negative integral values, $N = 0, 1, \ldots, N, \ldots$; the complete orthonormal set spanning $\mathfrak{H}$ will thus be the set

$$|0\rangle, \quad |i\rangle, \quad |i_i, i_2\rangle, \quad |i_i, i_2, \ldots, i_N\rangle, \quad \ldots, \quad (8.207)$$

where $|0\rangle$ is the vacuum state in which there are no particles present.

We now recall that in Chapter 4 we pointed out that the state described by the wavefunctions (4.206), (4.207), or a linear combination of these, such as (8.201) can be characterised either by stating that one particle is in the single-particle state φ_{i_1}, one particle in the single-particle state φ_{i_2}, ..., and, finally, one particle in the single-particle state φ_{i_N}, or by stating

---

[7] For the sake of simplicity we are assuming here that the φ_i are scalars; the generalisation to the case where they are spinors is straightforward — in that case the φ_i^* are the Hermitean conjugates of the φ_i.

[8] We have assumed here that i is a continuous parameter; if it is a discrete one, the Dirac delta-functions must be replaced by Kronecker deltas.

that that there are n_1 particles in the state φ_1, n_2 particles in the state φ_2, and so on. That is, instead of using the symbol $|i_i, i_2, \ldots, i_N\rangle$ for the wavefunction (8.201) we could have used the symbol $|n_1, n_2, \ldots\rangle$. We note, first of all, that whereas there are N arguments when we write $|i_i, i_2, \ldots, i_N\rangle$, the number of arguments of $|n_1, n_2, \ldots\rangle$ is infinite, corresponding to the, in general, infinite number of single-particle states. Secondly, we note that if we are dealing with a system of fermions the n_i can only take on the values 0 and 1, whereas they can take on any non-negative integral value, if we are dealing with a system of bosons. The order in which the different single-particle states are arranged in $|n_1, n_2, \ldots\rangle$ will turn out to be important in the fermion case, and we shall therefore assume that we have chosen it in some definite way and that we shall stick to that order of the single-particle states.

The advantage of using the $|n_1, n_2, \ldots\rangle$ notation is that there is no restriction on the total number of particles in the system, that is, $|n_1, n_2, \ldots\rangle$ can refer to any system of particles, independent of the total number of particles in it. In fact, the sum $\sum n_i$ can take on any value so that the Hilbert space $\mathfrak{H}$ is spanned by the set

$$|n_1, n_2, \ldots, n_i, \ldots\rangle. \tag{8.208}$$

We note, by the way, that $|0, 0, \ldots\rangle$ is now the vacuum state and that

$$\langle n_1', n_2', \ldots, n_i', \ldots | n_1, n_2, \ldots, n_i, \ldots\rangle = \delta_{n_1' n_1} \delta_{n_2' n_2} \cdots \delta_{n_i' n_i} \cdots. \tag{8.209}$$

So far we have not distinguished between boson and fermion systems, but we shall now assume that we are dealing with a system of bosons, and we shall introduce creation operators $\hat{a}^\dagger(i)$ which from a ket corresponding to a state in which there are n_i particles in the single-particle state φ_i will produce one in which there are $n_i + 1$ particles in that single-particle state. We define these operators by the relation

$$\hat{a}^\dagger(i) |n_1, \ldots, n_{i-1}, n_i, n_{i+1}, \ldots\rangle = \sqrt{n_i + 1} |n_1, \ldots, n_{i-1}, n_i + 1, n_{i+1}, \ldots\rangle. \tag{8.210}$$

First of all, using the definition (8.210) and (8.201) one can prove that the $\hat{a}^\dagger$ satisfy the commutation relation:

$$[\hat{a}^\dagger(i), \hat{a}^\dagger(j)]_- = 0, \tag{8.211}$$

where $[\hat{A}, \hat{B}]_-$ is the commutator of two operators $\hat{A}$ and $\hat{B}$, defined by Eq.(6.0110).

Secondly, it follows from the definition of the $\hat{a}^\dagger(i)$ that the $|n_1, n_2, \ldots, n_i, \ldots\rangle$ can all be obtained from the vacuum state through the repeated action of creation operators:

$$|n_1, n_2, \ldots, n_i, \ldots\rangle = \frac{\hat{a}^\dagger(1)^{n_1} \hat{a}^\dagger(2)^{n_2} \cdots \hat{a}^\dagger(i)^{n_i} \cdots}{\sqrt{n_1! \cdots n_i! \cdots}} |0, 0, \ldots, 0, \ldots\rangle. \tag{8.212}$$

We now consider the adjoint of Eq.(8.210):

$$\langle n_1, \ldots, n_{i-1}, n_i, n_{i+1}, \ldots | \widehat{a}(i) = \sqrt{n_i + 1} \langle n_1, \ldots, n_{i-1}, n_i + 1, n_{i+1}, \ldots |,$$
(8.213)

where $\widehat{a}(i)$ is the adjoint of $\widehat{a}^\dagger(i)$. First of all, it follows from this equation that the $\widehat{a}(i)$ satisfy the commutation relation

$$[\widehat{a}(i), \widehat{a}(j)]_- = 0,$$
(8.214)

To find out what the result is of $\widehat{a}(i)$ operating on a ket we consider the expansion of $\widehat{a}(i)|n_1, \ldots, n_i, \ldots\rangle$ in terms of the complete orthonormal set (8.208):

$$\widehat{a}(i)|n_1, \ldots, n_i, \ldots\rangle = \sum_{\{n'\}} |n_1', \ldots, n_i', \ldots\rangle$$
$$\times \langle n_1', \ldots, n_i', \ldots | \widehat{a}(i) |n_1, \ldots, n_i, \ldots\rangle. \quad (8.215)$$

From Eq.(8.213) it follows that the matrix element on the right-hand side of Eq.(8.215) satisfies the equation

$$\langle n_1', \ldots, n_i', \ldots | \widehat{a}(i) |n_1, \ldots, n_i, \ldots\rangle$$
$$= \sqrt{n_i' + 1} \langle n_1', \ldots, n_i' + 1, \ldots |n_1, \ldots, n_i, \ldots\rangle, \quad (8.216)$$

and from the orthonormality relation (8.209) of the set (8.208) it follows that this matrix element vanishes, unless $n_j' = n_j$ for all j, except $j = i$, while $n_i' = n_i - 1$; in the latter case the matrix element is equal $\sqrt{n_i}$. Hence it follows from Eq.(8.215) that

$$\widehat{a}(i)|n_1, \ldots, n_{i-1}, n_i, n_{i+1}, \ldots\rangle = \sqrt{n_i}\,|n_1, \ldots, n_{i-1}, n_i - 1, n_{i+1}, \ldots\rangle,$$
(8.217)

and we see that the $\widehat{a}(i)$ are annihilation operators: $\widehat{a}(i)$ removes a particle from the single-particle state φ_i.

Apart from Eqs.(8.211) and (8.214) the $\widehat{a}^\dagger(i)$ and $\widehat{a}(i)$ also satisfy the commutation relation

$$[\widehat{a}(i), \widehat{a}^\dagger(j)]_- = \delta_{ij}.$$
(8.218)

From Eqs.(8.210) and (8.217) it follows that we have

$$\widehat{a}^\dagger(i)\,\widehat{a}(i)\,|n_1, n_2, \ldots, n_i, \ldots\rangle = n_i\,|n_1, n_2, \ldots, n_i, \ldots\rangle.$$
(8.219)

The *number operator*

$$\widehat{n}(i) \equiv \widehat{a}^\dagger(i)\widehat{a}(i)$$
(8.220)

thus measures the number of particles in the single-particle state φ_i.

Consider now the case where the Hamiltonian is given by Eq.(8.103). One can show, by considering the matrix elements of the operator in the occupation number representation, that in terms of the annihilation and creation operators the Hamiltonian operator $\widehat{\mathcal{H}}$ can be written in the form

$$\widehat{\mathcal{H}} = \sum_{i,j} H_{i,j}\, \widehat{a}^{\dagger}(i)\widehat{a}(j) + \tfrac{1}{2} \sum_{i,j,k,l} V_{i,j,k,l}\widehat{a}^{\dagger}(i)\widehat{a}^{\dagger}(j)\widehat{a}(k)\widehat{a}(l), \qquad (8.221)$$

where

$$H_{i,j} = \langle i|\widehat{H}|j\rangle, \qquad (8.222)$$

with $\widehat{H}$ being the single-particle operator, and where we have assumed that the operators in the interaction term are all the same functions $\widehat{G}$ of their arguments so that

$$V_{i,j,k,l} = \langle i,j|\widehat{G}|k,l\rangle. \qquad (8.223)$$

In the case where the φ_i are the eigenfunctions of the single-particle operator $\widehat{H}$ with eigenvalues ε_i we have for the first sum on the right-hand side of Eq.(8.221) the expression

$$\sum_i \varepsilon_i\, \widehat{a}^{\dagger}(i)\widehat{a}(i). \qquad (8.224)$$

Often one uses for the φ_i the momentum eigenfunctions which are the eigenfunctions of the single-particle kinetic energy operator, $\widehat{p}^2/2m$, with m the particle mass. If, as is often the case, $\widehat{G}$ is a potential energy depending only on the distance apart of the two particles so that the interaction term is of the form $\tfrac{1}{2}\sum_{i,j} v(|\mathbf{r}_i - \mathbf{r}_j|)$, the Hamiltonian in terms of the $\widehat{a}^{\dagger}(i)$ and $\widehat{a}(i)$ will be of the form

$$\widehat{\mathcal{H}} = \sum_k \frac{(\hbar k)^2}{2m}\, \widehat{a}_k^{\dagger}\widehat{a}_k + \tfrac{1}{2} \sum_{k,k',q} v(q)\, \widehat{a}_k^{\dagger}\widehat{a}_{k'}^{\dagger}\widehat{a}_{k'+q}\widehat{a}_{k-q}, \qquad (8.225)$$

with $v(q)$ the Fourier transform of the potential energy,

$$v(q) = \frac{1}{V} \int v(r)\, e^{i(\mathbf{k}\cdot\mathbf{r})}\, d^3\mathbf{r}, \qquad (8.226)$$

where V is the volume of the system.

In its general form this problem posed by the Hamiltonian (8.225) is insoluble, but Bogolyubov[9] has shown how a simplified model can be solved. At low temperatures one expects that, by analogy with the case of a perfect boson gas (see § 4.7), a finite fraction of the bosons will be in the zero-momentum state so that $\langle\widehat{a}_0^{\dagger}\widehat{a}_0\rangle \equiv n_0$ will be of the same order of magnitude as the total number of particles, N, even though all the $\widehat{a}_k^{\dagger}\widehat{a}_k$ with $k \neq 0$ will be of the order of unity, that is, smaller than n_0 by a factor

[9] N.N.Bogolyubov, *J.Phys. U.S.S.R.* **11**, 23 (1947).

of the order unity. Put differently, in the commutor relation $\hat{a}_0 \hat{a}_0^\dagger - \hat{a}_0^\dagger \hat{a}_0 = 1$ we can neglect the right-hand side. We can therefore treat the operators $\hat{a}_0^\dagger$ and $\hat{a}_0$ to a good approximation as c-numbers, both equal to $\sqrt{n_0}$.

If we replace $\hat{a}_0^\dagger$ and $\hat{a}_0$ by $\sqrt{n_0}$ we get the Hamiltonian

$$
\begin{aligned}
\hat{\mathcal{H}}' = & {\sum_q}' \left(\frac{(\hbar q)^2}{2m} + n_0\, v(q) \right) \hat{a}_q^\dagger \hat{a}_q + \tfrac{1}{2} {\sum_q}'\, n_0\, v(q) \, [\hat{a}_q^\dagger \hat{a}_{-q}^\dagger + \hat{a}_q \hat{a}_{-q}] \\
& + \tfrac{1}{2} N^2 v(0) + {\sum_{q,q'}}'\, \sqrt{n_0}\, v(q) \, [\hat{a}_q^\dagger \hat{a}_{q-q'} \hat{a}_{q'} + \hat{a}_q^\dagger \hat{a}_{q'}^\dagger \hat{a}_{q+q'}] \\
& + \tfrac{1}{2} \sum_{q,q'} v(0) \, \hat{a}_q^\dagger \hat{a}_q \hat{a}_{q'}^\dagger \hat{a}_{q'} + \tfrac{1}{2} {\sum_{q,q',k}}'\, v(q) \hat{a}_q^\dagger \hat{a}_{q'}^\dagger \hat{a}_{q'+k} \hat{a}_{q-k},
\end{aligned}
\tag{8.227}
$$

where the prime on the summation sign indicates that the term with zero momentum has been omitted. If we now use the relation $N = n_0 + {\sum_q}' \hat{a}_q^\dagger \hat{a}_q$ and retain only terms in which at most two of the momenta differ from zero, we obtain the Bogolyubov Hamiltonian

$$
\hat{\mathcal{H}}_{\mathrm{Bog}} = \sum_q \frac{(\hbar q)^2}{2m} + \tfrac{1}{2} N^2\, v(0) + \tfrac{1}{2}\, N {\sum_q}'\, v(q)\, [2\hat{a}_q^\dagger \hat{a}_q + \hat{a}_q^\dagger \hat{a}_q^\dagger + \hat{a}_q \hat{a}_q].
\tag{8.228}
$$

Consider a canonical transformation to new boson creation and annihilation operators $\hat{b}_q^\dagger$ and $\hat{b}_q$, that is, a transformation such that the new operators satisfy the same commutation relations (8.211), (8.214), and (8.218) as the old ones:

$$
\hat{a}_q^\dagger = \alpha_q \hat{b}_q^\dagger + \beta_q \hat{b}_q, \qquad \hat{a}_q = \alpha_q \hat{b}_q + \beta_q \hat{b}_q^\dagger.
\tag{8.229}
$$

If we now determine the coefficients α_q and β_q such that the Hamiltonian is reduced to the form

$$
\hat{\mathcal{H}}_{\mathrm{Bog}} = \sum_q E_q \hat{b}_q^\dagger \hat{b}_q + \text{constant},
\tag{8.230}
$$

we find the quasi-particle energy E_q which is given by the equation

$$
E_q = \sqrt{\frac{N q^2 v(q)}{m} + \frac{q^4}{4m^2}},
\tag{8.231}
$$

which for small q leads to a linear energy spectrum,

$$
E_q \approx sq. \qquad s = \sqrt{\frac{N v(0)}{m}},
\tag{8.232}
$$

that is, a phonon spectrum.

8.3.   The Occupation Number Representation for Fermions

The case of a system of fermions is similar to, but rather more complicated than, the case of a system of bosons. The reason for this can be seen from expression (8.201) for the basic functions of the orthonormal set. Because of the appearance of the factor ε_P the order in which the states appear is important.[10]   This is the reason why we stated in the preceding section that we would fix the order in which the states occur in our wavefunctions.

We want to introduce creation operators $\hat{a}^\dagger(i)$ such that now their acting upon a function $|n_1, \ldots, n_i, \ldots\rangle$ produces zero, if $n_i = 1$, and produces $|n_1, \ldots, n_i + 1, \ldots\rangle$, if $n_i = 0$, and annihilation operators $\hat{a}(i)$ such that they produce $|n_1, \ldots, n_i - 1, \ldots\rangle$, if $n_i = 1$ and zero, if $n_i = 0$. This is accomplished, if we define them as follows:[11]

$$\hat{a}^\dagger(i) = (-1)^{\tau_i} \sqrt{1 - n_i} \, |n_1, \ldots, n_i, \ldots\rangle, \qquad (8.301)$$

$$\hat{a}(i) = (-1)^{\tau_i} \sqrt{n_i} \, |n_1, \ldots, n_i, \ldots\rangle, \qquad (8.302)$$

where

$$\tau_i = \sum_{j=1}^{i-1} n_j. \qquad (8.303)$$

One can prove that the $\hat{a}^\dagger(i)$ and $\hat{a}(i)$ satisfy the commutation relations[12]

$$[\hat{a}^\dagger(i), \hat{a}^\dagger(j)]_+ = 0, \qquad (8.304)$$

$$[\hat{a}(i), \hat{a}(j)]_+ = 0, \qquad (8.305)$$

$$[\hat{a}(i), \hat{a}^\dagger(j)]_+ = \delta_{ij}, \qquad (8.306)$$

where $[\hat{A}, \hat{B}]_+$ is the anticommutator of two operators $\hat{A}$ and $\hat{B}$, defined by the equation

$$[\hat{A}, \hat{B}]_+ = \hat{A}\hat{B} + \hat{B}\hat{A}. \qquad (8.307)$$

If we introduce once again the parameter γ from Chapter 4, which was defined by Eqs.(4.307),

$$\gamma_{\text{BE}} = 1, \qquad \gamma_{\text{FD}} = -1, \qquad (8.308)$$

we can combine Eqs.(8.210) and (8.301) for the creation operators, Eqs. (8.217) and (8.302) for the annihilation operators, and Eqs.(8.211) and

---

[10]Bear in mind that each state can appear at most only once in the case of fermions.

[11]The clumsy looking factors $(-1)^{\tau_i}$ — which are due to the fact that changing the order of states introduces minus signs — have been introduced in order that the commutation relations (8.304) to (8.306) remain simple.

[12]Although Eqs.(8.304) to (8.306) contain anticommutators rather than commutators, we shall for the sake of simplicity call them commutation relations.

(8.304), (8.224) and (8.305), and (8.218) and (8.306) for the commutation relations as follows:

$$\widehat{a}^\dagger(i) \;=\; \gamma^{\tau_i}\,\sqrt{1+\gamma n_i}\,\,|n_1,\dots,n_i,\dots\rangle, \qquad (8.309)$$

$$\widehat{a}(i) \;=\; \gamma^{\tau_i}\,\sqrt{n_i}\,\,|n_1,\dots,n_i,\dots\rangle, \qquad (8.310)$$

$$[\widehat{a}^\dagger(i),\,\widehat{a}^\dagger(j)]_{-\gamma} \;=\; 0, \qquad (8.311)$$

$$[\widehat{a}(i),\,\widehat{a}(j)]_{-\gamma} \;=\; 0, \qquad (8.312)$$

$$[\widehat{a}(i),\,\widehat{a}^\dagger(j)]_{-\gamma} \;=\; \delta_{ij}, \qquad (8.313)$$

where in Eqs.(8.311) to (8.313) the indices -1 and $+1$ denote, respectively, the commutator and the anticommutator.

One can show that Eqs.(8.221) and (8.225) remain valid for the case of a system of fermions.

8.4.   The Green Function Method in Statistical Mechanics

In many ways the basic problems in field theory and in statistical mechanics are similar, especially if we formulate them in the second quantisation language. One is in both cases often concerned with averages of quantum mechanical operators, but whereas in quantum field theory one usually considers averages over the ground state of the system — which is often referred to as the zero-temperature limit[13] — in statistical mechanics one is interested in ensemble averages. Another point of similarity is that one in both cases is concerned with systems possessing a large number of degrees of freedom which makes the application of ordinary perturbation theory difficult because the energy spectrum is practically a continuous one. However, again in both cases, diagrammatic methods involving selected classes of diagrams often are a great help. In those techniques Green functions, the so-called *propagators*, are used and in the present section we shall consider a particular class of Green functions which have been very useful in statistical mechanics.

We recall that Green functions occur in the theory of potential problems. For instance, in electromagnetism when we want to find the electromagnetic field produced by given current and charge distributions we want to find the field $\psi(r,t)$ produced by a source distribution $f(r,t)$, and to do this we must solve field equations of the form $\widehat{\Omega}\psi = f$, where $\widehat{\Omega}$ is a given differential operator.

---

[13] As pointed out by Kirzhnits (*Field Theoretical Methods in Many-Body Systems*, Pergamon Press, Oxford (1967), p.323) one has to be careful since often the ground state is degenerate so that the limit of the statistical mechanics expression contains several terms, each with their appropriate weight, whereas the field theoretical expression contains only a single term.

In this case the Green function $G(r, t; r', t')$ is defined as the field at the point r at time t produced by a unit strength point source at the point r' at time t'; this means that G satisfies the equation

$$\widehat{\Omega}G(r, t; r', t') = \delta(r - r')\delta(t - t').\tag{8.401}$$

If as an example we take the operator $\partial/\partial t - \alpha\nabla^2$, this equation reduces to the Fokker-Planck equation

$$\frac{\partial G}{\partial t} - \alpha\nabla^2 G = \delta(r - r')\delta(t - t'),\tag{8.402}$$

and we know that we then can interpret G as the probability for a particle undergoing Brownian motion to be at r at time t, if it were at r' at time t': G describes the propagation of the particle!

We now introduce three kinds of Green functions, similar to the ones used in field theory: the retarded, advanced, and causal Green functions:

$$\ll \widehat{A}(t); \widehat{B}(t') \gg_r = \frac{1}{i\hbar}\theta(t - t')\langle[\widehat{A}(t), \widehat{B}(t')]_{-\gamma}\rangle,\tag{8.403}$$

$$\ll \widehat{A}(t); \widehat{B}(t') \gg_a = \frac{-1}{i\hbar}\theta(t' - t)\langle[\widehat{A}(t), \widehat{B}(t')]_{-\gamma}\rangle,\tag{8.404}$$

$$\ll \widehat{A}(t); \widehat{B}(t') \gg_c = \frac{1}{i\hbar}\left[\theta(t - t')\langle\widehat{A}(t)\widehat{B}(t')\rangle + \gamma\,\theta(t' - t)\langle\widehat{B}(t')\widehat{A}(t)\rangle\right],\tag{8.405}$$

where γ is a disposable parameter which we shall choose to be either $+1$ or -1 according to circumstances, where $[\cdot\cdot, \cdot\cdot]_{-1}$ and $[\cdot\cdot, \cdot\cdot]_{+1}$ denote commutators and anticommutators, where $\theta(y)$ is again the Heaviside function,

$$\left.\begin{array}{rll}\theta(y) &= 0, & y < 0, \\ &= 1, & y > 0,\end{array}\right\}\tag{8.406}$$

where for the time being we assume that we are dealing with a system containing only one kind of particles, and where the averages $\langle\cdots\rangle$ are averages either over a grand canonical ensemble,

$$\langle\cdots\rangle = \frac{\mathrm{Tr}\,e^{-\beta(\widehat{\mathcal{H}} - \mu\widehat{N})}\cdots}{Z},\tag{8.407}$$

with Z the grand partition function,

$$Z = \mathrm{Tr}\,e^{-\beta(\widehat{\mathcal{H}} - \mu\widehat{N})}\tag{8.408}$$

and $\widehat{N}$ the number operator, or over a macrocanonical (petit) ensemble,

$$\langle\cdots\rangle = \frac{\mathrm{Tr}\,e^{-\beta\widehat{\mathcal{H}}}\cdots}{Z},\tag{8.409}$$

where now Z is the canonical partition function,

$$Z = \text{Tr}\, e^{-\beta \widehat{\mathcal{H}}}. \tag{8.410}$$

The $\widehat{A}(t)$ and $\widehat{B}(t)$ operators in Eqs.(8.403) to (8.405) are time-dependent operators in the Heisenberg representation, which are related to the time-independent operators in the Schrödinger representation through the equations

$$\widehat{A}(t) = e^{i(\widehat{\mathcal{H}} - \mu \widehat{N})t/\hbar}\, \widehat{A}\, e^{-i(\widehat{\mathcal{H}} - \mu \widehat{N})t/\hbar} \tag{8.411}$$

in the grand canonical case, and

$$\widehat{A}(t) = e^{i\widehat{\mathcal{H}}t/\hbar}\, \widehat{A}\, e^{-i\widehat{\mathcal{H}}t/\hbar} \tag{8.412}$$

in the petit canonical case.

We note that the Green functions (8.403) to (8.405) are not defined for $t = t'$ and that the retarded Green function vanishes for $t < t'$ and the advanced one for $t > t'$.

We notice also that the Green functions are made up of linear combinations of the following *correlation functions*:

$$F_{AB} = \langle \widehat{A}(t)\, \widehat{B}(t') \rangle, \qquad F_{BA} = \langle \widehat{B}(t')\, \widehat{A}(t) \rangle. \tag{8.413}$$

which are, in fact, the quantities which are more directly connected with any physical quantities which we might like to have information about. Note the slightly asymmetric definition of these two functions and also that, in contrast to the Green functions, they are defined for $t = t'$.

Using Eqs.(8.407), (8.408), and (8.411) — or (8.409), (8.410), and (8.412) — one can prove that the Green functions (8.403) to (8.405), like the correlation functions (8.413), depend on t and t' only in the combination $t - t'$.

If we use the equation of motion,

$$i\hbar\, \dot{\widehat{A}} = [\widehat{A}, \widehat{\mathcal{H}} - \mu \widehat{N}]_{-}, \tag{8.414}$$

or

$$i\hbar\, \dot{\widehat{A}} = [\widehat{A}, \widehat{\mathcal{H}}]_{-}, \tag{8.415}$$

for the Heisenberg operators and the relation

$$\dot{\theta}(t - t') = -\dot{\theta}(t' - t) = \delta(t - t'), \tag{8.416}$$

we find the following equation of motion for each of the three Green functions:

$$i\hbar \ll \dot{\widehat{A}}; \widehat{B} \gg = \delta(t - t')\langle [\widehat{A}, \widehat{B}]_{-\gamma} \rangle + \ll [\widehat{A}, \widehat{\mathcal{H}} - \mu \widehat{N}]_{-}; \widehat{B} \gg, \tag{8.417}$$

or

$$i\hbar \ll \dot{\widehat{A}}; \widehat{B} \gg \, = \, \delta(t - t') \langle [\widehat{A}, \widehat{B}]_{-\gamma} \rangle + \ll [\widehat{A}, \widehat{\mathcal{H}}]_{-}; \widehat{B} \gg, \qquad (8.418)$$

where a dot indicates differentiation with respect to t.

We note that the right-hand sides of Eqs.(8.417) and (8.418) contain higher-order Green functions.

We shall now for a moment restrict ourselves to the grand canonical case; it is straightforward to change the formulæ to apply to the petit canonical case.

If we introduce the Fourier transform, $\ll \widehat{A}; \widehat{B} \gg = \ll \widehat{A}; \widehat{B} \gg_E$, of the Green functions through the equations

$$\ll \widehat{A}(t); \widehat{B}(t') \gg \, = \, \int_{-\infty}^{+\infty} \ll \widehat{A}; \widehat{B} \gg_E e^{-iE(t-t')/\hbar} \, dE, \qquad (8.419)$$

$$\ll \widehat{A}; \widehat{B} \gg_E \, = \, \frac{1}{2\pi} \int_{-\infty}^{+\infty} \ll \widehat{A}(t); \widehat{B}(t') \gg e^{iE(t-t')/\hbar} \, d(t - t'), \qquad (8.420)$$

we can write Eq.(8.417) in the form

$$E \ll \widehat{A}; \widehat{B} \gg \, = \, \frac{1}{2\pi} \langle [\widehat{A}, \widehat{B}]_{-\gamma} \rangle + \ll [\widehat{A}, \widehat{\mathcal{H}} - \mu \widehat{N}]_{-}; \widehat{B} \gg . \qquad (8.421)$$

We define the Fourier transform of F_{BA} as follows:

$$F_{BA}(t, t') \, = \, \int_{-\infty}^{+\infty} J(\omega) e^{-i\omega} (t - t') \, d\omega. \qquad (8.422)$$

From the definitions of $F_{AB}(t, t')$ and $F_{BA}(t, t')$ it then follows that

$$F_{AB}(t, t') \, = \, \int_{-\infty}^{+\infty} J(\omega) e^{\beta \hbar \omega} e^{-i\omega(t-t')} \, d\omega. \qquad (8.423)$$

In order to express the retarded and advanced Green functions in terms of $J(\omega)$ we write the Heaviside function (8.406) as follows:

$$\left. \begin{array}{llll} \theta(y) & = & e^{-\varepsilon y} \quad (\varepsilon \rightarrow 0+), & y < 0; \\ & = & 1, & y > 0. \end{array} \right\} \qquad (8.424)$$

We then get

$$\ll \widehat{A}; \widehat{B} \gg_{\mathrm{r,a}} \, = \, \frac{1}{2\pi} \int_{-\infty}^{+\infty} \left(e^{\beta \hbar \omega} - \gamma \right) J(\omega) \frac{d\omega}{E - \omega \pm i\varepsilon}, \qquad (8.425)$$

where the plus (minus) sign refers to the retarded (advanced) Green function.

For later use we need the function $G(E)$ defined by the equation

$$G(E) = \frac{1}{2\pi} \int_{-\infty}^{+\infty} \left(e^{\beta \hbar \omega} - \gamma \right) J(\omega) \frac{d\omega}{E - \omega}, \qquad (8.426)$$

which can be shown to be an analytic function of E equal to $\ll \hat{A}; \hat{B} \gg_r$ in the upper and equal to $\ll \hat{A}; \hat{B} \gg_a$ in the lower half-plane with singularities on the real axis.

From the definition of $G(E)$ and the relation

$$\lim_{\varepsilon \to 0+} \frac{1}{x \pm i\varepsilon} = \frac{\mathcal{P}}{x} \mp \pi i \delta(x), \qquad (8.427)$$

where $\mathcal{P}$ indicates that on integrating we must take the principal part of the integral, we find the following relation between $J(\omega)$ and $G(E)$:[14]

$$J(\omega) = \frac{-i}{e^{\beta \hbar \omega} - \gamma} \lim_{\varepsilon \to 0+} \left[G(\omega + i\varepsilon) - G(\omega - i\varepsilon) \right]. \qquad (8.428)$$

The first application we want to consider is to a perfect gas. If we write the Hamiltonian in terms of creation and annihilation operators we have (compare Eqs.(8.224) and (8.220))

$$\mathcal{H} = \sum_i (\varepsilon_i - \mu) \hat{a}_i^\dagger \hat{a}_i, \qquad (8.429)$$

where the $\hat{a}_i^\dagger$ and $\hat{a}_i$ satisfy the commutation relations (8.311) to (8.313). We now choose γ in Eq.(8.403) and (8.404) as in Eqs.(8.311) to (8.313), that is, according to Eq.(8.308), and we consider the Green functions with $\hat{A} = \hat{a}_i$ and $\hat{B} = \hat{a}_i^\dagger$. From Eqs.(8.421) and (8.429) we then get

$$E \ll \hat{a}_i; \hat{a}_i^\dagger \gg = \frac{1}{2\pi} + (\varepsilon_i - \mu) \ll \hat{a}_i; \hat{a}_i^\dagger \gg, \qquad (8.430)$$

or

$$\ll \hat{a}_i; \hat{a}_i^\dagger \gg = \frac{1}{2\pi(E - \varepsilon_i + \mu)}. \qquad (8.431)$$

If we now use Eqs.(8.428), (8.427), (8.422), and (8.431) we find

$$\langle \hat{a}_i^\dagger(t') \hat{a}_i(t) \rangle = \frac{e^{-(\varepsilon_i - \mu)(t - t')/\hbar}}{e^{\beta(\varepsilon_i - \mu)} - \gamma}, \qquad (8.432)$$

which for $t = t'$ leads to the usual Bose-Einstein and Fermi-Dirac expressions for the average occupation number $\langle n_i \rangle = \langle \hat{a}_i^\dagger \hat{a}_i \rangle$. The case of a

---

[14]We have dropped here a term $C\delta(\omega)$ from the right-hand side of Eq.(8.428) which may appear in the case when $\gamma = +1$.

perfect gas is a very simple one and, of course, we have derived the expression for the average occupation numbers by much more simple means in earlier chapters.

In the remainder of this section we shall use the petit rather than the grand canonical description. In that case we see that the poles of the single-particle Green function $\ll \hat{a}_i; \hat{a}_i^\dagger \gg$ correspond to the energy levels of the system. Provided one takes care[15]   one can, in general, conclude that the poles of $\ll \hat{a}_i; \hat{a}_i^\dagger \gg$ give us the single-(quasi)-particle energies, especially the low-lying ones.

As an application of the Green function method we shall consider the case of a spin-$\frac{1}{2}$ Heisenberg ferromagnet. The Hamiltonian of our system is given by the equation

$$\mathcal{H} = -g\mu_B B \sum_f \hat{S}_f^z - 2 \sum_{f,g} I(f-g)(\hat{S}_f \cdot \hat{S}_g), \qquad (8.433)$$

where g is the Landé factor which we shall put equal to 2, μ_B is the Bohr magneton, B the strength of the magnetic induction, applied along the z-axis, $\hat{S}$ the spin operator vector for the spin on site f which in our case has the components[16]

$$\hat{S}^x = \tfrac{1}{2}\begin{pmatrix} 0 & 1 \\ 1 & 0 \end{pmatrix}, \quad \hat{S}^y = \tfrac{1}{2}\begin{pmatrix} 0 & -i \\ i & 0 \end{pmatrix}, \quad \hat{S}^z = \tfrac{1}{2}\begin{pmatrix} 1 & 0 \\ 0 & -1 \end{pmatrix}, \quad (8.434)$$

as far as site f is concerned, while it is the unit operator for the other sites, and the $I(f-g)$ are exchange integrals which are functions of $f-g$ and which we shall assume to satisfy the equations

$$I(0) = 0, \qquad I(f-g) = I(g-f). \qquad (8.435)$$

For the spin-$\frac{1}{2}$ case it is convenient to introduce instead of the $\hat{S}_f$ new operators by the equations

$$\hat{b}_f = \hat{S}_f^+ = \hat{S}_f^x + i\hat{S}_f^y, \qquad \hat{b}_f^\dagger = \hat{S}_f^- = \hat{S}_f^x - i\hat{S}_f^y, \qquad (8.436)$$

or

$$\hat{b}_f = \begin{pmatrix} 0 & 1 \\ 0 & 0 \end{pmatrix}, \qquad \hat{b}_f^\dagger = \begin{pmatrix} 0 & 0 \\ 1 & 0 \end{pmatrix}. \qquad (8.437)$$

---

[15]See the discussion in § 2.2 of D.ter Haar, *Lectures on Selected Topics in Statistical Mechanics*, Pergamon, Oxford (1977) and in D.ter Haar and W.E.Parry, *Phys. Lett.* **1**, 145 (1962).

[16]We have taken the spin operators to be dimensionless, rather than a multiple of $\hbar$ to simplify the equations. If we had taken for the spin operators $\hbar$ times the operators used here, we would in the first term have had to write $g\mu_B B/\hbar$ instead of $g\mu_B B$ and in the second term the exchange integrals would have been smaller by a factor $\hbar^2$.

As $\widehat{S}_f^2 = \frac{3}{4}$, only two of the components of $\widehat{S}_f$ are independent. From Eqs.(8.436) and (8.437) we find

$$\widehat{S}^x = \tfrac{1}{2}(\widehat{b}_f^\dagger + \widehat{b}_f), \quad \widehat{S}^y = \tfrac{1}{2}\mathrm{i}(\widehat{b}_f^\dagger - \widehat{b}_f), \quad \widehat{S}^z = \tfrac{1}{2}(1 - 2\widehat{n}_f), \qquad (8.438)$$

where

$$\widehat{n}_f = \widehat{b}_f^\dagger \widehat{b}_f = \begin{pmatrix} 0 & 0 \\ 0 & 1 \end{pmatrix}. \qquad (8.439)$$

The Hamiltonian in terms of the $\widehat{b}_f^\dagger$ and $\widehat{b}_f$ has the following form:

$$\widehat{\mathcal{H}} = -\mu_{\mathrm{B}} N B + 2\mu_{\mathrm{B}} B \sum_f \widehat{n}_f - 2 \sum_{f,g} I(f-g)\widehat{b}_f^\dagger \widehat{b}_g - \tfrac{1}{2} N \sum_f I(f)$$

$$+ 2\left(\sum_f I(f)\right) \sum_g \widehat{n}_g - 2 \sum_{f,g} I(f-g)\widehat{n}_f \widehat{n}_g, \qquad (8.441)$$

where N is the number of spins in the lattice.

We shall be interested in the Green function $\ll \widehat{b}_g; \widehat{b}_f^\dagger \gg$ so that we shall need the commutation relations satisfied by the $\widehat{b}_f^\dagger$ and $\widehat{b}_f$. These are

$$\left.\begin{array}{rcl} [\widehat{b}_f, \widehat{b}_g]_- = [\widehat{b}_f^\dagger, \widehat{b}_g^\dagger]_- = \widehat{b}_f^2 = \widehat{b}_f^{\dagger 2} = 0, \\[2mm] [\widehat{b}_f, \widehat{b}_g^\dagger]_- = [1 - 2\widehat{n}_f]\,\delta_{fg}. \end{array}\right\} \qquad (8.442)$$

From Eq.(8.421) with $\gamma = +1$ we now get

$$E \ll \widehat{b}_g; \widehat{b}_f^\dagger \gg = \frac{1}{2\pi}\delta_{fg}(1 - 2\langle\widehat{n}_f\rangle) + 2\left[\mu_{\mathrm{B}} B + \sum_f I(f)\right] \ll \widehat{b}_g; \widehat{b}_f^\dagger \gg$$

$$-2\sum_p I(p-g) \ll \widehat{b}_p; \widehat{b}_f^\dagger \gg$$

$$+4\sum_p I(p-g)\left\{\ll \widehat{n}_g \widehat{b}_p; \widehat{b}_f^\dagger \gg - \ll \widehat{n}_p \widehat{b}_g; \widehat{b}_f^\dagger \gg\right\}. \qquad (8.442)$$

We note the appearance of Green functions involving four rather than two $\widehat{b}_f^\dagger$ or $\widehat{b}_f$. To find equations for those we should use once again Eq.(8.421), but the resultant equations involve Green functions with six $\widehat{b}_f^\dagger$ or $\widehat{b}_f$, and so on. Sooner or later we must, therefore, in some way or other close the chain of equations. Since the $\widehat{n}_f$ and $\widehat{b}_f$ in the higher-order Green functions refer to different sites, as they are multiplied by an exchange integral for which Eq.(8.435) holds, we shall assume these operators to be uncorrelated to a first approximation, that is, we put

$$\ll \widehat{n}_f \widehat{b}_g; \widehat{b}_h^\dagger \gg \approx \langle\widehat{n}_f\rangle \ll \widehat{b}_g; \widehat{b}_h^\dagger \gg. \qquad (8.443)$$

The Hamiltonian (8.440) is translationally symmetric so that $\langle \hat{n}_f \rangle$ will be independent of f; we shall write

$$\langle \hat{n}_f \rangle = \bar{n}. \tag{8.444}$$

Using Eqs.(8.443) and (8.444) we can rewrite Eq.(8.442) as follows:

$$E \ll \hat{b}_g; \hat{b}_f^\dagger \gg = \frac{1 - 2\bar{n}}{2\pi} \delta_{fg} + \left[2\mu_B B + 2(1 - 2\bar{n}) \sum_f I(f) \right] \ll \hat{b}_g; \hat{b}_f^\dagger \gg$$
$$- 2(1 - 2\bar{n}) \sum_p I(p - g) \ll \hat{b}_p; \hat{b}_f^\dagger \gg . \tag{8.445}$$

To solve this equation we use once more the translational invariance and Fourier transform with respect to the lattice sites:

$$\ll \hat{b}_g; \hat{b}_f^\dagger \gg = \frac{1}{N} \sum_q e^{i([g - f] \cdot q)} G_q. \tag{8.446}$$

Using the fact that

$$\delta_{fg} = \frac{1}{N} \sum_q e^{i([g - f] \cdot q)}, \tag{8.447}$$

where the sum over q here and in Eq.(8.446) is over the first Brillouin zone, we get from Eq.(8.445)

$$G_q \left\{ E - 2\mu_B b - 2(1 - 2\bar{n}) \left[K(0) - K(q) \right] \right\} = \frac{1 - 2\bar{n}}{2\pi}, \tag{8.448}$$

with

$$K(q) = \sum_f I(f) e^{i(f \cdot q)}, \tag{8.449}$$

or

$$G_q = \frac{1 - 2\bar{n}}{2\pi (E - E_q)}, \tag{8.450}$$

with

$$E_q = 2\mu_B B + 2 \left[K(0) - K(q) \right]. \tag{8.451}$$

Bearing in mind what we said earlier about the poles of the single-particle Green functions, we may hope that for not too large values of q the E_q give us the energies of the low-lying elementary excitations. In fact, for small q they give us the spin-wave energies.

From Eqs.(8.450), (8.446), (8.428), (8.427), and (8.422) we now find

$$\langle \hat{b}_f^\dagger(t') \hat{b}_g(t) \rangle = \frac{1 - 2\bar{n}}{N} \sum_q \frac{e^{i([g - f] \cdot q) - i E_q (t - t')}}{e^{\beta E_q} - 1}. \tag{8.452}$$

Putting $t = t'$ and $g = f$, using Eqs.(8.439) and (8.444), changing from a summation over q to an integration,

$$\frac{1}{N} \sum_q \cdots \rightarrow \frac{v}{(2\pi)^3} \int d^3q \cdots, \tag{8.453}$$

where $v = V/N$, we get the following implicit equation for $\bar{n}$:

$$\frac{\bar{n}}{1 - 2\bar{n}} = \frac{v}{(2\pi)^3} \frac{d^3q}{e^{\beta E_q} - 1}. \tag{8.454}$$

The physical quantity usually measured is the magnetisation which is proportional to $\langle S^z \rangle$, and thus to $1 - 2\bar{n} \equiv \sigma$. Using the fact that the number of q-values in the first Brillouin zone is equal to N so that

$$1 = \frac{v}{(2\pi)^3} \int d^3q, \tag{8.455}$$

we get from Eq.(8.454)

$$\frac{1}{\sigma} = \frac{v}{(2\pi)^3} \int \coth \tfrac{1}{2}\beta E_q \, d^3q. \tag{8.456}$$

Equation (8.456) can be used to find the Curie temperature, as we shall do in a moment, or low-temperature and high-temperature expansions of the magnetisation (see Problems 11 and 12 at the end of this chapter).

To find out where the Curie point $T_c = 1/k\beta_c$ occurs in the approximation used to obtain Eq.(8.456) we must consider the case where there is no external magnetic field. Near, but below, T_c the quantity σ, and thus E_q, will be small. We can therefore expand the hyperbolic cotangent in Eq.(8.456) in a power series. If we define $\eta(q)$ by the relation

$$\eta(q) = 1 - \frac{K(q)}{K(0)}, \tag{8.457}$$

and the functions $F(n)$ by

$$F(n) = \frac{v}{(2\pi)^3} \int \eta^n(q) \, d^3q, \tag{8.458}$$

we find

$$\begin{aligned}
\frac{1}{\sigma} &= \frac{v}{(2\pi)^3} \int \left\{ \frac{2}{\sigma\beta K(0)\eta(q)} + \frac{\sigma\beta K(0)\eta(q)}{6} + \cdots \right\} \\
&= \frac{2F(-1)}{\sigma\beta K(0)} + \frac{\sigma\beta K(0)F(1)}{6} + \cdots .
\end{aligned} \tag{8.459}$$

By letting σ tend to zero, we find

$$\beta_c = \frac{2F(-1)}{K(0)}. \tag{8.460}$$

This result can be compared with the results from other models. If one uses the molecular field model[17] one finds

$$\beta_c = \frac{2}{K(0)}, \tag{8.461}$$

while Eq.(8.460) is exactly the same as the result obtained from the so-called spherical model.[18]

The functions $K(q)$, and thus $\eta(q)$, depend, of course, on the exchange integrals and, like the $F(n)$, on the crystal structure. For instance, in the case of nearest-neighbour interactions only with

$$\left.\begin{aligned}I(\boldsymbol{f}-\boldsymbol{g}) &= I, \quad \text{if } \boldsymbol{f} \text{ and } \boldsymbol{g} \text{ are nearest neighbours;} \\ &= 0, \quad \text{otherwise,}\end{aligned}\right\} \tag{8.462}$$

we have for the simple (sc), body-centred (bcc), and face-centred (fcc) cubic lattices:

$$\left.\begin{aligned}K(q) &= \tfrac{1}{3} K(0) \left[\cos q_x a + \cos q_y a + \cos q_z a\right], \quad \text{(sc)} \\ &= K(0) \cos q_x a \, \cos q_y a \, \cos q_z a, \quad \text{(bcc)} \\ &= \tfrac{1}{3} K(0) \left[\cos \tfrac{1}{2} q_x a \, \cos \tfrac{1}{2} q_y a + \cos \tfrac{1}{2} q_y a \, \cos \tfrac{1}{2} q_z a \right. \\ &\quad \left. + \cos \tfrac{1}{2} q_z a \, \cos \tfrac{1}{2} q_x a\right], \quad \text{(fcc)}\end{aligned}\right\} \tag{8.463}$$

where $a\sqrt{6/z}$ is the nearest neighbour distance with z the coordination number (the number of nearest neighbours; $z = 6$ (sc); $z = 8$ (bcc); $z = 12$ (fcc).

For the $F(n)$ we have

$$\left.\begin{aligned}&F(-1) = 1.56 \text{ (sc)}, = 1.393 \text{ (bcc)}, = 1.345 \text{ (fcc)}; \\ &F(1) = 1; \quad F(2) = \frac{z+1}{z}, \quad F(3) = \frac{z+3}{z}, \quad \cdots .\end{aligned}\right\} \tag{8.464}$$

For temperatures just below the Curie point, $\beta - \beta_c \ll \beta_c$, we get the following temperature dependence for σ:

$$\sigma = \sqrt{\frac{6}{F(-1)}} \sqrt{1 - \frac{T}{T_c}}. \tag{8.465}$$

This result is the same as the one following from the molecular field theory, but not in agreement with experimental data.

---

[17]For a discussion of various models of ferromagnetism we can, for instance, refer to R.A.Tahir-Kheli and D.ter Haar, *Phys. Rev.* **127**, 88 (1962); see also § 9.6.

[18]M.Lax, *Phys. Rev.* **97**, 629 (1955).

Problems

1. Prove that the transformation (8.107) leads to the Hamiltonian (8.109).
2. Prove Eqs.(8.210), (8.214), and (8.218).
3. Prove Eqs.(8.221) and (8.225) for the boson case.
4. Fill in the details of the calculations which lead from the Hamiltonian (8.225) to the quasi-particle energy spectrum (8.231).
5. Prove Eqs.(8.304) to (8.306).
6. Prove Eqs.(8.221) and (8.225) for the fermion case.
7. Use Eqs.(8.407) and (8.411) and the cyclic properties of the trace to prove that the double-time Green functions (8.403) to (8.405) depend on the two times only in the combination $t - t'$.
8. Prove that in the grand canonical case the Fourier transform $J(\omega)$ of $F_{BA}(t, t')$ satisfies the equation

$$J(\omega) = \frac{1}{Z} \sum_{\lambda, \nu} \langle \lambda | \widehat{B} | \nu \rangle \langle \nu | \widehat{A} | \lambda \rangle \, e^{-\beta(E_\lambda - \mu N_\lambda)} \, \delta(\omega - E_\lambda + E_\nu + \mu N_\lambda - \mu N_\nu),$$

(A)

where we are using a representation in which $\widehat{\mathcal{H}} - \mu \widehat{N}$ is diagonal,

$$\langle \lambda | \widehat{\mathcal{H}} - \mu \widehat{N} | \nu \rangle = \delta_{\lambda \nu} (E_\lambda - \mu N_\lambda).$$

and where Z is given by Eq.(8.408).

9. Use Eq.(A) from the previous problem to prove Eq.(8.423)
10. Prove that the Kubo formula from Problem 8 of Chapter 6 can be written in the form

$$\langle \widehat{G}(t) \rangle = \langle \widehat{G} \rangle_0 + 2\pi \, e^{i\omega t + \varepsilon t} \ll \widehat{G}; \widehat{V} \gg_{r, \gamma = +1},$$

where the perturbing Hamiltonian is $\widehat{V} \, e^{i\omega t + \varepsilon t} \ (\varepsilon \to 0+)$.

11. Consider the spin-$\frac{1}{2}$ Heisenberg ferromagnet at low temperatures with $B = 0$ and with the exchange integrals satisfying Eq.(8.462). By expanding $\eta(q)$ in powers of q find an expansion of the right-hand side of Eq.(8.454) in powers of β^{-1}, and hence an expansion for σ in powers of T.

12. Consider the spin-$\frac{1}{2}$ Heisenberg ferromagnet at high temperatures, now with $B \neq 0$, and with the exchange integrals satisfying Eq.(8.462). By writing

$$\coth \frac{1}{2} \beta E_q = \frac{1 + t_0 t_1}{t_0 + t_1},$$

where

$$t_0 = \tanh \beta \mu_B B; \qquad t_1 = \tanh \beta \sigma K(0) \eta(q),$$

expand $\coth \frac{1}{2}\beta E_q$ in powers of t_1. By then expanding t_1 in powers of β find an expansion of the susceptibility χ in powers of β; show that for $\beta \ll \beta_c$ it can be written approximately in the Curie-Weiss form

$$\chi = \frac{\mu_B^2}{k(T - \Theta)},$$

and determine Θ.

13. When studying ferromagnetic resonance we must take into account the finite shape of the sample on which resonance experiments are performed and add a term $\frac{1}{2}(N_x \widehat{M}_x^2 + N_y \widehat{M}_y^2 + N_z \widehat{M}_z^2)$ to the Hamiltonian, where $\widehat{M}$ is the magnetisation vector:

$$\widehat{M} = 2\mu_B \sum_f \widehat{S}_{vf}.$$

In ferromagnetic resonance experiments there will be a radio-frequency field B_1 with components $b \cos \omega t$, $b \sin \omega t$, 0 at right angles to the constant field B along the z-axis. The perturbing Hamiltonian $\widehat{\mathcal{H}}'$ will be of the form

$$\widehat{\mathcal{H}}' = -2\mu_B \sum_f (S_f \cdot B_1) = \widehat{V} e^{i\omega t} + \widehat{V}^* e^{-i\omega t},$$

where

$$\widehat{V} = -n\mu_B \sum_f \widehat{S}_f^-.$$

If $\delta \widehat{M}$ is the additional magnetisation in the xy-plane produced by the radio-frequency field we can write

$$\delta \widehat{M}_\pm = \delta(\widehat{M}_x \pm i\widehat{M}_y) = N\chi_\pm b\, e^{\pm i\omega t},$$

and using the Kubo formula from Problem 10 we find for the susceptibilities χ_+ and χ_- the equations

$$\chi_-^* = \chi_+ = -4\pi\mu_B^2 \sum_f \ll \widehat{S}_g^+; \widehat{S}_f^- \gg .$$

Writing

$$\chi_\pm = \chi' \pm \chi'',$$

and using the fact that the energy, W, absorbed per unit time is given by the equation

$$W = (B_1 \cdot \dot{\delta M}) = -\tfrac{1}{2}iNb^2\omega(\chi_+ - \chi_-) = Nb^2\omega\chi'',$$

it follows that χ'' determines the energy absorbed in ferromagnetic resonance experiments. It also follows that to determine χ'' we must find $\ll \hat{b}_g; \hat{b}_f^\dagger \gg$. If we apply Eq.(8.421) to this Green function, we see that we need to find also the Green function $\ll \hat{b}_g^\dagger; \hat{b}_f^\dagger \gg$. Write down the equations of motion for these two Green functions and use the decoupling approximation (8.443) as well as

$$\ll \hat{n}_f \hat{b}_g^\dagger; \hat{b}_h^\dagger \gg \approx \langle \hat{n}_f \rangle \ll \hat{b}_g^\dagger; \hat{b}_h^\dagger \gg$$

to determine χ''.

14. Consider a antiferromagnet with two sublattices such that each lattice site of the one sublattice is a nearest neighbour of lattice sites of the other sublattice and with interactions only between nearest neighbours. If we denote the sites of the one sublattice by f and those of the other sublattice by g, the Hamiltonian of this system will be of the form

$$\hat{\mathcal{H}} = \sum_{f,g} I\left(\hat{S}_f \cdot \hat{S}_g\right) - 2\mu_{\mathrm{B}}B\left[\sum_f \hat{S}_f^z + \sum_g \hat{S}_g^z\right],$$

where we have used Eq.(8.462) for the exchange integrals, except that we have taken the nearest-neighbour exchange integrals to be equal to $-I$ rather than I to take into account that we are dealing with an antiferromagnet.

Consider the Green functions $\ll \hat{b}_f; \hat{b}_h^\dagger \gg$ and $\ll \hat{b}_g; \hat{b}_h^\dagger \gg$ where h may refer either to the f or to the g sublattice. In solving the equations of motion for these two Green functions and using decouplings similar to Eq.(8.443), bear in mind that $\langle \hat{n}_f \rangle$ will not necessarily be equal to $\langle \hat{n}_g \rangle$. Find the sublattice magnetisations and the Néel temperature, assuming that in the absence of a magnetic field the magnetisations of the two sublattices will be equal and opposite

Bibliographical Notes

Section 8.1. For a general discussion of many-body systems see, for instance,

1. D.ter Haar, *Introduction to the Physics of Many-Body Systems*, Interscience, New York (1958).
2. D.ter Haar, *Contemporary Phys.* 1, 112 (1959).
3. R.D.Mattuck, *A Guide to Feynman Diagrams in the Many-Body Problem*, McGraw-Hill, London (1967), Chapter 1.
4. W.E.Parry, *The Many-Body Problem*, Clarendon Press, Oxford (1973), Chapter 1.

For a discussion of the problem of finding suitable transformations to reduce the Hamiltonian to a simpler form see Ref.1 and also:

5. D.Bohm and D.Pines, *Phys. Rev.* **92**, 609 (1953).
6. S.Tomonaga, *Progr. Theor. Phys.*, **13**, 467, 482 (1955).
7. G.J.Yevick and J.K.Percus, *Phys. Rev.* **101**, 1186 (1956).
8. J.K.Percus and G.J.Yevick, *Phys. Rev.* **101**, 1192 (1956).
9. D.ter Haar, *Repts. Progr. Phys.* **20**, 130 (1957).

Sections 8.2 and 8.3. For a discussion of the occupation number representation see, for instance, Ref.3, Chapter 7, and also

10. A.A.Abrikosov, L.P.Gor'kov, and I.Ye.Dzyaloshinskii, *Quantum Field Theoretical Methods in Statistical Physics*, Pergamon Press, Oxford (1965), § 3.
11. R.Balescu, *Equilibrium and Nonequilibrium Statistical Mechanics*, John Wiley, New York (1977), § 1.5
12. D.ter Haar, *Lectures on Selected Topics in Statistical Mechanics*, Pergamon Press, Oxford (1977), Chapter 1.
13. J.C.Inkson, *Many-Body Theory of Solids*, Plenum Press, New York (1984), Chapter 3.
14. R.Balian, *From Microphysics to Macrophysics*, Springer, Berlin (1992), § 10.2.

Section 8.4. Bogolyubov and Tyablikov were the first to use the equation of motion method:

15. N.N.Bogolyubov and S.V.Tyablikov, *Sov. Phys. Doklady* **4**, 604 (1959).

For a general discussion of temperature-dependent double-time Green functions see

16. D.N.Zubarev, *Sov. Phys. Uspekhi* **3**, 920 (1960).
17. V.L.Bonch-Bruevich and S.V.Tyablikov, *The Green Function Method in Statistical Mechanics*, North-Holland, Amsterdam (1962).
18. D.ter Haar, in *Fluctuations, Relaxation and Resonance in Magnetic Systems*, Oliver and Boyd, Edinburgh (1962), p.119.
  and also Ref.4 and Ref.12, Chapter 2.

For the Green function theory of the Heisenberg ferromagnet see Refs.15, 12, 18, and

19. R.A.Tahir-Kheli and D.ter Haar, *Phys. Rev.* **127**, 88 (1962).
20. R.A.Tahir-Kheli and D.ter Haar, *Phys. Rev.* **127**, 95 (1962).
21. A.C.Hewson and D.ter Haar, *Physica* **30**, 271 (1964).

For the Green function theory of the ferromagnetic resonance see Refs.12, 18, and

22. S.V.Tyablikov, *Sov. Phys. Solid State* **2**, 332, 1805 (1960).
23. S.V.Tyablikov and Fu-Cho, *Sov. Phys. Solid State* **3**, 102 (1961).

For the Green function theory of antiferromagnetism see Ref.12 and

24. Fu-Cho, *Sov. Phys. Solid State* **5**, 128, 321 (1962).
25. M.E.Lines, *Phys. Rev.* **131**, 540 (1963).
26. M.E.Lines, *Phys. Rev.* **133**, A841 (1964).
27. A.C.Hewson and D.ter Haar, *Physica* **30**, 890 (1964).
28. A.C.Hewson, D.ter Haar, and M.E.Lines, *Phys. Rev.* **137**, A1465 (1965).

CHAPTER 9

PHASE TRANSITIONS

9.1.   Introduction

Many physical systems show *phase transitions*; these are characterised by singularities in one of the thermodynamic potentials, such as the free energy, and hence by discontinuities in their derivatives. Ehrenfest[1] classified phase transitions according to whether the first, second, ... derivative of the free energy showed a discontinuity, and called them, respectively, first-, second-, third-, ... order transitions. As the first derivative of the relevant thermodynamic potential is discontinuous in the case of a first-order phase transition, such a phase transition will usually be accompanied by a *latent heat*. Nowadays it is realised that not only discontinuities, but also divergences at the phase transition point are important, and all higher-order phase transitions are grouped together as *critical* or *continuous* phase transitions.[2]

Typical first-order phase transitions are melting and condensation. In a typical *phase diagram* (Fig.9.1) we see regions corresponding to the solid phase (I), the liquid phase (II), and the vapour (or gas) phase (III). The sublimation curve is the line separating regions I and III, the melting curve the line separating regions I and II, and the evaporation or condensation curve, the line separating regions II and III. These three lines meet in the triple point. The evaporation line ends in the critical point; beyond that point one can move continuously, without going through a phase transition, from the vapour to the liquid. Whenever one passes through one of the phase separation lines, except in the critical point, a first-order phase transition occurs. In § 7.2 we have already discussed some aspects of the gas-liquid diagram when we dealt with the van der Waals equation of state. In the next three sections we shall consider the same subject from a slightly more general point of view, first considering a simplified model introduced by Wergeland in which the gas is considered to consist of an assembly of droplets of different sizes, then briefly considering Mayer's original approach, and fi-

---

[1] P.Ehrenfest, *Proc. Kon. Ned. Akad. Wet.* **36**, 153 (1933).

[2] The term "continuous" phase transitions was introduced by Landau (*Phys. Zs. Soviet Un.* **11**, 26 (1937)) in his classic paper on the theory of phase transitions.

nally, discussing some general ideas due to Yang and Lee. Unfortunately, it seems as if we are still far from having a complete theory of condensation or of melting. The situation is more satisfactory when we come to continuous phase transitions. There are several models which can be treated either completely rigorously or with rather satisfactory approximate methods. We shall turn to those in §§ 9.5 to 9.9. Among the phase transitions which can be treated by those methods we mention ferromagnetism and antiferromagnetism, order-disorder transitions in solids — which can often be described by the Ising model of ferromagnetism — ferroelectricity, phase separations, and structural phase transitions.

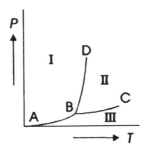

Fig.9.1. In this phase diagram C is the critical point, AB the sublimation line, BD the melting line, BC, ending in the point C, the condensation line, I is the solid region, II the liquid region, and III the vapour or gas region.

Let us consider for a moment the general problem and let us assume for the time being that the system consists of only one kind of particles. The generalisation to systems containing several kinds of particles is straightforward. Let us, as in § 7.1 assume that the forces between the particles are binary, central forces which can be derived from a potential energy function $\phi(r)$ where r is the distance apart of the two interacting particles. All equilibrium properties of the system can be derived from the grand potential q which for classical systems is given by Eq.(7.101),

$$e^q = \sum_\nu \frac{e^{\alpha\nu}}{\nu!} \int_\Gamma e^{-\beta\varepsilon}\, d\Omega, \qquad (9.101)$$

where we shall, as in § 7.1, use for the element of extension in phase, $d\Omega$, the formula

$$d\Omega = \prod_{i=1}^{\nu} \frac{d^3p_i\, d^3r_i}{h^3}. \qquad (9.102)$$

The first question which arises is how from an equation in which all functions occurring in it are continuous one can arrive at isotherms, such as the van der Waals isotherms, which have discontinuities. The answer, first given by Kramers in his discussion of the Heisenberg ferromagnet,[3] is that, in fact, there are no discontinuities as long as the number of particles, N, in the system is finite, but that such discontinuities appear in the limit as $N \rightarrow \infty$. Actually, one must take at the same time the limit as the volume V goes to infinity, in such a way that the ratio N/V, that is, the number density, stays fixed. This limit is called the *thermodynamic limit*. It has been shown by van Hove[4] that in the thermodynamic limit q/N becomes a function of the number density $n(= N/V)$, or the specific volume $v(= 1/n)$ and the temperature T and that the pressure P, which is given by Eq.(5.0931),

$$P = kT \left(\frac{\partial(q/N)}{\partial v} \right)_T, \tag{9.103}$$

is a strictly non-negative quantity so that the slope $(\partial P/\partial v)_T$ of the isotherms is *never* positive, although it can be zero corresponding, as we saw in Chapter 7, to the coexistence of two or more phases.[5]

Of course, if the slope is zero for a range of v-values this implies that there are discontinuities in the second derivative of P. The exact vanishing of the slope and the appearance of the discontinuities appear only in the thermodynamic limit and in Problem 2 at the end of the present chapter it is shown that in the case of the Einstein condensation the discontinuities and the exact vanishing of the slope of the line of coexisting phases only occur in this limit.

Apart from condensation where we are looking at the pressure P as a function of the volume V, much attention in the literature on phase transitions is paid to magnetic systems where the external parameter is the magnetic field B while the magnetisation M plays the role of the pressure. Considering magnetic systems with a constant volume, we have a complete one-to-one correspondence between the two cases. If we use macrocanonical petit ensembles Eq.(5.0314) and (5.0317) are replaced by

$$M = -\frac{\partial \psi}{\partial B}, \tag{9.104}$$

and

$$M = \frac{1}{\beta} \frac{\partial \ln Z_\Gamma}{\partial B}, \tag{9.105}$$

---

[3] H.A.Kramers, *Commun. Kamerlingh Onnes Lab.* **22**, Suppl.22, 1 (1936); *Collected Scientific Papers*, North-Holland, Amsterdam (1956), p.607. On page 20 of this paper Kramers notes that, in fact, in the limit as the number of particles goes to infinity a certain quantity which determines the free energy consists of two analytically different parts. The first explicit statement of the consequences of such behaviour for isotherms can be found in Kahn's Utrecht thesis in 1938.

[4] L.van Hove, *Physica* **15**, 951 (1949).

[5] It is easy to prove that $(\partial P/\partial v)_T$ is always negative for finite V: see Problem 1 at the end of the present chapter.

whereas if we use canonical grand ensembles Eq.(5.0931) is replaced by

$$M = kT \frac{\partial q}{\partial B}.$$

9.2. The Liquid Drop Model of Condensation

In 1935 Becker and Döring discussed condensation from a thermodynamic and kinetic point of view. The main points of their theory are the following. As long as for a given temperature the pressure is less than the saturated vapour pressure most of the gas will consist of single molecules and only a small fraction will be present as small droplets. The number of droplets of a given size decreases extremely rapidly with increasing size. The reason for this is that the smaller the droplet the larger the relative influence of the surface tension. As long as the presure is below the saturation pressure the tendency of all drops will be to evaporate rather than to grow, and the larger-size drops are present only due to fluctuations. However, as soon as the pressure exceeds the saturation pressure there will be a tendency for most drops to grow, and thus condensation will set in.

Several authors have proposed statistical counterparts of Becker and Döring's theory. In the present section we shall consider a slightly altered version of a theory by Wergeland, whose main idea was to treat the system as consisting of non-interacting droplets. In our treatment we shall use grand ensemble theory rather than the petit ensembles used by Wergeland.

The grand potential q of our system of non-interacting droplets will satisfy the equation

$$q = \sum_l q_l, \tag{9.201}$$

where q_l is the grand potential of the subsystem consisting of only those drops which contain l atoms (or molecules). As we are considering the drops to be non-interacting we are dealing in each subsystems with a perfect Boltzmann gas and we can use Eq.(6.0606) for the q_l:

$$q_l = e^{\alpha_l - \beta f_l}, \tag{9.202}$$

where f_l is the (free) energy of a single drop and α_l/β the thermal potential per drop, all corresponding to drops containing l atoms. Since a free exchange of atoms between the drops must be possible, we have the equilibrium condition

$$\alpha_l = l\alpha. \tag{9.203}$$

To calculate f_l we represent the subsystem by a macrocanonical ensemble and write (compare Eq.(5.0204))

$$e^{-\beta f_l} = \frac{1}{l!} \int_\Gamma e^{-\beta \varepsilon} \, d\Omega. \tag{9.204}$$

With Wergeland we assume that each drop containing l atoms occupies a volume v_l in which the l atoms move independently in a smoothed-out constant negative potential, $-\chi_l$, where χ_l and v_l are the following functions of l:

$$\chi_l = \frac{l-1}{l}\chi \qquad v_l = v_1 l^{\frac{l-2}{l-1}}, \tag{9.205}$$

which guarantee that χ_l is nearly constant and v_l nearly proportional to l for large values of l, as should be the case.

Integration over the momenta in Eq.(9.204) gives a factor v_0^{-l} where v_0 is again given by the Eq.(4.736),

$$v_0 = \left(\frac{\beta h^2}{2\pi m}\right)^{3/2}. \tag{9.206}$$

Using the fact that each drop has the total volume V at its disposal, but the atoms in the drop only the volume v_l, while each atom contributes a factor $e^{\beta \chi_l}$ we find

$$e^{-\beta f_l} = \frac{v_0^{-l}}{l!} e^{l\beta \chi_l} \int_V d^3 r_1 \prod_{i=2}^{l} \int_{v_l} d^3 r_i$$

$$= \frac{V}{v_1} \frac{e^{-\beta \chi}}{l!} \left(\frac{v_1}{v_0} e^{\beta \chi}\right)^l l^{l-2}; \tag{9.207}$$

in getting the last equality we have used Eqs.(9.205). Hence we finally get for q:

$$q = \frac{V}{v_1} e^{-\beta \chi} \sum_l \frac{l^{l-2}}{l!} z^l, \tag{9.208}$$

with

$$z = \frac{v_1}{v_0} e^{\alpha + \beta \chi}. \tag{9.209}$$

For the pressure, P, and the average number of atoms, N, we now get the equations

$$P = \frac{kTq}{V} = \frac{kT}{v_1} e^{-\beta \chi} \sum_l \frac{l^{l-2}}{l!} z^l, \tag{9.210}$$

$$N = \frac{\partial q}{\partial \alpha} = \frac{V}{v_1} e^{-\beta \chi} \sum_l \frac{l^{l-1}}{l!} z^l. \tag{9.211}$$

Equations (9.210) and (9.211) together gives us the equation of state in parameter form. Problem 3 at the end of this chapter shows how by eliminating z one obtains the equation

$$P = \frac{NkT}{V} \left(1 - \frac{N v_1}{2V} e^{\beta \chi}\right). \tag{9.212}$$

Introducing again the specific volume $v(= V/N)$, we have from Eq.(9.211)

$$\frac{1}{v} = \frac{e^{-\beta\chi}}{v_1} \sum_l \frac{l^{l-1}}{l!} z^l. \tag{9.213}$$

The series on the right-hand side of this equation converges for $z \leqslant 1/e$, but $z = e^{-1}$ is a singularity of the function represented by the series. The same point is also a singularity of the series on the right-hand side of Eq.(9.210), and we see that the equation of state given by Eq.(9.212) will hold as long as $v < v_c$ where v_c is given by the equation[6]

$$\frac{1}{v_c} = \frac{e^{-\beta\chi}}{v_1}. \tag{9.214}$$

As soon as $v < v_c$, however, one must proceed with more care. One can show that the pressure is constant for $v < v_c$:

$$P = P_c = \frac{kT}{2v_1} e^{-\beta\chi}, \quad v < v_c. \tag{9.215}$$

In fact, Eq.(9.215) holds exactly only in the limit as $N \to \infty$; for finite N there are corrections of the order of $1/N$.[7]

To prove Eq.(9.215) we note that Eqs.(9.210) and (9.211) are of the form

$$q = Vf(z), \tag{9.216}$$

$$N = Vzf'(z), \quad f'(z) = \frac{df}{dz}. \tag{9.217}$$

It is convenient to introduce again the specific volume v and a function $g(z)$,

$$g(z) = vf(z), \tag{9.218}$$

so that Eqs.(9.216) and (9.217) take the form

$$q = Ng(z), \tag{9.219}$$

$$1 = zg'(z). \tag{9.220}$$

The functions $g(z)$ and $g'(z)$ possess the following properties: (i) they have a singularity on the real axis at $z = z_0(= e^{-1})$; (ii) $zg'(z)$ is monotonically increasing on the real axis between $z = 0$ and $z = z_0$; (iii) they are both finite at $z = z_0$, and the singularity at that point is a branch point, so that we can write

$$g(z) = g(z_0) + \sum_{n=1}^{\infty} a_n(z - z_0)^n + (z - z_0)^\alpha \sum_{n=1}^{\infty} b_n(z - z_0)^n, \tag{9.221}$$

---

[6] From the definition of the parameter u_0 in Problem 3 at the end of this chapter it follows that $u_0 = 1$ for $z = e^{-1}$ and hence that the series on the right-hand side of Eq.(9.213) is equal to 1 in that point.

[7] Strictly speaking, the corrections are of order $\ln N/N$; see Eq.(9.228).

$$zg'(z) = z_0 g'(z_0) + \sum_{n=1}^{\infty} c_n (z - z_0)^n + (z - z_0)^{\alpha-1} \sum_{n=1}^{\infty} d_n (z - z_0)^n, \quad (9.222)$$

where α is positive and non-integral.

We now follow a trick suggested by Kramers[8] and write

$$e^q = \frac{1}{2\pi i} \oint e^{N g(z)} d\ln\left(z g'(z) - 1\right), \quad (9.223)$$

where the path of integration excludes z_0, but includes the zeroes of Eq.(9.220).

As Eq.(9.214) can be written in the form

$$\frac{v}{v_c} = z_0 g'(z_0), \quad (9.224)$$

the cases $v > v_c$ and $v < v_c$ correspond to $z_0 g'(z_0) > 1$ and $z_0 g'(z_0) < 1$. In the first case Eq.(9.223) leads straightforwardly to (9.219), where z is given by Eq.(9.220) as before. However, in the second case Eq. (9.220) no longer holds, but Eq.(9.223) may be regarded as the definition of q, giving the analytical continuation of the expression given by Eq.(9.219). The integral in Eq.(9.223) can be evaluated by the method of steepest descents. The main contribution comes from the neighbourhood of $z = z_0$ and we can thus use Eqs.(9.221) and (9.222).

We write

$$e^{q - N g(z_0)} = \frac{1}{2\pi i} \oint \frac{g'(z) + z g''(z)}{z g'(z) - 1} e^{-Ns} dz, \quad (9.225)$$

where

$$s = g(z_0) - g(z). \quad (9.226)$$

Using Eqs.(9.221) and (9.222), introducing s as a new integration variable, and retaining only the first term of the series expansion, we find

$$q = N g(z_0) + \ln \frac{A}{2\pi i} \oint e^{-Ns} s^{\alpha} ds, \quad (9.227)$$

where A is a numerical constant, independent of N, and where the contour can be taken to start from $s = \infty - i\varepsilon$ ($\varepsilon > 0$), to encircle the origin, and to go to $s = \infty + i\varepsilon$. The integral leads to a Γ-function and the final result is

$$q = N g(z_0) + \mathcal{O}(\ln N), \quad (9.228)$$

whence follows Eq.(9.215) in the limit as $N \to \infty$.

Combining Eqs.(9.212) and (9.215) we can write the equation of state in the form

$$P = \frac{NkT}{V} \left(1 - \frac{NkT}{4 P_c V}\right), \quad (9.229)$$

---

[8] H.A.Kramers, *Commun. Kamerlingh Onnes Lab.* **22**, Suppl.22, 1 (1936); *Collected Scientific Papers*, North-Holland, Amsterdam (1956), p.607; compare also the derivation of Eqs.(7.128).

where now only measurable quantities appear. Wergeland has shown that Eq.(9.229) represents the observational data for nitrogen and argon with good accuracy in the neighbourhood of the saturation pressure where the deviations from the perfect gas law are by no means small.

In order to get some idea of the physical processes involved in the condensation it is instructive to consider the numbers of droplets of various sizes at temperatures in the neighbourhood of the condensation temperature T_0. For a given volume $V(= Nv)$ this temperature is given by the equation (compare Eq.(9.214); $\beta_0 = 1/kT_0$)

$$e^{\beta_0 X} = \frac{v}{v_1}. \tag{9.230}$$

If $\langle m_l \rangle$ be the average number of droplets containing l atoms, we find from grand ensemble theory (compare Eq.(6.0611))

$$\langle m_l \rangle = \frac{-1}{\beta} \frac{\partial q}{\partial \varepsilon_l}, \tag{9.231}$$

where ε_l is the energy of such a drop. Using Eqs.(9.201), (9.202), and (9.207) to (9.209) we find

$$\langle m_l \rangle = \frac{V}{v_1} e^{-\beta X} \frac{l^{l-2}}{l!} z^l, \tag{9.232}$$

or,

$$\langle m_l \rangle = N \frac{l^{l-2}}{l!} \eta^{l-1} e^{-l\eta}, \tag{9.233}$$

where η is a parameter measuring the degree of saturation, given by the equation

$$\eta = \frac{N v_1}{V} e^{\beta X}, \tag{9.234}$$

which is equal to 1 at the condensation point. At infinite dilution, $v \to \infty$, we have $\eta \to 0$, and hence from Eq.(9.233)

$$\langle m_l \rangle = \delta_{l1} N, \tag{9.235}$$

that is, every atom is single!

From Eqs.(9.232) and (9.210) we see that the pressure is the sum of the pressures of the subsystems which are all perfect gases,

$$P = \frac{kT}{V} \sum_l \langle m_l \rangle, \tag{9.236}$$

which is, of course, a consequence of our assumption of non-interacting droplets (compare also Eq.(9.201)).

On the other hand, at the condensation point $\langle m_l \rangle$ is proportional to[9] $l^{-5/2}$ and, although the decrease in $\langle m_l \rangle$ with increasing l is relatively slow, even then only the smallest drops are present in appreciable amounts.

Let us now consider temperatures just above T_0. From Eqs.(9.230) and (9.234) we find

$$\eta = \left(\frac{v_1}{v}\right)^{\Delta T/T}, \tag{9.237}$$

where

$$\Delta T = T - T_0. \tag{9.238}$$

Substituting Eq.(9.237) into Eq.(9.233) we find

$$\langle m_l \rangle = N \frac{l^{l-2}}{l!} \left(\frac{v_1}{v}\right)^{(l-1)\Delta T/T} \exp\left[-l\left(\frac{v_1}{v}\right)^{\Delta T/T}\right], \tag{9.239}$$

which shows that $\langle m_l \rangle$ is only appreciable — even for the smallest l — in a temperature region of the order of $\Delta T \sim T_0/l$. The fact that there are no appreciable numbers of drops at temperatures just above T_0 corresponds to the fact that right up to the saturation point no condensation can be observed.

The liquid drop model is only an approximate theory and fails to account for the existence of a critical temperature: at any temperature condensation will occur. This is a consequence of the fact that the coefficients in the power series on the right-hand side of Eq.(9.208) are independent of the temperature. This breakdown of the theory is not surprising, since the theory is essential concerned with the behaviour near the saturation point, where it gives, indeed, a clear picture of what is happening. Far from the saturation point, however, the theory will break down, since there the approximate formulæ (9.205) must be applied for small values of l and they will no longer hold.

9.3.   Mayer's Theory of Condensation

In the present section we shall briefly consider the theory of condensation developed by J.E.Mayer. We have mentioned earlier that all properties of our system should be derivable from the grand potential q which is given by Eq.(9.101):

$$e^q = \sum_\nu \frac{e^{\alpha\nu}}{\nu!} \int_\Gamma e^{-\beta\varepsilon}\, d\Omega, \tag{9.301}$$

with the element of extension in phase, $d\Omega$, given by the formula

$$d\Omega = \prod_{i=1}^{\nu} \frac{d^3 p_i\, d^3 r_i}{h^3}, \tag{9.302}$$

---

[9] We use the equation for the factorial given in Problem 5 of Ch.4.

and the energy given by Eq.(7.106),

$$\varepsilon = \sum_{i=1}^{\nu} \left(\frac{p_i^2}{2m} + F(r_i) \right) + \sum_{i<j} \phi(r_{ij}). \tag{9.303}$$

We saw in § 7.1 that Eq.(9.101) leads to Eq.(7.120),

$$q = \frac{V}{v_0} \sum_{n=1}^{\infty} b_n e^{n\alpha}, \tag{9.304}$$

for the grand potential q, and to Eq.(7.122) and (7.123),

$$P = \frac{kT}{v_0} \sum_{n=1}^{\infty} b_n e^{n\alpha} \tag{9.305}$$

and

$$N = \frac{V}{v_0} \sum_{n=1}^{\infty} n b_n e^{n\alpha}, \tag{9.306}$$

for the pressure P and the average number of atoms, N. In Eqs.(9.304) to (9.306) the b_n are the so-called *cluster integrals* defined by Eqs.(7.112) and (7.116) to (7.118). We should remark here that the b_n are functions of the temperature which also depend on the volume. However, the volume-dependence comes into play only when the specific volume becomes of the order of the molecular volume, v_m, which is of the order of d^3, where d is the range of the intermolecular forces. In fact, Mayer assumed that one can replace the $b_n(v)$ by their asymptotic values $b_n(\infty)$.

We now note that Eqs.(9.304) and (9.306) are of the same form as Eqs.(9.216) and (9.217) where the function $f(z)$ is in the present case given by the formula

$$f(z) = \frac{1}{v_0} \sum_{n=1}^{\infty} b_n z^n. \tag{9.307}$$

We can therefore use the same arguments as in the preceding section to prove that in the limit as $N \to \infty$ the isotherms consists of two analytically different parts, provided the function $f(z)$ given by Eq.(9.307) satisfies the conditions (i), (ii), and (iii) given on p.325. As the b_n are functions of the temperature, it is possible that for temperatures above a critical value T_c the function $f(z)$ is regular for all finite values of z but that for temperatures below T_c a singularity occurs, say at $z = z_0$. If that is the case, the isotherms will give P as a monotonically decreasing function of v for $T > T_c$, but for $T < T_c$ the isotherms will consist of two different parts joining at $v = v_c$, where

$$\frac{1}{v_c} = z_0 f(z_0). \tag{9.308}$$

As long as $v > v_c$, the pressure P is a monotonically decreasing function of v, given by Eq.(9.305). However, for $v < v_c$ we have, apart from terms of the order $1/N$,

$$P = \frac{1}{\beta} f(z_0). \tag{9.309}$$

We must remark here that the derivation of Eqs.(9.304) and (9.306) is general only as long as we may neglect the molecular volume, that is, as long as the replacement of the volume-dependent b_n by their asymptotic value is a good approximation. This means that we are really restricting the equation of state to specific volumes larger than the molecular volume. As soon as we are dealing with specific volumes of the order of, or smaller than, v_m the derivation of Eq.(9.304) is no longer valid, and especially it is no longer correct to assume that the b_n are independent of the volume.[10] It is thus not surprising that the theories discussed in the present and the preceding section fail to account for the third part of the isotherm which corresponds to the pure liquid.

As long as we do not make any specific assumptions about the T-dependence of the b_n we cannot reach any quantitative conclusions about the existence and the value of the critical temperature. Such assumptions were made by Mayer and we refer to his work for a more detailed discussion of his theory.

In the preceding section we saw that in Wergeland's liquid drop model of an imperfect gas in the gas phase, right down to the condensation temperature, only the smallest drops were present, and also that in his model the gas is essentially a perfect gas in which the constituents are the drops of various sizes. We shall now show that the situation is very similar in Mayer's theory.

We first of all note the similarity between Eqs.(9.208), (9.210), and (9.211), on the one hand, and Eqs.(9.304), (9.305), and (9.306), on the other hand. This suggests that we can introduce again quantities m_n, now to be defined by the formula

$$m_n = \frac{V}{v_0} b_n e^{n\alpha}, \tag{9.310}$$

and interpret them as the number of clusters containing n atoms.[11] We note that it follows from Eqs.(9.305) and (9.310) that Eq.(9.236) will hold again.

To find the behaviour of the m_n for large n we must study the behaviour of the b_n. Consider Eqs.(9.305) and (9.306) which we shall write in the form

$$P(z) = \frac{kT}{v_0} \sum b_n z^n \tag{9.311}$$

---

[10]Similar considerations apply to the discussion of § 9.2.

[11]One should not take this interpretation too literally, since the b_n are not necessarily always positive.

and

$$\frac{v_0 N}{V} = \frac{v_0}{v} \equiv x(z) = \sum n b_n z^n. \tag{9.312}$$

The inverse of Eq.(9.312) we write in the form

$$z(x) = x\, e^{-\varphi(x)}, \tag{9.313}$$

and we can then write Eq.(9.311) as

$$
\begin{aligned}
P(z) &= \frac{kT}{v_0} \int_0^z \frac{dz}{z\, x(z)} \\
&= \frac{kT}{v_0} \int_0^x x^2 \frac{d\ln z}{d\ln x}\, dx \\
&= \frac{kT}{v_0} \int_0^x x^2 \left(\frac{1}{x} - \varphi'(x)\right), dx \\
&= \frac{kT}{v_0} \left[x - \sum_{k=1}^\infty \frac{k}{k+1} \beta_k x^{k+1} \right] \\
&= \frac{kT}{v} \left[1 - \sum_{k=1}^\infty \frac{k}{k+1} \beta_k x^k \right],
\end{aligned}
\tag{9.314}
$$

where we have used a series expansion of $\varphi(x)$,

$$\varphi(x) = \sum_k \beta_k x^k. \tag{9.315}$$

which is justified since we have $z \approx x$, if $z \ll 1$.

We note in passing that Eq.(9.314) gives us the virial expansion for the pressure, since $x = v_0/v$.[12]

In order to find the behaviour of the b_n we must express the variable x in terms of z, that is, we must find the solution of Eq.(9.313). We use the following theorem: If $z(x) = x/F(x)$, we can write the inverse function $x(z)$ in the form

$$x(z) = \sum_{j=1}^\infty \frac{z^j}{j!} \left[\frac{d^{j-1}}{d\xi^{j-1}} \{f(\xi)\}^j \right]_{\xi=0}. \tag{9.316}$$

We leave the proof of this theorem as an exercise (see Problem 5 at the end of the present chapter). We note that the expression within the square

---

[12]We see that the virial coefficients are proportional to the β_k which we have introduced in a purely formal way through the series expansion (9.315). From our way of introducing these quantities it is extremely difficult to see their physical meaning. They are the so-called *irreducible cluster integrals*. We do not want to enter into a detailed discussion of the β_k and refer to R.K.Pathria, *Statistical Mechanics*, Pergamon Press, Oxford (1972), §§ 9.2 and 9.4.

brackets on the right-hand side of Eq.(9.316) is $(j-1)!$ times the coefficient of ξ^{j-1} in the Taylor expansion of $(f(\xi))^{j}$.

Using this theorem to express x in terms of z and comparing the resulting expression for the pressure with Eq.(9.311) we find the following relation between the b_n and the coefficients β_k in the expansion (9.315):

$$b_n = \frac{1}{n^2} \sum_{\{p_k\}} \prod_{k=1}^{n-1} \frac{(n\beta_k)^{p_k}}{p_k!}, \tag{9.317}$$

where the sum over the p_k is over all combinations such that

$$\sum_{k=1}^{n-1} kp_k = n - 1. \tag{9.318}$$

We now bear in mind that we are interested in the behaviour of the m_n, or the b_n, for large values of n. We can then expect that the sum over the p_k in Eq.(9.317) can essentially be reduced to its largest term. Using the Lagrangian multiplier method of § 1.5 we find the p_k corresponding to that term by maximising the expression

$$\ln b_n \approx -2\ln n + \sum_{k=1}^{n-1} p_k \{\ln(n\beta_k) - \ln p_k + 1\}, \tag{9.319}$$

with the p_k being restricted by the condition (9.318).

This leads to the following equation for the p_k we are looking for:

$$p_k^{\max} = n\beta_k\, e^{-k\gamma}, \tag{9.320}$$

where γ is a Lagrangian multiplier which can be determined by using condition (9.318).

Substituting expression (9.320) into Eq.(9.319) we find that the b_n are given by the equation

$$b_n \approx cb_0^n, \tag{9.321}$$

with c a constant and b_0 given by the equation

$$b_0 = \exp\left(\gamma + \sum_k \beta_k\, e^{-k\gamma}\right). \tag{9.322}$$

The $\approx$-sign in Eq.(9.321) refers to the fact that this equations is, in fact, true only asymptotically. If we assume that it is true for all values, rather than only for large values, of n Eqs.(9.311) and (9.312) become

$$P = \frac{ckT}{v_0} \sum (b_0 z)^n \tag{9.323}$$

and

$$\frac{1}{v} = \frac{c}{v_0} \sum (b_0 z)^n, \tag{9.324}$$

and Eq.(9.310) becomes

$$m_n = \frac{cV}{v_0} (b_0 z)^n. \tag{9.325}$$

We note, first of all, that the sums in Eqs.(9.324) and (9.325) converge as long as $z < 1/b_0$, but diverge when $z > 1/b_0$. In fact, these equations are very similar to Eqs.(9.210) and (9.213), especially if we use the Stirling formula $l! \approx (l/e)^l$ in the latter. Secondly, we note that for $z < 1/b_0$ the m_n are very small for large n: as in the case of Wergeland's model, there are hardly any large clusters (drops) for those temperatures for which the series for P and $1/v$ converge. We can carry out an analysis similar to the one given in the preceding section and the result is again that the isotherms will consist of a horizontal part for specific volumes smaller than a critical value and a monotonically decreasing part for larger values of the specific volume. Of course, we have to bear in mind that these conclusions are only approximately correct for a real gas, since in deriving Eqs.(9.324) and (9.325) we have made several simplifying assumptions.

9.4.   Yang and Lee's Theory of Phase Transitions

We start again from Eq.(7.101) for the grand potential q and we shall assume that the interparticle potential $\phi(r)$ is given by the equations (cf. Eqs.(7.201))

$$\left.\begin{array}{ll} \phi(r) = \infty, & \text{if } r \leqslant a; \\ \phi(r) = 0, & \text{if } r \geqslant b; \\ \phi(r) > -\infty : \end{array}\right\} \tag{9.401}$$

each particle is a hard sphere with a finite range.[13]   This means that each particle occupies a finite volume and that we can have in the volume V at most a finite number of particles, M, which is the maximum number of particles which can be accommodated in the volume so that the sum in Eq.(7.101) is a finite one:

$$e^q = \sum_{\nu=0}^{M} \frac{e^{\alpha\nu}}{\nu!} \int_{\Gamma} e^{-\beta\epsilon} \, d\Omega. \tag{9.402}$$

As in §7.1 we can integrate over the momenta, and we get, instead of Eq.(7.108), the equation

$$e^q = \sum_{\nu=0}^{M} \frac{y^\nu}{\nu!} Q_\nu, \tag{9.403}$$

---

[13]The finite-range requirement is needed for some of the proofs of the various properties of the grand potential and its derivatives.

where, to simplify the notation, we have introduced a variable y through the definition

$$y = \frac{e^{\alpha}}{v_0},$$ (9.404)

with v_0 again given by Eq.(7.109),

$$v_0 = \left(\frac{\beta h^2}{2\pi m} \right)^{3/2},$$ (9.405)

and the configurational partition function, which we now denote by Q_{ν}, given by Eq.(7.111):

$$Q_{\nu} = \int \cdots \int \prod_{i<j} e^{-\beta\phi(r_{ij})} d^3 r_1 \cdots d^3 r_{\nu}.$$ (9.406)

We can write Eq.(9.403) for the grand partition function in the form

$$e^q = \prod_{i=1}^{M} \left(1 - \frac{y}{y_i} \right),$$ (9.407)

where the y_i are the roots of the equation

$$e^{q(y,V,\beta)} = 0.$$ (9.408)

From Eq.(9.408) it follows that the roots are functions of the volume and the temperature:

$$y_i = y_i(V, \beta).$$ (9.409)

As the right-hand side of Eq.(9.403) is a polynomial of the M-th degree, there are M roots. Since all coefficients in the sum in Eq.(9.403) are positive, none of the roots can be positive real. If we let the volume V tend to infinity, the number of roots will also tend to infinity and we can speak of a distribution function $g(y, \beta)$ of the roots, such that $g(y) \, d\xi \, d\eta$ is the number of roots with real parts lying between ξ and $\xi + d\xi$ and imaginary parts lying between η and $\eta + d\eta$ ($y \equiv \xi + i\eta$); as the roots are functions of both the volume and the temperature, g is still a function of the temperature. From Eq.(9.407) we see that the roots y_i completely determine the state of our system, and in the thermodynamic limit, as $V \rightarrow \infty$, the state of the system will be determined by $g(y, \beta)$.

We remind ourselves that the pressure P and the density ρ are given by Eqs.(5.0926) and (5.0928) (cf. Eqs.(7.121) and (7.123))

$$\frac{P}{kT} = \frac{q}{V},$$ (9.410)

and

$$\rho \equiv \frac{N}{V} = \frac{\partial q/V}{\partial \ln y}.$$ (9.411)

In the thermodynamic limit we are thus interested in the quantities

$$F(y) \equiv \lim_{V \to \infty} \frac{q}{V},$$

(9.412)

and

$$G(y) \equiv \lim_{V \to \infty} \frac{\partial q/V}{\partial \ln y}.$$

(9.413)

Without proof we now state two important theorems which were proved by Yang and Lee:[14]

First Theorem of Yang and Lee. The function $F(y)$ exists for all real, positive y and it is a continuous, monotonically increasing function of y.

Moreover, P/kT as function of y is for finite V an analytic function of y in the whole of the complex plane, except in the points $y = y_i$. This means that ρ as function of y, as well as all higher derivatives of q/V with respect to y, are analytic everywhere on the real positive y-axis, since there are no roots on that axis. As a result, as long as V is finite, there can be no singularities in any of the thermodynamic functions.

We now consider a region $\mathcal{R}$ of the complex y plane, such that in the limit as $V \to \infty$ it does not contain any zeros y_i.

Second Theorem of Yang and Lee. The quantity q/V converges uniformly for all y in $\mathcal{R}$ to the function $F(y)$ as $V \to \infty$.

This means that $F(y)$ will be analytic throughout the region $\mathcal{R}$ and that, if $\mathcal{R}$ contains part of the positive real axis, the derivatives of $F(y)$ will also be analytic throughout $\mathcal{R}$ and, moreover, the taking of the derivative and the taking of the limit as $V \to \infty$ are interchangeable. This means that for all real positive $y \in \mathcal{R}$ the equation of state is given by the equations

$$\frac{P}{kT} = F(y),$$

(9.414)

and

$$\frac{N}{V} = \frac{\partial F(y)}{\partial \ln y}.$$

(9.415)

We shall now consider what happens in the limit as $V \to \infty$. Let us first look at the situation where we can find a region $\mathcal{R}$ such that it contains the whole of the real positive y axis (Fig.9.2a). In that case Eqs.(9.414) and (9.415) give us the equation of state for all real positive values of y. From Yang and Lee's first theorem it then follows that P as function of y, and hence of α, is monotonically increasing (see Fig.9.2b), and that the same applies to N/V (see Fig.9.2c). Combining these two functions we get a behaviour of P as function of $v \equiv N/V$ as shown in Fig.9.2d: the isotherm

---

[14]For the proofs of these theorems we refer to their paper (*Phys. Rev.* **87**, 404 (1952)).

is smooth and the quantity $\partial P/\partial v$ is everywhere negative. There are no phase transitions and we are clearly at a temperature which lies above the critical temperature.

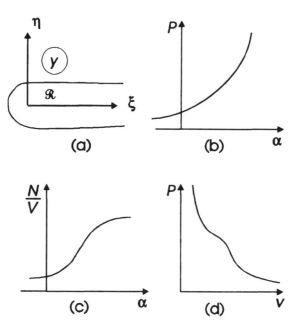

Fig.9.2. The region $\mathcal{R}$ without roots in the y-plane at a temperature above the critical temperature (a); the pressure P as function of α (b); the density N/V as function of α (c); and the isotherm (d).

We mentioned earlier that the roots y_i are functions of the temperature. We just considered a temperature above the critical temperature which meant that there was a region $\mathcal{R}$ such that it contained no roots and contained the whole of the real positive y axis. When the temperature is lowered the distribution of the roots $g(y)$ will change and it may happen that it will become such that it is impossible to find any finite region near a point y_1 of the real positive y axis such that it does not contain any roots. Let us consider the case where there are two such points, y_1 and y_2. We then may have regions, $\mathcal{R}_1$, $\mathcal{R}_2$, and $\mathcal{R}_3$, such that they do not contain any roots and each of them contains portions of the real positive y axis (Fig.9.3a).

Along the real positive axis within each region $\mathcal{R}_i$ the equation of state is given by Eqs.(9.414) and (9.415). It follows from the first theorem that in the points y_1 and y_2 the pressure will still be a continuous function of α, but the same will no longer be true of all its derivatives. If in these two points its first derivative has a discontinuity we have the situation depicted in Fig.9.3: in the two points y_1 and y_2 a first-order phase transition takes place and the isotherm will look like the one shown in Fig.9.3d. If, on the other hand, the first derivative is still continuous, but the second derivative is discontinuous we have what Ehrenfest calls a second-order phase transition.[15]

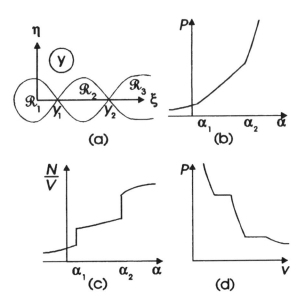

Fig.9.3. Regions $\mathcal{R}_1$, $\mathcal{R}_2$, and $\mathcal{R}_3$, without roots in the y-plane at a temperature below the critical temperature (a); the pressure P as function of α (b); the density N/V as function of α (c); and the isotherm (d).

We see that for the kind of systems considered by Yang and Lee, that is, for systems with an interatomic potential given by Eq.(9.401) the distribution $g(y)$ of the roots in the limit as $V \to \infty$ will determine the behaviour of the system, especially the occurrence of phase transitions. One might hope that this will be true in general. From the theory of complex functions it follows that if we have a function $f(y)$ which is defined in the complex y plane and if there is a point $y = y_0$ such that there are zeros of $f(y)$ in any region, however small, around that point, this point will be a singularity of $f(y)$. If therefore, we have a function defining the pressure P as function of some variable y and if there exists such a point y_0 on the positive real axis

---

[15]There is also the possibility that the first derivative, albeit continuous, becomes infinite in the point y_1. In that case we have a phase transition like the one shown by liquid helium in its λ point.

we expect that at that point there will be a phase transition. In that connection we may draw attention to the fact that in the two cases where we were able to prove the existence of a phase transition — the Einstein condensation in § 4.7 and Wergeland's liquid drop model in § 9.2 — we found, indeed, singularities on the real positive axis, and the behaviour of the function $g(z)$ given by Eq.(9.221) determined the order of the transition.

In the limit as $V \to \infty$ we can write Eqs.(9.407) and (9.410) in the form

$$q(y) = \int \ln\left(1 - \frac{y}{s}\right) \lim_{V \to \infty} Vg(s)\, d\xi\, d\eta, \qquad (9.416)$$

where $\lim_{V \to \infty} Vg(s)$ is the distribution of the zeros, where the integration is over the whole of the complex s plane, and where ξ and η are the real and imaginary parts of s; and

$$\frac{P}{kT} = \int g(s) \ln\left(1 - \frac{y}{s}\right)\, d\xi\, d\eta. \qquad (9.417)$$

We see once again that if we know the distribution of the zeros $g(s)$ we can find the equation of state, but we also see that in principle we should be able to find $g(s)$ from a knowledge of the equation of state. There are very few cases where $g(s)$ is known:[16] the one-dimensional Ising ferromagnet,[17] a gas of hard rods, a gas with very weak, very long-range repulsive forces, a Weiss-field ferromagnet, and two-dimensional Ising ferromagnets. Lee and Yang[18] have shown that the problem of an Ising ferromagnet is mathematically equivalent to that of a lattice gas, that is, a system where its constituents are constrained to be situated on lattice sites and where the potential energy u for the interaction between two constituents is given by the equation

$$
\left.
\begin{array}{lll}
u = \infty, & \text{if the two atoms occupy the same lattice site;} \\
u = -2\epsilon, & \text{if the two atoms are nearest neighbours;} \\
u = 0, & \text{otherwise.}
\end{array}
\right\} \qquad (9.418)
$$

---

[16]Hemmer and Hiis Hauge (*Phys. Rev.* **133A**, 1010 (1964)) and Stephenson (*J. Phys.* **A20**, 4513 (1987)) give most of those cases; we refer to them for references to the original papers.

[17]The Ising model is defined as follows. Consider a lattice such that nearest neighbours of one site are not themselves nearest neighbours of each other. Now assign to each lattice site i a variable σ_i which can take on two values, $+1$ or -1. The Hamiltonian H_{Ising} of the system is now defined by the equation

$$H_{\text{Ising}} = -\tfrac{1}{2} J \sum_{\{j,k\}} \sigma_j \sigma_k,$$

where the summation is over all the nearest-neighbour pairs in the lattice. We shall discuss the Ising model of ferromagnetism in more detail in §§ 9.5 to 9.9.

[18]T.D.Lee and C.N.Yang, *Phys. Rev.* **87**, 410 (1952.

They also proved that for an even more general model where the interaction potential satisfies the relations

$$u = \infty, \quad \text{if the two atoms occupy the same lattice site;}$$
$$u \leqslant 0, \quad \text{otherwise,}$$

$\left. \right\}$ (9.419)

the roots of the partition function lie on a circle in the complex y plane
with its centre at $y = 0$. This result is independent of the range of
the potential, the dimensionality of the lattice, and the structure or
size of the lattice. In the case of a one-dimensional Ising model the
distribution of the roots can be found quite easily (see Problem 17 at
the end of the present chapter). In that case, there is a gap in the
distribution which includes the intersection of the unit circle with the
real positive axis: there is in that case no phase transition, as is well
known.

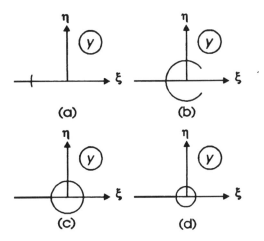

Fig.9.4. Distribution of zeros for a van der Waals gas;
(a): $T \gg T_c$; (b): $T \gtrsim T_c$; (c): $T = T_c$; (d): $T < T_c$.

Another way of approaching the question of the distribution of the
zeros is to study the equation of state, if it is known. We noticed
earlier that $g(s)$ can be found, once we know the equation of state.
This approach has been used by Hemmer and Hiis Hauge who studied
the case of the van der Waals equation of state (1.305). The calculations
are quite cumbersome, but it is interesting to look at their results which
we show in Fig.9.4. In this case we know the critical temperature, T_c,
above which there is no phase transition and below which there is just
a single phase transition. We see that at very high temperatures, the
zeros are practically confined to the real negative axis (Fig.9.4a), when
the temperature approaches T_c, the distribution of zeros has a part

which begins to approach the real positive axis (fig.9.4b), closing at $T = T_c$ (Fig.9.4c). When the temperature is lowered even further, the radius of the circle of zeros diminishes (Fig.9.4d).

9.5.   The Ising Model of Ferromagnetism

In the preceding sections we have looked at first-order phase transitions, especially the gas-liquid phase transition, although we have seen that Yang and Lee's theory in principle also covers higher-order phase transitions. In the remaining sections of this chapter we shall look at a broad class of phase transitions which are often described as cooperative phenomena. The simplest model describing the systems undergoing such phase transitions is the so-called Ising model,[19] introduced in 1925 to describe ferromagnetism. We have already mentioned in the previous section that the Ising model is mathematically equivalent to a lattice gas, that is, a system where the constituents are constrained to lie on lattice sites and where the interaction has a hard core so that no two constituents can lie on the same lattice site. In the limit as the lattice spacing goes to zero the lattice gas resembles an ordinary gas.

To simplify our discussion we shall consider only such lattices that the nearest neighbours of one lattice site are not themselves nearest neighbours of each other. Such lattices include the square lattice in two dimensions and the simple cubic and body-centred cubic lattices in three dimensions, but they do not include the triangular lattice in two dimensions or the face-centred cubic lattice in three dimensions. This restriction is not essential and can easily be lifted, but it simplifies the discussion. If we introduce this simplification we can split the lattice into two sublattices which we shall call the α and the β lattices, such that each α site has only β sites as nearest neighbours and vice versa.

Before introducing the Ising model we shall briefly discuss the problem of binary alloys. If one adds to a metal A a quantity of another metal B, it sometimes happens that the crystal structure of the resulting alloy is the same as that of the pure metal A, except that some lattice sites are occupied by B atoms rather than by A atoms. Such an alloy is called a *solid solution* or, since it can be obtained by a simple one for one substitution, a *substitutional solid solution*. X-ray diffraction experiments showed that under certain conditions an ordered structure was attained. This ordered structure was manifested by the occurrence of the so-called *superstructure lines*. The origin of these superstructure lines can be seen from Fig.9.5. In Figs.9.5a and c we depict possible situations in a crystal, the first one in a state of disorder and the second one in a state of complete order. Imagine an X-ray beam of wavelength λ incident on the crystal at an angle θ such that the path length difference PQR between two rays reflected by two successive planes is equal to $\frac{1}{2}\lambda$ (Fig.9.5b). If the reflecting properties

---

[19]E.Ising, *Z. Physik* **31**, 253 (1925).

of the two planes are the same — as they will be when there is a state of disorder — no line will appear on the photographic plate. If, however, a state of order exists, it is possible that a line will be produced, provided the scattering properties of the A and B atoms are sufficiently different.

Let us now consider an alloy for which the ordered arrangement of Fig. 9.5c is energetically more favourable. Of course, if this were not the case the ordered state would be very unlikely ever to occur. If, then, the ordered state is the state of lowest energy, there will be a tendency for A atoms to be surrounded by B atoms and *vice versa* rather than a tendency to form clusters of A or of B atoms. At very low temperatures we may expect a state of order to exist. We shall call the lattice sites occupied by the A atoms α sites and those occupied by B atoms β sites. An A atom on an α site or a B atom on a β site will be called a *right* atom and an A atom on a β site or a *b* atom on an α site a *wrong* atom.[20]

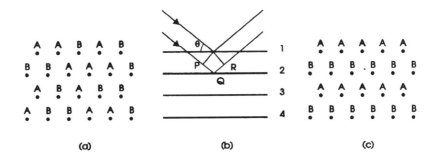

(a)   (b)   (c)

Fig.9.5. The origin of superstructure lines; (a): a disordered state of an AB alloy; (b): incidence of an X-ray beam on the crystal ; (c): an ordered state of the AB alloy.

At sufficiently low temperatures the crystal will be in an ordered state. If we then raise the temperature, there will be a chance that an A atom on an α site will swap places with a B atom on a β site thus leading to two atoms changing from right to wrong, and a certain amount of disorder will ensue. If we want to discuss the degree of order or disorder in the crystal quantitatively we must distinguish between two kinds of order. The first

[20]If, as will be the case for a two-dimensional square lattice or for a simple or body-centred cubic lattice, there are as many α- as β sites and if there are as many A as B atoms — otherwise a completely ordered state would be impossible — there would be a second ordered state with all A atoms on β sites and all B atoms on α sites. The transition from the one ordered state to the other would, however, involve the rearrangement of every single atom and is therefore practically impossible to occur. We shall not consider this second ordered state.

kind, called *long-range order* or *order at distance*, measures how large a fraction of A atoms is situated on α sites. The second kind, called *short-range order* or *local order*, measures how well on average A atoms are surrounded by B atoms. At the absolute zero there will be perfect long-range order and perfect short-range order, but at finite temperatures both will be destroyed. There will be, on the one hand, A atoms on β sites and B atoms on α sites and, on the other hand, there will be nearest neighbours of the same kind. As long as the temperature is not too high, the tendency of atoms to surround themselves with atoms of the other kind will counteract the thermal movement of the atoms in the lattice. At temperatures below a certain critical temperature there will be more right atoms than wrong ones and the tendency for unlike nearest neighbours will maintain a certain amount of long-range order. However, an A atom on a β site makes it energetically more favourable to have B atoms on the neighbouring α sites, thus lowering the potential barrier which must be overcome in order that these α sites can be occupied by B atoms. We see thus that the higher the long-range disorder, the greater will be the ease with which this disorder can be further increased. This "avalanche" effect will finally lead to a complete and abrupt disappearance of long-range order at a certain critical temperature T_c. By analogy with ferromagnetism this temperature is often called the *Curie temperature*. Above this temperature there will be as many right as wrong atoms, but short-range order will still persist to a certain degree.

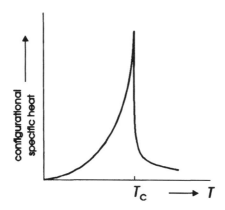

Fig.9.6. The configurational specific heat.

Since it requires energy to produce disorder, the disordering effect can be seen in the specific heat. The part of the specific heat which is connected with the ordering of the crystal is called the *configurational specific heat.* A qualitative sketch of its behaviour as function of the temperature is shown in Fig.9.6. The steep rise in the neighbourhood of T_c is due to the "avalanche" effect described in the preceding paragraph. That the configurational specific heat does not vanish above T_c is due to the fact that there still remains a certain amount of short-range order.

The order-disorder transformation in alloys is one example of the large class of *cooperative phenomena*. Other examples are ferromagnetism, antiferromagnetism, and phase separation in binary solutions. Characteristic of them is the cooperation of certain subsystems to form units which stick together notwithstanding the disruptive influence of thermal agitation. and the fact that the ability of sticking together depends on the degree to which the subsystems have already cooperated. We shall now introduce a model to describe such systems.

To simplify our considerations we shall restrict ourselves to either a one-dimensional chain, or a two-dimensional square lattice, or a simple cubic lattice; the generalisation to other lattices is in most cases straightforward, although sometimes it can be cumbersome. Now consider a lattice with a spin situated on each of its sites and let the spin be capable of just two orientations which, for the sake of simplicity, we shall call the up- and the down-orientation. Moreover, we shall assume that these spins interact only with their nearest neighbours. In a ferromagnet the energy will be lowest if all spins are parallel. We assign to each spin a parameter σ which is equal to $+1$ when the spin is up and equal to -1 when the spin is down. We assume that if a nearest-neighbour pair changes from a parallel to an antiparallel alignment the energy of the ferromagnet is increased by an amount J. In that case the total energy, E, of the ferromagnet will, apart possibly from an additive constant, be of the form

$$E = -\tfrac{1}{2} J \sum_{\{j,k\}} \sigma_j \sigma_k, \tag{9.501}$$

where $\sum\limits_{\{j,k\}}$ here and henceforth indicates summation over all nearest-neighbour pairs in the lattice. A ferromagnet characterised by the energy (9.501) is called an *Ising ferromagnet* since Ising introduced this model to describe ferromagnetism in 1925.

In an antiferromagnet the tendency will be to have antiparallel alignment, and if we take J to be always a positive quantity, the energy will be given by the equation[21]

$$E = \tfrac{1}{2} J \sum_{\{j,k\}} \sigma_j' \sigma_k'. \tag{9.502}$$

In the case of a binary alloy we shall also have an energy given by Eq.(9.502). This can be seen as follows. Let each A atom correspond

---

[21]This is the point where the restriction of nearest neighbours of one site not being nearest nearest neighbours of one another comes into play as a moment's reflection will make clear.

to a σ' equal to $+1$ and each B atom to a σ' equal to -1. If v_{AA}, v_{BB}, and v_{AB} are the energies associated with an AA, a BB, and an AB pair, respectively, the total energy will, apart from a possible constant, be given by Eq.(9.502) if we write

$$J = \tfrac{1}{2}(v_{AA} + v_{BB}) - v_{AB}. \tag{9.503}$$

From the definition of the v's it follows that the total energy is given by the expression

$$E = v_{AA}Q'_{AA} + v_{BB}Q'_{BB} + v_{AB}Q'_{AB} + \text{constant}, \tag{9.504}$$

where $Q'_{AA}, \ldots$ are the number of AA, $\ldots$ pairs in the lattice. If Q is the total number of pairs, we can rewrite Eq.(9.504) in the following form:

$$E = \tfrac{1}{2}Q(v_{AA} + v_{BB}) - JQ'_{AB} + \text{constant}, \tag{9.505}$$

where we have used Eq.(9.503).

Since $\sigma'_j \sigma'_k$ equals $+1$ for an AA or a BB pair and -1 for an AB pair, we have

$$\sum_{\{j,k\}} \sigma'_j \sigma'_k = Q'_{AA} + Q'_{BB} - Q'_{AB} = Q - 2Q'_{AB}, \tag{9.506}$$

whence Eq.(9.502) follows through a suitable choice of the constant in the energy.

In the case of a binary alloy Eq.(9.502) can be reduced to Eq.(9.501) by the following procedure. Choose one lattice site and call it an α-site; then call all its nearest neighbours β-sites, and so on. In this way we have divided the lattice into two sublattices such that each site of the first lattice, the α-sites, say, have sites of the other lattice, β-sites, as nearest neighbours and *vice versa*. Let there be N lattice sites so that $\tfrac{1}{2}N$ are α-sites and $\tfrac{1}{2}N$ β-sites,[22] and so that there are $\tfrac{1}{2}N$ A atoms with $\sigma'_i = +1$ and $\tfrac{1}{2}N$ B atoms with $\sigma'_i = -1$. In the state of complete order all A atoms will be on α-sites and all B atoms on β-sites. Introducing the σ_i by the equations

$$\left. \begin{array}{ll} \sigma_i = \sigma'_i, & \text{if } i \text{ is an } \alpha-\text{site} \\ \sigma_i = -\sigma'_i, & \text{if } i \text{ is a } \beta-\text{site}, \end{array} \right\} \tag{9.507}$$

we see that a state of complete order is described by all σ_i being equal to $+1$.[23]   Using Eq.(9.507) to change from the σ'_i to the σ_i we see that Eq.(9.502) reduces to Eq.(9.501) which means that we can reduce the discussion of binary alloys or of antiferromagnets to that of ferromagnets. We

---

[22]We have carefully chosen to consider only such lattices that the number of α-sites and the number of β-sites are equal.

[23]The state where all σ_i are equal to -1 is, of course, also a state of complete order, but we shall not consider this case; compare footnote 20.

remind ourselves that Lee and Yang proved the equivalence of an Ising ferromagnet, that is, a ferromagnet characterised by the energy (9.501), and a
lattice gas which is a model which in the limit as the lattice constant gives
a good description of a real gas where the constituents interact through a
delta-function potential.

If the spins on the lattice have a magnetic moment, μ, which we assume
to be along the spin direction and if the system is in a magnetic field B
which we assume to be in the up-spin direction, there will be an extra term
in the energy and instead of Eq.(9.501) we have

$$E' = -\tfrac{1}{2}J \sum_{\{j,k\}} \sigma_j\sigma_k - I \sum_{\{j\}} \sigma_j, \qquad (9.508)$$

where $I = \mu B$ and where $\sum_{\{j\}}$ here and elsewhere indicates a sum over all
spins in the system.

We mentioned earlier that there are two kinds of order: long-range order
which measures how well α-sites are occupied by A atoms — or β-sites
by B atoms — and short-range order which measures how well A atoms
are surrounded by B atoms and vice versa. We shall now introduce two
parameters which measure these two kinds of order. Let r_α and w_α denote,
respectively, the fraction of α-sites occupied by A or by B atoms, and
likewise let r_β and w_β denote the fraction of β-sites occupied by B or by A
atoms. We see easily that there are three relations between these quantities
for the lattices we are considering which have equal numbers of A and B
atoms and of α- and β-sites:

$$r_\alpha = 1 - w_\alpha = r_\beta = 1 - w_\beta, \qquad (9.509)$$

so that only one of them is an independent quantity.

If $N_{A\beta}$ is the number of A atoms on β-sites we have

$$\sum_{\{j\}} \sigma_j = N - 4N_{A\beta}, \qquad (9.510)$$

where we have used the fact that for complete order $\sum_{\{j\}} \sigma_j = N$ and
that for every A atom on an α-site which we interchange with a B atom
on a β-site two of the σ_j change from $+1$ to -1. We also have the relation

$$\tfrac{1}{2}N - N_{A\beta} = \tfrac{1}{2}Nr_\alpha, \qquad (9.511)$$

and combining Eqs.(9.510) and (9.511) we find

$$2r_\alpha = 1 + \frac{1}{N} \sum_{\{j\}} \sigma_j. \qquad (9.512)$$

We now introduce the long-range order parameter R by the equation

$$R = 2r_\alpha - 1 = \frac{1}{N} \sum_{\{j\}} \sigma_j. \qquad (9.513)$$

We see that R ranges from a value of 1 for perfect order, when all σ_j are equal to $+1$, to 0 for perfect disorder, when there are as many σ_j equal to $+1$ as to -1.

To introduce the short-range order parameter ϱ we consider a pair of nearest-neighbour sites which will always consist of one α- and one β- site. The total number of such nearest-neighbour pairs, Q, satisfies the equation[24]

$$Q = \tfrac{1}{2}zN, \tag{9.514}$$

where z is the coordination number, that is, the number of nearest neighbours per site; z equals 2, 4, and 8, respectively, for the one-dimensional chain, the square lattice, and the simple cubic lattice. Moreover, if Q_{AA}, Q_{AB}, Q_{BA}, and Q_{BB} now denote, respectively, the number of pairs with an A (A, B, B) atom on the α-site and an A (B, A, B) atom on the β-site — so that the earlier $Q'_{AA}, \ldots$ are related to the present ones by $Q'_{AA} = Q_{AA}$, $Q'_{BB} = Q_{BB}$, $2Q'_{AB} = Q_{AB} + Q_{BA}$ — we have

$$Q_{AA} + Q_{BB} + Q_{AB} + Q_{BA} = Q, \tag{9.515}$$

and

$$Q_{AA} = Q_{BB}. \tag{9.516}$$

We now define ϱ by the equation

$$\varrho Q = Q_{AB} + Q_{BA} - Q_{AA} - Q_{BB}, \tag{9.517}$$

and using Eq.(9.506) and the relation between Q'_{AB}, Q_{AB} and Q'_{BA} we find

$$\varrho = \frac{1}{Q} \sum_{\{j,k\}} \sigma_j \sigma_k, \tag{9.518}$$

so that using Eqs.(9.513) and (9.517) we can write the energy of the ferromagnet in a magnetic field in terms of the long-range and short-range order parameters:

$$E = -\tfrac{1}{2}JQ\varrho - INR. \tag{9.519}$$

We see that like the long-range order parameter the short-range order parameter equals $+1$ for perfect order and 0 for perfect disorder.

Introducing the quantities

$$K = \frac{J}{2kT}, \tag{9.520}$$

and

$$C = \frac{I}{kT}, \tag{9.521}$$

---

[24]We have neglected here end or surface effects.

we can write the partition function Z of the ferromagnet in the form

$$Z = \sum_{\{\sigma\}} \exp\left[K \sum_{\{j,k\}} \sigma_j \sigma_k + C \sum_{\{j\}} \sigma_j\right], \tag{9.522}$$

where the summation is over all possible combinations of σ_j values.

Treating J (or K) and I (or C) as external parameters we see that knowing the partition function we can find the long- and short-range order from the equations

$$R = \frac{1}{N} \frac{\partial \ln Z}{\partial C}, \tag{9.523}$$

and

$$\varrho = \frac{1}{Q} \frac{\partial \ln Z}{\partial K}. \tag{9.524}$$

From Eq.(9.523) it follows that in the case of a ferromagnet the long-range order is related to the magnetisation.

This can also be seen directly, as follows. If we are dealing with a ferromagnet, the magnetisation M will be given by the relation

$$M = (N_{\text{up}} - N_{\text{down}})\mu, \tag{9.525}$$

or, in terms of the r_α, r_β, w_α, and w_β,

$$M = (r_\alpha + r_\beta - w_\alpha - w_\beta)\mu, \tag{9.526}$$

In Eq.(9.525) N_{up} and N_{down} are, respectively, the total number of up- and down-spins. Using Eqs.(9.509) we find

$$M = 2\mu(2r_\alpha - 1) = 2\mu R, \tag{9.527}$$

which proves our statement.

In the case of a binary alloy there is no term with I in the energy so that Eq.(9.523) has to be changed to

$$R = \frac{1}{N} \left(\frac{\partial \ln Z}{\partial C}\right)_{C=0}. \tag{9.528}$$

9.6.   The Mean-Field Approximation

In § 9.9 we shall discuss some exact solutions of the Ising problem, but before doing that we want to discuss some approximations. The first approach we shall discuss, the so-called mean-field, molecular field, or Bragg-Williams approximation — called the zeroth approximation by Pathria[25] — is the one suggested originally by Gorsky in 1928 and developed by

---

[25] R.K.Pathria, *Statistical Mechanics*, Pergamon, Oxford (1972), § 12.7.

Bragg and Williams in the 1930s. The simplifying assumptions made in this approach are:

1. The thermodynamic behaviour of the system depends on the σ_j only through its dependence on R;

2. The average energy Φ needed to exchange an A atom on an α-site with a B atom on a β-site will be proportional to R:

$$\Phi = R\Phi_0, \tag{9.601}$$

where Φ_0 is the energy needed for this exchange in the state of perfect order.

In our model we have

$$\Phi_0 = 2zJ. \tag{9.602}$$

It is now a matter of straightforward calculations to find the energy, specific heat, and order parameters as functions of the temperature. We shall consider the case where there is no magnetic field present. If we change the number of A atoms on α-sites from $N_{A\alpha}$ to $N_{A\alpha} + dN_{A\alpha}$ we have from Eqs.(9.511) and (9.513)

$$dN_{A\alpha} = \tfrac{1}{2}N\,dr_\alpha = \tfrac{1}{4}N\,dR. \tag{9.603}$$

Since according to our second assumption each move will cost an energy $R\Phi_0$, the change in energy will be given by the equation

$$dE = -R\Phi_0\,dN_{A\alpha} = \tfrac{1}{4}N\Phi_0 R\,dR, \tag{9.604}$$

whence

$$E = E_0\left(1 - R^2\right) + E_1, \tag{9.605}$$

where

$$E_0 = \tfrac{1}{8}N\Phi_0 = \tfrac{1}{4}zJN, \tag{9.606}$$

and where E_1 is the energy of the state of perfect order. If we take the energy zero to correspond to Eq.(9.519), we have $E_1 = -E_0$, but if we want the energy in the perfectly ordered state to be zero, we must put $E_1 = 0$, as we shall do in this section.

If $W(R)$ is the number of ways of arranging the atoms over the lattice sites in a way consistent with a given value of R, the entropy S will be given by the equation

$$S = k\ln W(R), \tag{9.607}$$

where k is again Boltzmann's constant.

Since we must arrange $\tfrac{1}{2}Nr_\alpha$ A atoms over $\tfrac{1}{2}N$ α-sites and $\tfrac{1}{2}Nw_\beta$ A atoms over $\tfrac{1}{2}N$ β-sites, $W(R)$ will be given by the equation

$$W(R) = \begin{pmatrix} \tfrac{1}{2}N \\ \tfrac{1}{2}Nr_\alpha \end{pmatrix} \begin{pmatrix} \tfrac{1}{2}N \\ \tfrac{1}{2}Nw_\beta \end{pmatrix}. \tag{9.608}$$

Using Eqs.(9.509) and (9.513) to express r_α and w_β in terms of R and using the Stirling formula for the factorial,

$$\ln x! = x \ln x - x, \tag{9.609}$$

we get, after a straightforward calculation,

$$S = kN \left[\ln 2 - \tfrac{1}{2}(1 + R)\ln(1 + R) - \tfrac{1}{2}(1 - R)\ln(1 - R)\right]. \tag{9.610}$$

In order to find the equilibrium value of R at a given temperature we must minimise the free energy with respect to R. The free energy $F\ (= E - TS)$ is given by the equation

$$
\begin{aligned}
F = {} & E_1 + E_0(1 - R^2) \\
& - kNT \left[\ln 2 - \tfrac{1}{2}(1 + R)\ln(1 + R) - \tfrac{1}{2}(1 - R)\ln(1 - R)\right]. \tag{9.611}
\end{aligned}
$$

From the minimisation condition,

$$\frac{\partial F}{\partial R} = 0, \tag{9.612}$$

we then get for R the relation

$$R = \tanh \frac{R\Phi_0}{4kT}, \tag{9.613}$$

an equation which must be solved numerically or graphically.

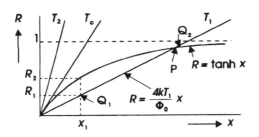

Fig.9.7. The graphical way of determining the long-range order parameter R.

In order to see how the second method works we introduce an auxiliary quantity X through the equation

$$X = \frac{\Phi_0}{4kT}, \tag{9.614}$$

so that we can write Eq.(9.613) in the form

$$R = \tanh X. \qquad (9.615)$$

We must now solve R and X from Eqs.(9.614) and (9.615). First we draw attention to an essential difference between these two equations: Eq.(9.614) being the definition of X holds whether or not we are dealing with an equilibrium situation whereas Eq.(9.615) only holds if we are dealing with an equilibrium situation.

The graphical method of solving for X and R consists in drawing in an X-R-plane the two curves corresponding to Eqs.(9.614) and (9.615) (Fig.9.7). The intersection of these two curves gives us the equilibrium value of R. From Fig.9.7 we see, first of all, that $R = 0$ is always a possible solution; this follows also directly from Eq.(9.613). However, if the temperature is sufficiently low (T_1 in Fig.9.7), so that the slope of the straight line $R = (4kT/\Phi_0)X$ is sufficiently small, there is also another solution, corresponding to the point P in Fig.9.7. We shall show presently that P corresponds to stable equilibrium while the origin corresponds to unstable equilibrium. When the temperature is increased there comes a moment when P is also at the origin. The corresponding temperature is the *critical temperature*, T_c. For temperatures above T_c (T_2 in Fig.9.7) the only solution is $R = 0$ and there will therefore not be any long-range order above T_c. The value of T_c can be calculated from the condition that for this temperature the straight line represented by Eq.(9.614) should be tangent in the origin to the curve represented by Eq.(9.615), that is,

$$\frac{\Phi_0}{4kT_c} = \left(\frac{dR}{dX}\right)_{X=0} = 1, \qquad (9.616)$$

or

$$T_c = \frac{\Phi_0}{4k}. \qquad (9.617)$$

We next show that P corresponds to stable equilibrium. We mentioned earlier that Eq.(9.614) holds even for a non-equilibrium situation. The point Q_1 (see Fig.9.7) thus corresponds to such a non-equilibrium situation. If the temperature and the ordering energy Φ are kept fixed at the same value as that corresponding to Q_1, X is fixed at the value X_1. The equilibrium value of R corresponding to X_1 is equal to R_2 (see Fig.9.7) and the system will thus move towards a *higher* value of R, that is, the representative point will move towards P and not in the direction of the origin. Similar reasoning can be applied to a point Q_2 corresponding to a value of the long-range order parameter larger than its equilibrium value. Another way of proving that P corresponds to stable equilibrium is discussed in Problem 9 at the end of the present chapter.

Once R has been determined by the graphical method as a function of the temperature the configurational energy follows from Eq.(9.605) and the configurational specific heat from the equation

$$c_v = \frac{dE}{dT}. \qquad (9.618)$$

The general behaviour of R, E, and c_v is shown in Fig.9.8, where we have chosen $E_1 = 0$.

Comparing Eqs.(9.605) and (9.519) we see that the Bragg-Williams approximation corresponds to the following approximation for the short-range order in terms of the long-range order:

$$\varrho = R^2, \tag{9.619}$$

or

$$\frac{1}{Q} \sum_{\{j,k\}} \sigma_j \sigma_k = \frac{1}{N^2} \left(\sum_{\{j\}} \sigma_j \right) \left(\sum_{\{k\}} \sigma_k \right). \tag{9.620}$$

Substituting this expression into Eq.(9.501) for the energy we get

$$E = -\left(\frac{\frac{1}{2}JQ}{N^2} \sum_{\{k\}} \sigma_k \right) \left(\sum_{\{j\}} \sigma_j \right), \tag{9.621}$$

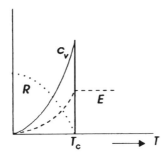

Fig.9.8. The long-range order parameter R, the configurational energy E, and the configurational specific heat c_v in the Bragg-Williams (mean-field) approximation as functions of the temperature.

and comparing this with Eq.(9.508) we see that the Bragg-Williams approximation corresponds to assuming that the spins are situated in a magnetic field B_{eff} given by the equation

$$B_{\text{eff}} = \frac{zJ}{4\mu} \left(\frac{1}{N} \sum_{\{j\}} \sigma_j \right), \tag{9.622}$$

which shows that each spin is subject to a "mean field", proportional to the average value of all the spins in the lattice. Hence the name *mean-field approximation* for this particular approximation. Another way of looking at it is that the approximation is equivalent to assuming that every spin in the lattice is a nearest neighbour of every other spin — of course, with a suitably renormalized interaction constant.

9.7.   The Quasi-Chemical Approximation

In the present section we shall discuss the so-called *quasi-chemical approximation* which is the simplest approximation to take short-range order into account. We shall again consider the case where there is no magnetic field, but we now take into account the fact that there are two parameters, R and ϱ, which characterise the state of the system.

Let $W(R, \varrho)$ be the number of ways in which a certain combination of values of R and ϱ can be realised. The free energy will then be given by the equation

$$F = -kT \ln W(R, \varrho) + E(\varrho). \tag{9.701}$$

The equilibrium value of ϱ for a given value of R is obtained by minimising F with respect to ϱ,

$$\frac{\partial F}{\partial \varrho} = 0, \tag{9.702}$$

and this will give us ϱ as a function of R.

To find the equilibrium value of R we proceed as follows. Let $W(R)$ be again the number of ways in which a situation corresponding to a given value of R can be realised; it will again be given by Eq.(9.608). Introduce now an energy $E'(R)$ by the equation[26]

$$F(R) = -kT \ln W(R) + E'(R). \tag{9.703}$$

This equation is similar to the one leading to Eq.(9.611), and once we know $E'(R)$ we can proceed in the same way as in the previous section, obtaining R by minimising F and the critical point from the equations

$$R = 0, \qquad \frac{\partial F}{\partial R} = 0, \qquad \frac{\partial^2 F}{\partial R^2} = 0. \tag{9.704}$$

From Eq.(9.703) it follows that the partition function Z is given by the equation

$$Z = W(R) \, e^{-\beta E'(R)}. \tag{9.704}$$

On the other hand, we also have

$$Z = \sum_{\{R\}} e^{-\beta E}, \tag{9.705}$$

where the summation is over all configurations with the same value of R and where E is obtained from $E(\varrho)$ by substituting for ϱ its equilibrium

---

[26]It follows from Eq.(9.707) that $E'(R)$ is, in fact, a free energy.

value obtained from Eq.(9.702). We also know that at equilibrium the energy can be expressed in terms of the partition function as follows

$$E = -\frac{\partial \ln Z}{\partial \beta}, \tag{9.706}$$

whence we have

$$E = \frac{\sum_{\{R\}} E(\varrho) e^{-\beta E(\varrho)}}{\sum_{\{R\}} e^{-\beta E(\varrho)}}, \tag{9.707}$$

where again ϱ is a function of R through the use of Eq.(9.702).

Combining Eqs.(9.705) and (9.707) we have

$$E = \frac{\partial \beta E'}{\partial \beta}, \tag{9.708}$$

or

$$E'(R) = kT \int_{\beta=0}^{\beta} E(\varrho) \, d\beta, \tag{9.709}$$

where we have made use of the fact that $E'(R) \to 0$ as $\beta \to 0$.

Substituting Eq.(9.709) for $E'(R)$ into Eq.(9.703) we have

$$F(R) = -kT \ln W(R) + kT \int_{0}^{\beta} E(\varrho) \, d\beta. \tag{9.710}$$

We can now proceed to evaluate $F(R)$. All equations up to this point have been exact, but in obtaining an expression for $W(R, \varrho)$ we shall make some approximations. In the previous section we assumed in writing down Eq.(9.611) for the free energy with the number of configurations $W(R)$ given by Eq.(9.9608) that all configurations with the same value of R had the same weight. We now shall assume that we may treat all pairs in the lattice, that is the quantities Q_{AA}, Q_{AB}, Q_{BA}, and Q_{BB}, as independent quantities.[27] This means that we can write $W(R, \varrho)$ as the product of a factor depending solely on R and a factor giving the number of ways Q can be written as the sum of Q_{AA}, Q_{AB}, Q_{BA}, and Q_{BB}, or

$$W(R, \varrho) = W_1(R) \frac{Q!}{Q_{AA}! Q_{AB}! Q_{BA}! Q_{BB}!}, \tag{9.711}$$

where $W_1(R)$ is again given by Eq.(9.608).

Using Eqs.(9.515) to (9.517) it follows immediately that

$$Q_{AA} = Q_{BB} = \tfrac{1}{4} Q(1 - \varrho), \tag{9.712}$$

---

[27]The name "quasi-chemical" approximation is due to the fact that we are treating here the pairs as independent chemical bonds.

while it follows from the fact that $Q_{AA} + Q_{AB}$ is the number of pairs in the lattice for which the α-site is correctly occupied that we have for Q_{AB} and Q_{BA}

$$\left. \begin{array}{l} Q_{AB} = \frac{1}{4}Q\,(1 + 2R + \varrho), \\ Q_{BA} = \frac{1}{4}Q\,(1 - 2R + \varrho). \end{array} \right\} \tag{9.713}$$

Using Eqs.(9.701), (9.711), and (9.519) (with $I = 0$), as well as the Stirling formula (9.609) for the factorial, we get the following expression for the free energy:

$$F = -kT \left[\ln W_1(R) + Q \ln Q - Q_{AA} \ln Q_{AA} - Q_{AB} \ln Q_{AB} \right.$$
$$\left. - Q_{BA} \ln Q_{BA} - Q_{BB} \ln Q_{BB} \right] - \frac{1}{2}JQ\varrho. \tag{9.714}$$

Using Eqs.(9.712) and (9.713) to express F in terms of ϱ and applying Eq.(9.702), we find the following equation to determine the short-range order parameter:

$$\ln \frac{Q_{AA}\,Q_{BB}}{Q_{AB}\,Q_{BA}} = -\frac{2J}{kT}, \tag{9.715}$$

or

$$\frac{Q_{AA}\,Q_{BB}}{Q_{AB}\,Q_{BA}} = x^2, \tag{9.716}$$

where we have introduced x through the equation

$$x = e^{-2K}, \tag{9.717}$$

with K defined by Eq.(9.520).

Once again using Eqs.(9.712) and (9.713) we find the following expression for ϱ:

$$\varrho = \frac{1 + x^2 - 2x\sqrt{1 - R^2 + R^2 x^2}}{1 - x^2}, \tag{9.718}$$

which for temperatures above T_c where $R = 0$ reduces to

$$\varrho = \frac{1 - x}{1 + x}. \tag{9.719}$$

Using Eq.(9.709) for $E'(R)$, Eq.(9.519) (with $I = 0$) for E, and Eq.(9.718) for ϱ, we find after integration

$$-\frac{E'(R)}{\frac{1}{2}QkT} = 2 \ln \tfrac{1}{2}(u + 1) + (R - 1) \ln \frac{u - R}{1 - R}$$
$$- (R + 1) \ln \frac{u + R}{1 + R} + \frac{J}{kT}. \tag{9.720}$$

and hence, for the free energy

$$F(R) = \tfrac{1}{2}kNT \left[(1+R)\ln(1+R) + (1-R)\ln(1-R) \right.$$

$$- 2\ln 2 + \frac{z}{2} \left\{ (1+R)\ln \frac{u+R}{1+R} \right.$$

$$\left. \left. + (1-R)\ln \frac{u-R}{1-R} - 2\ln \frac{u+1}{2} \right\} \right], \qquad (9.721)$$

where

$$u = \frac{1}{x}\sqrt{1 - R^2 + R^2 x^2}. \qquad (9.722)$$

The equilibrium value of R follows from minimising F and the result is the following equation for R:

$$\left(1 - \tfrac{1}{2}z\right)\ln \frac{1+R}{1-R} + \tfrac{1}{2}z \ln \frac{u+R}{u-R} = 0. \qquad (9.723)$$

The critical temperature follows from Eqs.(9.704) and the result is

$$x_c = 1 - \frac{2}{z}, \qquad (9.724)$$

where

$$x_c = e^{-2J/kT_c}. \qquad (9.725)$$

9.8.   Critical Phenomena

There is a large class of phase transitions in which the two phases between which the transition takes place have different symmetries. This is, of course, true for the first-order transition between a solid and a liquid, but it is also true for the continuous phase transitions between different crystal structures; in those transitions there is no latent heat and the energy, and the volume, are continuous functions, but some thermodynamic function must show a discontinuity since a particular symmetry is either present or not. It is those transitions which were studied in particular by Landau in his classic paper of 1937.[28] The important assumption made by Landau is that in the vicinity of T_c one can expand all thermodynamic functions in power series in the long-range order parameter R, taking into account the symmetry of the system. If we apply this to the Helmholtz free energy for the systems considered in § 9.5, we can write

$$F = F_0 + AR^2 + BR^4, \qquad (9.801)$$

[28]L.D.Landau, *Phys.Zs.Soviet Un.* **11**, 26 (1937); *Collected Papers*, Pergamon Press, Oxford (1965) p.193.

where $B > 0$ because the free energy must be bounded from below, where we have included only even terms because of the symmetry,[29]   and where we have included only quadratic and quartic terms because one can prove that the conclusions we shall be reaching about the critical behaviour are independent of whether or not we include higher-order terms as long as $B > 0$.

The coefficient A in Eq.(9.801) will, in general be a function of the temperature. In Fig.9.9 we have sketched the behaviour of F for two different values of A. We see that if $A > 0$ the equilibrium state of the system corresponds to $R = 0$: there is no long-range order. If, on the other hand, $A < 0$, the equilibrium state of the system corresponds to a non-zero value of R: there is long-range order. It follows that $T = T_c$ corresponds to $A = 0$ and, provided A is not singular at that point, we can expand A in a power series in $T - T_c$:

$$A = a(T - T_c) + \ldots, \qquad (9.802)$$

where a is positive. Hence we get, in the immediate vicinity of T_c,

$$F = F_0 + a(T - T_c)R^2 + BR^4, \qquad (9.803)$$

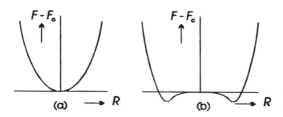

Fig.9.9. The Helmholtz free energy F as function of the long-range order parameter R for $A > 0$ (a) and $A < 0$ (b).

The equilibrium value of the long-range order parameter follows from minimising the free energy, $\partial F / \partial R = 0$:

$$2a(T - T_c)R + 4BR^3 = 0. \qquad (9.803)$$

This equation always has the solution

$$R = 0, \qquad (9.804)$$

[29]Strictly speaking, we considered only positive values of R in our earlier discussions, but we pointed out that there is a symmetry between the states where we exchange all A and B atoms, without renaming the α- and β-sites.

and that is the only solution for $T > T_c$, while for $T < T_c$ there is another solution,

$$R = (a/2B)^{1/2}(T_c - T)^{1/2}, \qquad (9.805)$$

which is for those temperatures the equilibrium solution.

We see that the long-range order parameter is identically equal to zero above T_c, as in the mean-field approximation. This is not surprising since it follows from Eq.(9.611) that for small values of R the free energy in the mean-field approximation can be written in the form

$$F = E_1 + E_0 - kNT \ln 2 - E_0 R^2 + \tfrac{1}{2}NkT\left[R^2 + \tfrac{1}{6}R^4\right], \qquad (9.806)$$

or, if we use Eqs.(9.606) and (9.617),

$$F = F_0 + \tfrac{1}{2}Nk(T - T_c)R^2 + \tfrac{1}{12}NkTR^4. \qquad (9.807)$$

We see thus that Landau's theory gives the same results as the mean-field approximation in the vicinity of the critical point.

The specific heat below the transition temperature is given by Eq.(9.618) and from that equation and Eqs.(9.605) and (9.805) we find to lowest order in $T_c - T$

$$c_v = \tfrac{3}{2}Nk. \qquad (9.808)$$

From a theoretical point of view the behaviour of the system close to the transition temperature is of great interest. Special attention has been paid to the so-called *critical exponents* or *critical indices*, that is, the exponents of the powers of $T - T_c$ of the first term in a power series expansion near the critical point of the various thermodynamics quantities of interest. For instance, in the case of the Ising ferromagnet we see from Eq.(9.805) that if we describe the Ising ferromagnet by the mean-field approximation the critical exponent for the magnetisation — or the long-range order parameter R — is equal to $\tfrac{1}{2}$ while the critical exponent of the specific heat is 0 in the same approximation according to Eq.(9.808).

The importance of the critical exponents is that whereas the value of the transition temperature depends on the details of the interactions (see.g., Eq.(9.617) or bear in mind that in Landau's theory the transition temperature is determined by the vanishing of the system-dependent function $A(T)$) the critical exponents have values which depend solely on the model used, which serves as a model for a large class of physical systems. This has the advantage that when studying a particular physical system one can choose a model which also describes a much simpler physical system for which one can evaluate the critical exponents. In this way one can determine those exponents for the original system. We should warn, however, against attaching too much importance to the critical exponents: we have just seen that in the mean-field approximation the critical exponent for the specific heat is equal to zero, whereas we see from Fig.9.8 that this exponent describes the behaviour of the specific heat only in the immediate vicinity of T_c without giving any indication about the behaviour of the specific heat at a finite distance from T_c. One might think that this is because the mean-field approximation is too rough a model, but that is not the case: for many purposes the mean-field approximation is an extremely good guide to the behaviour of the system, expecially below T_c.

There is another reason why a great deal of attention has been paid to the behaviour of physical systems at and near the critical temperature. The reason is that at the critical temperature the system is scale invariant, that is, the system behaves the same at whatever scale we look at it. To see the effect of scaling consider Fig.9.10 which shows a two-dimensional square lattice. In the case of the two-dimensional Ising ferromagnet there will be one spin on each lattice site which can be in one of two directions, corresponding to $\sigma_i = +1$ or $\sigma_i = -1$ (see § 9.5).

We can now look at the effect of scaling as follows. Divide the lattice in blocks of nine spins each (one such block has been outlined in Fig.9.10). Take the average spin value in that block and "renormalise" it such that we replace it by $+1$ if the average is positive, and by -1 if the average is negative. We then are led to a scaled lattice which has still the same symmetry structure as the old one and where again on each "lattice site" there is a spin with a spin-value equal to either $+1$ or -1. Repeat this and see what happens (see Problems 12 and 13 at the end of this chapter).

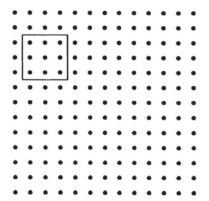

Fig.9.10. The two-dimensional square lattice.

The result is the following. If we start from a situation corresponding to a temperature below T_c, that is, a situation where the long-range order parameter is different from zero so that there are more up- than down-spins (Problem 13) the chances for blocks to have a positive average will be greater than to have a negative average and therefore there will be more of them; renormalising them to a spin value of $+1$ increases the magnetisation, that is, the long-range order and the situation will increasingly, with each scaling step, look like the situation at $T = 0$, that is, a situation where there is perfect order.

On the other hand, if we start from a situation where there are as many up- as down-spins, the situation will be perpetuated and, in fact, the situation will in that case look more and more like the situation at $T = \infty$. In order to see that one must use a larger lattice than is envisaged in Problems 12 and 13. This has been done by numerical methods and the situation is, in fact, as we have described it.[30] The situation at T_c itself is such that it does not change under scaling: it is scale invariant.

One can look at these results also from the point of view of correlation lengths. We define the two-spin correlation function $G^{(2)}$ by the following equation:[31]

$$G_{ij}^{(2)} = \langle (\sigma_i - \langle \sigma_i \rangle) \rangle \langle (\sigma_j - \langle \sigma_j \rangle) \rangle, \qquad (9.809)$$

where the pointed brackets indicate thermal or ensemble averages. In the case of a translationally invariant lattice, such as the two-dimensional square lattice, this correlation function will depend only on the distance between the two spins, $r_i - r_j \equiv r_{ij}$. For distances between the spins large as compared to the lattice constant, $G^{(2)}$ will depend only on the absolute magnitude, r, of r_{ij} and for temperatures either below or above T_c is is found that the dependence is exponential:

$$G^{(2)} \propto e^{-r/\xi}, \qquad (9.810)$$

where ξ is the so-called *correlation length*.

At the critical temperature itself Eq.(9.810) no longer holds, but we have

$$G^{(2)} \propto \frac{1}{r^{d-2+\eta}}, \qquad (9.811)$$

where d is the dimensionality of the lattice and where η is another critical exponent. In terms of a correlation length, the power-law relation (9.811) means that the correlation length at T_c is infinite.

The correlation length essentially measures the extent to which short-range order prevails. If it is finite, this means that over distances larger than ξ from a given spin another spin no longer feels the influence of the first spin. Hence we can understand that when we carry out the scaling exercise after a while only the long-range order will be of importance and hence the system will behave as if it were at either $T = 0$, if there was long-range order to start with, or $T = \infty$, if there was no long-range order at the start. At $T = T_c$, however, scaling cannot get rid of the short-range order, and the system behaves at all times in the same way: it is scale invariant.

---

[30]See, for instance, J.M.Yeomans, *Statistical Mechanics of Phase Transitions*, Oxford University Press (1992), Figs.1.8, 1.9, and 1.10.

[31]In the definition of $G^{(2)}$ we have taken into account that below T_c the average value of the spin variable σ_i is non-vanishing so that, in fact, $G^{(2)}$ measures the correlation in the fluctuations of the σ_i.

This scale invariance of the system at T_c is an important reason for the interest in continuous phase transitions since there are many systems in physics which also show scale invariance: instances are the inertial range of hydrodynamic turbulence, the clustering of galaxies, and polymers.

We do not intend to go into a detailed discussion of critical exponents or of the scaling transformations which enable one to derive the values of some of them, but we shall briefly discuss the so-called *renormalisation group* methods. For details we must refer to the literature given in the Bibliographical Notes at the end of the chapter.

We must now formalise the rough and ready procedure we described earlier. We start with the example of the one-dimensional Ising ferro-magnet (Fig.9.11). We start off with a spin on each of the lattice sites. The energy of the system is given by Eq.(9.508),

$$E = -\tfrac{1}{2} J \sum_{\{j,k\}} \sigma_j \sigma_k - I \sum_{\{j\}} \sigma_j, \tag{9.812}$$

(a)

$$\cdots \ 1 \quad 2 \quad 3 \quad 4 \quad 5 \quad 6 \quad 7 \quad 8 \quad 9 \quad 10 \quad 11 \ \cdots$$

(b)

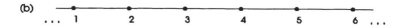

$$\cdots \ 1 \qquad 2 \qquad 3 \qquad 4 \qquad 5 \qquad 6 \ \cdots$$

Fig.9.11. The one-dimensional Ising model: (a) before and (b) after renormalisation.

where we have assumed that the system is placed in an external magnetic field in the direction of the up-spins. We shall rewrite this equation slightly (i) to bring out the special nature of the one-dimensional lattice, (ii) to distinguish the even- and the odd-numbered spins, since the former will be got rid of in the first renormalisation step (see Fig.9.11), and (iii) to introduce an extra term which will become essential in later steps of the renormalisation process, as we shall see presently:

$$E = -\tfrac{1}{2} J \sum_{j=\ldots,2,4,6,\ldots} \sigma_j \left(\sigma_{j+1} + \sigma_{j-1} \right)$$

$$- I \sum_{j=\ldots,2,4,6,\ldots} \left(\sigma_j + \tfrac{1}{2}(\sigma_{j+1} + \sigma_{j-1}) \right) - \sum_{j=\ldots,2,4,6,\ldots} 2A, \tag{9.813}$$

where A is a constant.

This means that we can write the partition function in the form

$$Z = \sum_{\sigma_i = \pm 1} \prod_{j=\ldots,2,4,6,\ldots} e^{K \sigma_j \left(\sigma_{j+1} + \sigma_{j-1} \right) + C \left(\sigma_j + \tfrac{1}{2} \left(\sigma_{j+1} + \sigma_{j-1} \right) \right) + 2D},$$

$$\tag{9.814}$$

where K and C are given by Eqs.(9.520) and (9.521) and where

$$D = \frac{A}{kT}. \tag{9.815}$$

The first renormalisation step now consists in getting rid of the even-numbered spins by summing over the corresponding σ_j. This is straightforward and the result is the "renormalised" partition function

$$Z' = \sum_{\sigma_i=\pm 1}' \prod_{j=\ldots,2,4,6,\ldots} \left[e^{K\left(\sigma_{j+1}+\sigma_{j-1}\right)+C+\frac{1}{2}C\left(\sigma_{j+1}+\sigma_{j-1}\right)+2D} \right.$$
$$\left. + e^{-K\left(\sigma_{j+1}+\sigma_{j-1}\right)-C+\frac{1}{2}C\left(\sigma_{j+1}+\sigma_{j-1}\right)+2D} \right], \tag{9.816}$$

or, after renumbering the remaining spins consecutively (see Fig.9.11b),

$$Z' = \sum_{\sigma_i=\pm 1} \prod_j \left[e^{\left(K+\frac{1}{2}C\right)\left(\sigma_{j+1}+\sigma_{j-1}\right)+C+2D} \right.$$
$$\left. + e^{\left(-K+\frac{1}{2}C\right)\left(\sigma_{j+1}+\sigma_{j-1}\right)-C+2D} \right], \tag{9.817}$$

where the prime on the summation sign in Eq.(9.816) indicates that the summation is over the odd-numbered spins only.

We now require that Eq.(9.817) has the same form as the original partition function, but with renormalised constants, that is, we we require that Z' has the form

$$Z' = \sum_{\sigma_i=\pm 1} \prod_j e^{K'\sigma_j\sigma_{j+1}+C'\sigma_j+D'}. \tag{9.818}$$

This leads to the recursion formulæ

$$e^{4K'} = \frac{\cosh(2K+C)\cosh(2K-C)}{\cosh^2 C}, \tag{9.819}$$

$$e^{2C'} = \frac{e^{2C}\cosh(2K+C)}{\cosh(2K-C)}, \tag{9.820}$$

$$e^{4D'} = e^{8D}\cosh(2K+C)\cosh(2K-C)\cosh^2 C. \tag{9.821}$$

In the three-dimensional temperature, magnetic field, exchange coupling constant space Eqs.(9.819) to (9.821) map a trajectory which leads from one point to another after each renormalisation step. Scale invariance occurs when the renormalised point in this parameter space is the same as the original point. For the one-dimensional Ising ferromagnet this occurs only at zero temperature. There is no finite critical temperature in this case, that is, there is no phase transition at any finite temperature.

The one-dimensional Ising model is particular in that the renormalisation procedure leads to the same form of the partition function after

each step. In general this is not the case if we use a simple form of
the Hamiltonian; the exact renormalisation procedure will lead to extra couplings as can be seen from the example of the two-dimensional
Ising ferromagnet on a square lattice. We shall briefly consider this
case.

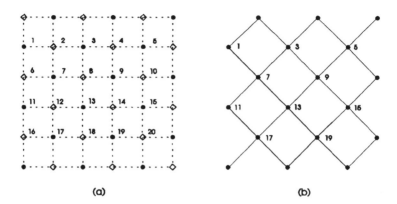

(a)   (b)

Fig.9.12. The two-dimensional Ising ferromagnet on
a square lattice: (a) before and (b) after renormalisation.

In Fig.9.12a we show the original lattice. As in the case of the one-
dimensional Ising ferromagnet we want to scale by eliminating the
even-numbered spins. We shall not consider the case where there is
a magnetic field present so that we put $I = 0$ in Eq.(9.812). In order
to eliminate the even-numbered spins we proceed as before and, first
of all, write the energy of the lattice in such a way that we gather together all terms referring to each of the even-numbered spins (compare
Eq.(9.813):

$$E = -\tfrac{1}{2} J \left\{ \ldots + \sigma_8 \Big[\sigma_3 + \sigma_7 + \sigma_9 + \sigma_{13}\Big] + \sigma_{12}\Big[\sigma_7 + \sigma_{11} \right.$$
$$\left. + \sigma_{13} + \sigma_{17}\Big] + \sigma_{14}\Big[\sigma_9 + \sigma_{13} + \sigma_{15} + \sigma_{19}\Big] + \ldots \right\}. \quad (9.822)$$

Substituting expression (9.822) into the expression for the partition
function and summing over the even-numbered spins we get for the new
partition function

$$Z' = \sum_{\sigma_i = \pm 1}' \ldots \times \left\{ e^{K\left(\sigma_3 + \sigma_7 + \sigma_9 + \sigma_{13}\right)} + e^{-K\left(\sigma_3 + \sigma_7 + \sigma_9 + \sigma_{13}\right)} \right\}$$
$$\times \left\{ e^{K\left(\sigma_7 + \sigma_{11} + \sigma_{13} + \sigma_{17}\right)} + e^{-K\left(\sigma_7 + \sigma_{11} + \sigma_{13} + \sigma_{17}\right)} \right\} \times \ldots \quad (9.823)$$

where the prime on the summation sign indicates that the summation is over the odd-numbered spins only.

We now want the partition function to look again the same as before the summation over the even-numbered spins, that is, we want to find a constant K' such that we can write

$$e^{K\left(\sigma_3+\sigma_7+\sigma_9+\sigma_{13}\right)} + e^{-K\left(\sigma_3+\sigma_7+\sigma_9+\sigma_{13}\right)}$$
$$= f(K)\,e^{K'\left(\sigma_3\sigma_7+\sigma_7\sigma_{13}+\sigma_{13}\sigma_9+\sigma_9\sigma_3\right)}, \qquad (9.824)$$

independent of the values of σ_3, σ_7, σ_{13}, and σ_9. If that were possible we would have exactly the same partition function for the scaled lattice of Fig.12b[32] as for the original lattice of Fig.12a, but with a renormalised constant K' instead of K. However, it is not possible and the simplest possibility is one involving three new constants K_1, K_2, and K_3 such that[33]

$$e^{K\left(\sigma_3+\sigma_7+\sigma_9+\sigma_{13}\right)} + e^{-K\left(\sigma_3+\sigma_7+\sigma_9+\sigma_{13}\right)}$$
$$= f(K)\,e^{\frac{1}{2}K_1\left(\sigma_3\sigma_7+\sigma_7\sigma_{13}+\sigma_{13}\sigma_9+\sigma_9\sigma_3\right)}$$
$$\times\, e^{K_2\left(\sigma_3\sigma_{13}+\sigma_7\sigma_9\right)+K_3\left(\sigma_3\sigma_7\sigma_{13}\sigma_9\right)}, \qquad (9.825)$$

where K_1, K_2, and K_3 satisfy the equations

$$\left.\begin{aligned}
K_1 &= \tfrac{1}{4}\,\ln\cosh(4K), \\
K_2 &= \tfrac{1}{8}\,\ln\cosh(4K), \\
K_3 &= \tfrac{1}{8}\,\ln\cosh(4K) - \tfrac{1}{2}\,\ln\cosh(2K), \\
f(K) &= 2\left[\cosh(2K)\right]^{1/2}\left[\cosh(4K)\right]^{1/8}.
\end{aligned}\right\} \qquad (9.826)$$

This means that the partition function $Z(K,N)$ of the original lattice of, say, N spins which is a function of the single constant K now, after summing over all even-numbered spins, takes the form

$$Z(K,N) = \sum_{\sigma_i=\pm 1} e^{K\sum'_{ij}\sigma_i\sigma_j}, \qquad (9.827)$$

$$= \left[f(K)\right]^{N/2} \sum_{\sigma_i=\pm 1} e^{K_1\sum'_{ij}\sigma_i\sigma_j+K_2\sum''_{kl}\sigma_k\sigma_l}$$
$$\times\, e^{K_3\sum'''_{mnpq}\sigma_m\sigma_n\sigma_p\sigma_q}, \qquad (9.828)$$

where the sums over all spins are over N spins in Eq.(9.827) and over $\frac{1}{2}N$ spins (the odd-numbered ones) in Eq.(9.828), where the primed

---

[32] Note that in this case the square lattice has been turned over an angle of $\pi/4$.

[33] The factor $\frac{1}{2}$ in front of K_1 is inserted because each pair of nearest neighbours occurs twice when we sum over the whole of the lattice: the pair $\sigma_7\sigma_{13}$, for instance, comes from the squares centered around lattice sites 8 and 12.

sums are over all nearest-neighbour pairs in the relevant lattice — that of Fig.12a in Eq.(9.827) and that of Fig.12b in Eq.(9.828), — where the doubly primed sum is over all pairs of next-nearest neighbours, such as spins 7 and 9 or spins 9 and 19, and where the triply primed sum is over all four nearest-neighbour spins around a square, such as spins 9, 15, 19, and 15.

We see that we have not succeeded in our attempt to reproduce a renormalised partition function of the same form. The next reduction in the number of spins, that is, the number of degrees of freedom, would further complicate the partition function and the renormalisation procedure clearly has to be supplemented by an approximation. The simplest would be to put $K_2 = K_3 = 0$. In that case K_1 is simply related to K by the first of Eqs.(9.826). This is similar to the result of the one-dimensional Ising ferromagnet and would not lead to a phase transition. The next step would be to put $K_3 = 0$, but incorporate the effect of next-nearest neighbour interactions by altering the nearest-neighbour interactions, that is, by trying to find a constant $K'(K_1, K_2)$ such that

$$K_1 {\sum_{ij}}' \sigma_i \sigma_j + K_2 {\sum_{kl}}'' \sigma_k \sigma_l \ = \ K'(K_1, K_2) {\sum_{ij}}' \sigma_i \sigma_j. \qquad (9.829)$$

By considering the state where all spins are up-spins one finds

$$K' \ = \ K_1 + K_2, \qquad (9.830)$$

and this leads to the prediction of a phase transition (see Problem 16 at the end of this chapter).

Let us now for a moment consider the general procedure. The energy of the system after the n-th scaling step, $E^{(n)}$, will be a function of several parameters — in the case of the one-dimensional Ising ferromagnet these were K, C, and D — so that we can represent it by a vector, $L^{(n)}$, say, in parameter space. The next scaling step will transform this vector into a new vector, $L^{(n+1)}$. Representing the scaling process, including the renormalisation of the parameters using recursion formulæ such as Eqs.(9.819) to (9.821), by an operator $\widehat{\Omega}$ we can write the scaling operation in the symbolic form

$$\widehat{\Omega} L^{(n)} \ = \ L^{(n+1)}, \qquad (9.831)$$

which defines a mapping in the parameter space.

In parameter space there will, in general, be some *fixed points*, that is, points such that they will be mapped onto themselves. They will thus satisfy the equation

$$\widehat{\Omega} L \ = \ L. \qquad (9.832)$$

In these fixed points the scaling process does not change the Hamiltonian and we see that those are the situation where we have scale invariance.

There are three kinds of fixed points: attractive, repulsive, and mixed fixed points. In the case of an attractive fixed point, any point in its immediate vicinity will move towards it — be attracted to it — if we repeat the scaling operation (9.831). On the other hand in the case of a repulsive fixed point, any point in its immediate vicinity will move away from it — be repelled by it — if we repeat the scaling operation (9.831). The interesting fixed points, however, are the mixed points where for some points in their immediate vicinity there is attraction and for other points there is repulsion. In Fig.9.13 we have schematically sketched the situation corresponding to the cd cd two-dimensional

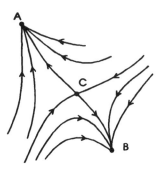

Fig.9.13. The fixed points of the two-dimensional Ising model on a square lattice: A and B are the low- and the high-temperature attractive fixed points and C is the mixed fixed point corresponding to the transition point.

Ising model on a square lattice. We discussed earlier that Problems 12 and 13 show that the system will tend to move to either the situation at $T = 0$ or the situation at $T = \infty$. There are thus two attractive fixed points, the high- and the low-temperature points (A and B in Fig.9.13). The third fixed point is a mixed point and corresponds to $T = T_c$, and we see that we have, indeed, scale invariance at $T = T_c$, but not at any other finite temperature. The situation there is clearly complicated since we can have either attraction or repulsion, and we would expect that the regions of attractions and those of repulsion will be separated from one another. In the case of the two-dimensional Ising model that is, indeed, the case and there is a *critical surface* which separates them. The behaviour on this critical surface determines the critical exponents; for a detailed discussion of the use of the renormalisation group method to obtain information about the critical behaviour we refer to the literature.[34]

---

[34]See, for instance, J.J.Binney, N.J.Dowrick, A.J.Fisher, and M.E.J.Newman, *The Theory of Critical Phenomena*, Oxford University Press (1992), Chapter 5.

9.9. Some Exact Results

The partition function of the Ising ferromagnet is given by Eq.(9.522):

$$Z = \sum_{\{\sigma\}} \exp\Big[K \sum_{\{j,k\}} \sigma_j\sigma_k + C \sum_{\{j\}} \sigma_j\Big], \qquad (9.901)$$

where the summation is over all possible combinations of σ_j values.

Let us first consider a linear chain. The probability $P(\sigma_1,\sigma_2,\ldots,\sigma_{n-1})$ that the spins $\sigma_1,\sigma_2,\ldots,\sigma_{n-1}$ of a chain of $n-1$ members will have given values (for instance $\sigma_1 = -1$, $\sigma_2 = +1$, $\sigma_3 = +1$, $\sigma_4 = -1$, $\ldots$, $\sigma_{n-1} = +1$) will be proportional to the Boltzmann factor $e^{-E/kT}$, since the a priori probability (or weight) for any arrangement is the same. Using Eq.(9.812) for the energy and Eqs.(9.520) and (9.521) we then find that

$$P(\sigma_1,\sigma_2,\ldots,\sigma_{n-1}) = p_{n-1}\, e^{K(\sigma_1\sigma_2+\sigma_2\sigma_3+\cdots+\sigma_{n-2}\sigma_{n-1})+C(\sigma_1+\cdots+\sigma_{n-1})},$$
$$(9.902)$$

where p_{n-1} is a normalising constant. Including one more spin in the chain (see Fig.9.14) we have

$$P(\sigma_1,\sigma_2,\ldots,\sigma_n) = \frac{p_n}{p_{n-1}}\, P(\sigma_1,\sigma_2,\ldots,\sigma_{n-1})\, e^{K\sigma_{n-1}\sigma_n+C\sigma_n}. \quad (9.903)$$

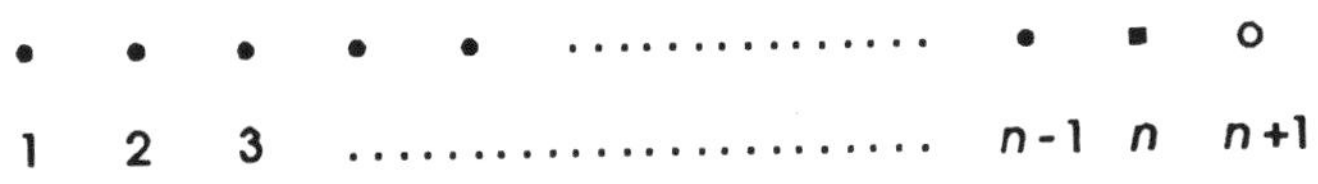

Fig.9.14. Adding one more spin to a linear chain.

From Eqs.(9.902) and (9.903) we can obtain the probabilities $P(\sigma_{n-1})$ and $P(\sigma_{n-1},\sigma_n)$ that, respectively, σ_{n-1} and both σ_{n-1} and σ_n have given values, independent of the values of the $n-2$ other spins. For those probabilities we have

$$P(\sigma_{n-1}) = \sum_{\sigma_1=\pm1} \cdots \sum_{\sigma_{n-2}=\pm1} P(\sigma_1,\sigma_2,\ldots,\sigma_{n-1}), \qquad (9.904)$$

and

$$P(\sigma_{n-1},\sigma_n) = \sum_{\sigma_1=\pm1} \cdots \sum_{\sigma_{n-2}=\pm1} \frac{p_n}{p_{n-1}} P(\sigma_1,\sigma_2,\ldots,\sigma_{n-1})\, e^{K\sigma_{n-1}\sigma_n+C\sigma_n},$$
$$(9.905)$$

or

$$\lambda\, P(\sigma_{n-1},\sigma_n) = P(\sigma_{n-1})\, e^{K\sigma_{n-1}\sigma_n+C\sigma_n}, \qquad (9.906)$$

with $\lambda = p_{n-1}/p_n$.

Summing both sides of Eq.(9.906) over the two possible values of σ_{n-1} we get the probability $P(\sigma_n)$ that σ_n has a given value irrespective of the values of the $n-1$ preceding spins, that is,

$$\lambda P(\sigma_n) = \sum_{\sigma_{n-1}=\pm 1} P(\sigma_{n-1}) e^{K\sigma_{n-1}\sigma_n + C\sigma_n}. \tag{9.907}$$

If our chain is sufficiently long $P(\sigma_n)$ and $P(\sigma_{n-1})$ should be the same functions of their argument. We can considering the $P(\sigma)$ to be one by two matrices.[35] Equation (9.907) is then a mtrix equation and, introducing the new spinor $A(\sigma)$ through the relation

$$A(\sigma) = P(\sigma) e^{-\frac{1}{2}C\sigma}, \tag{9.908}$$

we can write it in the form

$$\lambda A(\sigma) = \sum_{\sigma'=\pm 1} \mathcal{T}(\sigma,\sigma') A(\sigma'), \tag{9.909}$$

where $\mathcal{T}(\sigma,\sigma')$ is a two by two matrix whose elements are given by the equation[36]

$$\mathcal{T}(\sigma,\sigma') = e^{K\sigma\sigma' + \frac{1}{2}C(\sigma+\sigma')}. \tag{9.910}$$

Written out in all detail Eq.(9.909) looks as follows:

$$\lambda \begin{pmatrix} A(+1) \\ A(-1) \end{pmatrix} = \begin{pmatrix} e^{K+C} & e^{-K} \\ e^{-K} & e^{K-C} \end{pmatrix} \begin{pmatrix} A(+1) \\ A(-1) \end{pmatrix}. \tag{9.911}$$

Equation (9.909) has the form of a matrix eigenvalue equation. Let us examine the importance of the eigenvalues of $\mathcal{T}$. We denote the orthonormalised eigenvectors of $\mathcal{T}$ by $A_1(\sigma)$ and $A_2(\sigma)$ and the corresponding eigenvalues by λ_1 and λ_2. Since $\mathcal{T}(\sigma,\sigma')$ is a function of both σ and σ' we can expand it in the form

$$\mathcal{T}(\sigma,\sigma') = \sum_{i,j=1}^{2} c_{ij} A_i(\sigma) A_j(\sigma'), \tag{9.912}$$

where the c_{ij} are given by the equations

$$c_{ij} = \sum_{\sigma=\pm 1} \sum_{\sigma'=\pm 1} \mathcal{T}(\sigma,\sigma') A_i(\sigma) A_j(\sigma'). \tag{9.913}$$

---

[35]Such matrices are called *spinors* because of their occurrence in the Dirac theory of the spinning electron.

[36]The matrix $\mathcal{T}$ is the so-called *transfer matrix*.

Since the $A_i(\sigma)$ are the orthonormalised eigenvectors of $\mathcal{T}$ we have

$$\sum_{\sigma'=\pm 1} \mathcal{T}(\sigma, \sigma') A_j(\sigma') = \lambda_j A_j(\sigma), \qquad (9.914)$$

and

$$\sum_{\sigma=\pm 1} A_i(\sigma) A_j(\sigma) = \delta_{ij}, \qquad (9.915)$$

where δ_{ij} is the Kronecker symbol.

Using these relations we find from Eq.(9.913) that $c_{ij} = \lambda_i \delta_{ij}$ and hence that

$$\mathcal{T}(\sigma, \sigma') = \lambda_1 A_1(\sigma) A_1(\sigma') + \lambda_2 A_2(\sigma) A_2(\sigma'). \qquad (9.916)$$

It now easily follows from the orthonormality relation (9.915) and Eq.(9.916) that we have

$$\sum_{\sigma_2=\pm 1} \mathcal{T}(\sigma_1, \sigma_2) \mathcal{T}(\sigma_2, \sigma_3) = \lambda_1^2 A_1(\sigma_1) A_1(\sigma_3) + \lambda_2^2 A_2(\sigma_1) A_2(\sigma_3), \qquad (9.917)$$

$$\sum_{\sigma_2=\pm 1} \sum_{\sigma_3=\pm 1} \mathcal{T}(\sigma_1, \sigma_2) \mathcal{T}(\sigma_2, \sigma_3) \mathcal{T}(\sigma_3, \sigma_4)$$
$$= \lambda_1^3 A_1(\sigma_1) A_1(\sigma_4) + \lambda_2^3 A_2(\sigma_1) A_2(\sigma_4), \qquad (9.918)$$

and so on, until

$$\sum_{\sigma_2,\ldots,\sigma_N=\pm 1} \mathcal{T}(\sigma_1, \sigma_2) \mathcal{T}(\sigma_2, \sigma_3) \cdots \mathcal{T}(\sigma_N, \sigma_{N+1})$$
$$= \lambda_1^N A_1(\sigma_1) A_1(\sigma_{N+1}) + \lambda_2^N A_2(\sigma_1) A_2(\sigma_{N+1}). \qquad (9.919)$$

Assuming that $\sigma_1 = \sigma_{N+1}$, which corresponds either to closing the chain to form a ring or to imposing a periodicity condition on the chain, we can sum Eq.(9.919) over σ_1 and obtain

$$\sum_{\sigma_1,\ldots,\sigma_N=\pm 1} \mathcal{T}(\sigma_1, \sigma_2) \mathcal{T}(\sigma_2, \sigma_3) \cdots \mathcal{T}(\sigma_N, \sigma_1) \left(= \mathrm{Tr}\, \mathcal{T}^N \right) = \lambda_1^N + \lambda_2^N. \qquad (9.920)$$

From Eqs.(9.911) and (9.901) it follows that the left-hand side of Eq.(9.920) is just the partition function Z of a chain of N spins.

We have thus proven that the partition function of the one-dimensional Ising ferromagnet of N spins in a magnetic field is given by the equation

$$Z = \lambda_1^N + \lambda_2^N, \qquad (9.921)$$

where the λ_i are the eigenvalues of the matrix (9.911). If N is sufficiently large and if $|\lambda_2| < |\lambda_1|$ we can write Eq.(9.921) in the form[37]

$$Z = \lambda_1^N. \tag{9.922}$$

The eigenvalues of $\mathfrak{T}$ are found from the equation

$$\begin{vmatrix} e^{K+C} - \lambda & e^{-K} \\ e^{-K} & e^{K-C} - \lambda \end{vmatrix} = 0, \tag{9.923}$$

or

$$\lambda_{1,2} = e^K \cosh C \pm \sqrt{e^{2K} \sinh^2 C + e^{-2K}}. \tag{9.924}$$

In the case when there is no magnetic field present, so that $C = 0$, we find for Z

$$Z = (2 \cosh K)^N, \tag{9.925}$$

a result which could also have been obtained from directly summing the partition function of the one-dimensional Ising ferromagnet.

Using Eqs.(9.922), (9.924), and (9.528) we see that the long-range order is always equal to zero. This corresponds to the well known fact that a one-dimensional lattice will not show ferromagnetism.[38] The reason is that one wrong spin will upset completely any tendency for long-range order since there is no way for the later spins to find out whether or not they are in accordance with the earlier parts of the chain. The situation is different in the two-dimensional case since then each spin is connected with all the other spins in the lattice in a multitude of ways and not only through a single nearest neigbour.[39]

The reason for treating the rather trivial one-dimensional case so extensively is that we now can immediately use the same method for higher-dimensional ferromagnets. In those cases the calculation of the partition function can also be reduced to finding the largest eigenvalue of a matrix. However, the matrices are no longer simple and it requires more powerful methods to obtain either an exact solution — in the cases where this is possible — or an approximate one. We shall confine our discussion to the two-dimensional Ising ferromagnet on a square lattice for the case without a magnetic field.

Let us remind ourselves what we did to find the partition function of the linear chain. We built up the chain by adding spins to it one by

---

[37]The various thermodynamic quantities which one calculates using the partition function involve $\ln Z$ and in going over from Eq.(9.921) to (9.922) one neglects terms of order $(\lambda_2/\lambda_1)^N$ as compared to the main term $N \ln \lambda_1$.

[38]This is true only as long as the interaction is short-range. If the interaction has an infinite range as in the Kac-Hemmer-Uhlenbeck model of a van der Waals gas (see §7.2) there can be a phase transition at a finite temperature.

[39]We have assumed here that we are dealing with the Ising model with nearest-neighbour interactions only. As long as the interactions between the spins are short-range, the same argument holds *mutatis mutandis*.

one. If the chain is sufficiently long, the addition of one more spin does not alter the physical situation and we get an equation relating $P(\sigma_n)$ to $P(\sigma_{n-1})$, or, since these two probabilities should for a long chain be the same functions of their arguments, a matrix equation for $P(\sigma)$. The largest eigenvalue of the (symmetrised) matrix then gives us the partition function.

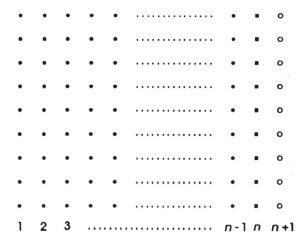

Fig.9.15. Adding one more column to a square lattice.

In the two-dimensional case the step to be repeated is the addition of one more column (Fig.9.15), while in the three-dimensional case one adds successive planes of spins. Let us consider a two-dimensional lattice where each column conatins a spins and where we add columns until the lattice contains $a \times b \ (= N)$ spins. Denoting the spins of the nth column by $\sigma_i \ (i = 1, 2, \ldots, a)$ and those of the $n - 1$st column by $\sigma_i' \ (i = 1, 2, \ldots, a)$ we get instead of Eq.(9.907) the equation

$$\rho P(\sigma_i) = \sum_{\sigma_i' = \pm 1} \mathcal{K}(\sigma_i, \sigma_i') \, P(\sigma_i'), \tag{9.926}$$

where ρ appears due to normalisation, where $P(\sigma_i')$ and $P(\sigma_i)$ are the probabilities that the a spins in the $n - 1$st and the nth columns have given values, and where $\mathcal{K}(\sigma_i, \sigma_i')$ is given by the equation

$$\mathcal{K}(\sigma_i, \sigma_i') = \exp\left[K \sum_{i=1}^{a-1} \sigma_i' \sigma_{i+1}' + K \sum_{i=1}^{a} \sigma_i \sigma_i' \right]. \tag{9.927}$$

Introducing $A(\sigma_i)$ by the equations

$$A(\sigma_i) = P(\sigma_i) \, e^{\frac{1}{2} K \sum \sigma_i \sigma_{i+1}}, \tag{9.928}$$

we have brought Eq.(9.926) into symmetric form:

$$\rho A(\sigma_i) = \sum_{\sigma_i' = \pm 1} \mathcal{T}(\sigma_i, \sigma_i') \, A(\sigma_i'), \qquad (9.929)$$

with

$$\mathcal{T}(\sigma_i, \sigma_i') = e^{K\left[\sum \sigma_i \sigma_i' + \frac{1}{2} \sum \sigma_i \sigma_{i+1} + \sum \sigma_i' \sigma_{i+1}'\right]}. \qquad (9.930)$$

The matrix $\mathcal{T}$ is now of order 2^a and if we expand it in terms of its eigenvectors we can again show that the partition function Z is determined by the eigenvalues ρ_j of $\mathcal{T}$ in the following way:

$$Z \,(= \operatorname{Tr} \mathcal{T}^b \,) = \sum_{j=1}^{2^a} \rho_j^b, \qquad (9.931)$$

where we have once again imposed a periodicity condition $(\sigma_i^{(1)} = \sigma_i^{(b+1)})$ on the lattice.

If b is sufficiently large only the largest eigenvalue will contribute and we have

$$Z = \rho_{\max}^b. \qquad (9.932)$$

In 1944 Onsager succeeded in finding an exact expression for $\rho_{\max}$ but, unfortunately, the calculation is rather long and complicated so that we restrict ourselves by referring to the literature given in the Bibliographical Notes at the end of this chapter.

In concluding this section we want to indicate how one can determine the exact position of the transition temperature without having the complete solution. If we introduce a variable K^* through the equation

$$\sinh 2K \, \sinh 2K^* = 1, \qquad (9.933)$$

and put

$$K^* = \frac{J}{2kT^*}, \qquad (9.934)$$

we have associated with each temperature T another temperature T^*. We see that as T increases from 0 to ∞, T^* decreases from ∞ to 0.

For the case of the two-dimensional Ising ferromagnet on a square lattice without a magnetic field one can show that[40]

$$\frac{Z(T)}{2^{N/2}(\cosh 2K)^N} = \frac{Z(T^*)}{2^{N/2}(\cosh 2K^*)^N}, \qquad (9.935)$$

which relates the partition function at temperature T to its value at temperature T^*. This means that critical temperatures must occur in

---

[40] For a proof see, for instance, D.ter Haar, *Elements of Statistical Mechanics*, Rinehart, New York (1954), § 12.7.

pairs, or, if there is just one critical temperature, T_c, it must occur for a value of K_c which satisfies the equation

$$\sinh 2K_c = 1. \tag{9.936}$$

Problems

1. Use Eqs.(5.0926), (5.0928), and (5.1001) and the definition $v = V/\langle N \rangle$ to prove that $(\partial P/\partial v)_{V,T}$ is always negative for finite V.

2. Use grand ensemble theory to consider the Einstein condensation.[41]

   (i) In § 4.7 we shifted the energy levels so that the lowest level, ε_0, was equal to zero. Show that the zero-point pressure neglected in this way is of the order of $N^{-2/3}$.

   (ii) Consider the region where $v < v_c$. Assuming that

   $$-\alpha \ll -\alpha + \beta \varepsilon_1, \tag{A}$$

   prove that $-\alpha = \mathcal{O}(N^{-1})$ so that, indeed, (A) is satisfied. Use this result to find the slope of that part of the isotherm where $v < v_c$.

   (iii) Consider the region where $|v - v_c| < \mathcal{O}(N^{-1/3})$. Using the formulæ in footnote 24 on p.120 prove that

   $$N = \frac{1}{-\alpha} + \frac{Nv}{v_c}\left[\frac{v_0}{v_c} - 3.54(-\alpha)^{1/2}\right].$$

   Use this result to find $\partial^n P/\partial v^n$ for $v = v_c$.

3. Prove Eq.(9.212). Hint: Introduce three parameters ξ, η, and ζ by the equations

   $$\xi = \sum_{l=1}^{\infty} \frac{l^{l-2}}{l!} z^l, \qquad \eta = z\frac{d\xi}{dz}, \qquad \zeta = z\frac{d\eta}{dz}.$$

   and use the integral representation

   $$\frac{l^l}{l!} = \frac{1}{2\pi i} \oint \frac{du}{u^{l+1}} e^{lu}$$

   to express ξ, η, and ζ in terms of a parameter u_0 which satisfies the equation

   $$u_0 = z e^{u_0}.$$

4. Prove Eq.(9.233). Hint: Use the parameters η and u_0 introduced in the previous problem and the relations between them.

---

[41]See D.ter Haar, *Proc. Roy. Soc. (London)* **A212**, 552 (1952).

5. Prove Eq.(9.316).

6. Prove Eqs.(9.320) and (9.321).

7. Assume that in the virial expansion (9.314) one can at the critical temperature and density neglect all terms, except the ones corresponding to $k = 1$ and $k = 2$. Prove that this leads to the relation[42]

$$P_c v_c / kT_c = \tfrac{1}{3}.$$

8. Determine the critical temperature of an Ising ferromagnet in the mean-field approximation from the equations

$$R = 0, \qquad \frac{\partial F}{\partial R} = 0, \qquad \frac{\partial^2 F}{\partial R^2} = 0.$$

9. By considering $\partial^2 F / \partial R^2$ prove that the point P in Fig.9.7 corresponds to stable equilibrium, that is, that P for temperatures below T_c corresponds to a minimum and the origin to a maximum.

10. Prove Eqs.(9.720) and (9.721).

11. Prove that the expressions for the long-range and short-range order parameters obtained in the quasi-chemical approximation tend to those from the mean-field approximation in the limit as $z \to \infty$. Hint: Introduce a variable δ through the equation $R = \tanh \delta$; if Φ_0 is kept constant it follows that as $z \to \infty$, we must have $J \to 0$.

12. Consider the two-dimensional lattice of Fig.9.10. Put spins on it in a random way such that the average spin value is zero (corresponding to a vanishing long-range order, that is, temperatures above T_c). Take the lattice large enough that the scaling procedure outlined in §9.8 can be applied a few times, that is, take at least a lattice of 27 by 27 when you can do the scaling exercise twice, but preferably an 81 by 81 lattice when it can be done three times. To occupy the lattice sites randomly use the decimals of π and put an up spin on the lattice site if the decimal is even and a down spin if the decimal is odd.[43] For the second thousand

---

[42]R.K.Pathria, *Statistical Mechanics*, Pergamon Press, Oxford (1972), p.439.

[43]To 1000 decimals π is equal to: 3.14159 26535 89793 23846 26433 83279 50288
41971 69399 37510 58209 74944 59230 78164 06286 20899 86280 34825 34211
70679 82148 08651 32823 06647 09384 46095 50592 23172 53594 08128 48111
74502 84102 70193 85211 05559 64462 29489 54930 38196 44288 10975 66593
34461 28475 64823 37867 83165 27120 19091 45648 56692 34603 48610 45432
66482 13393 60726 02491 41273 72458 70066 06315 58817 48815 20920 96282
92540 91715 36436 78925 90360 01133 95305 48820 46652 13841 46951 94151
16094 33057 27036 57595 91953 09218 61173 81932 61179 31051 18548 07446
23799 62749 56735 18857 52724 89122 79381 83011 94912 98336 73362 44065
66430 86021 39494 63952 24737 19070 21798 60943 70277 05392 17176 29317
67523 84674 81846 76694 05132 00056 81271 45263 56082 77857 71342 75778

decimals repeat the first thousand.[44] Now carry out the scaling exercise outlined in § 9.8, that is, look at blocks of 3×3 spins and replace them by an up spin (down spin), if the average spin in the block is up (down). Repeat this three time for the case where you started with an 81 by 81 lattice.

13. Repeat the exercise of the previous problem, but now with a starting situation where there are twice as many up- as down-spins. Put an up-spin on the lattice site for each decimal of π which is equal to 3, 6, or 9 and a down-spin for each decimal which is 1, 2, 4, 5, 7, or 8; neglect the decimals which are zero.

14. Prove that the recursion formulæ (9.819) to (9.821) follow from the requirement that the renormalised partition function (9.817) for the one-dimensional Ising ferromagnet has the form (9.818).

15. Prove Eqs.(9.826) by considering all possible combinations of spin values in the square of the spins 3, 7, 13, and 9.

16. Prove Eq.(9.830) and find the recursion formula to which it leads. Find the non-trivial fixed point for that recursion relation and give the value of the corresponding transition temperature.

17. Consider the one-dimensional Ising ferromagnet in the case when there is no magnetic field. Use Eq.(9.921) to find the zeros of the partition function.

18. Consider a lattice gas with the interactions given by Eq.(9.418).[45]

   (i) Discuss how the grand partition function of the lattice gas is related to the canonical partition function of the Ising ferromagnet.

  (ii) Use the results of (i) to discuss the behaviour of the lattice gas in the mean-field approximation.

 (iii) Use the results of (i) and of the previous problem to confirm Yang and Lee's theorem that the zeros of the grand partition function of the lattice gas lie on a circle. Show that there is no phase transition at any finite temperature and discuss what happens as $T \to 0$.

19. Consider the two-dimensional Ising ferromagnet in a magnetic field. Use

---

96091 73637 17872 14684 40901 22495 34301 46549 58537 10507 92279 68925
89235 42019 95611 21290 21960 86403 44181 59813 62977 47713 09960 51870
72113 49999 99837 29780 49951 05973 17328 16096 31859 50244 59455 34690
83026 42522 30825 33446 85035 26193 11881 71010 00313 78387 52886 58753
32083 81420 61717 76691 47303 59825 34904 28755 46873 11595 62863 88235
37875 93751 95778 18577 80532 17122 68066 13001 92787 66111 95909 21642
01989.

[44]This means, of course, that even if the decimals of π would represent a good random sequence — and this is not at all certain — our extended sequence is certainly not truly random. However, this should not matter for our exercise.

[45]See R.K.Pathria, *Statistical Mechanics*, Pergamon Press, Oxford (1972), §§ 12.6, 7, 9C, and 10.

Ritz's variational method in the form

$$\rho_{\max} = \underset{\text{all} A's}{\text{Max}} \frac{\displaystyle\sum_{\sigma_i,\sigma_i'=\pm1} \mathfrak{T}(\sigma_i,\sigma_i')A(\sigma_i)A(\sigma_i')}{\displaystyle\sum_{\sigma_i=\pm1} A^2(\sigma_i)}, \tag{A}$$

to find equations for $\rho_{\max}$ from Eq.(9.931).

In the case where there is a magnetic field present $\mathfrak{T}$ will be given by the equation

$$\mathfrak{T}(\sigma_i,\sigma_i') = e^{K\left[\sum \sigma_i\sigma_i' + \frac{1}{2}\sum \sigma_i\sigma_{i+1} + \sum \sigma_i'\sigma_{i+1}'\right] + \frac{1}{2}C\left[\sum \sigma_i + \sum \sigma_i'\right]}, \tag{B}$$

rather than by Eq.(9.930).

Restrict the possible trial functions A to those of the form

$$A(\sigma_i) = e^{a\left[H(K,C)\xi + B(K,C)\eta\right]}, \tag{C}$$

with

$$\xi = \frac{1}{a}\sum_{i=1}^{a-1}\sigma_i\sigma_{i+1}, \qquad \eta = \frac{1}{a}\sum_{i=1}^{a}\sigma_i. \tag{D}$$

Hint:First derive the following equation for the ath power root, $\lambda_{\max}$, of $\rho_{\max}$:

$$\lambda_{\max} = \underset{H,B}{\text{Max}} \frac{\chi(H,B)}{\psi(H,B)}, \tag{E}$$

where χ and ψ are partition functions, respectively, of a lattice consisting of two parallel, interacting chains and of a one-dimensional chain. Both of those can be evaluated using the method which led to Eq.(9.921).

Find $\lambda_{\max}$ for temperatures above the transition temperature by putting $B = 0$.[46]

Find the transition temperature in this approximation by solving the simultaneous equations

$$\frac{\partial^2\lambda}{\partial B^2} = 0, \qquad \frac{\partial\lambda}{\partial B} = 0, \qquad B = 0.$$

---

[46]In the region below the transition temperature the equations become complicated and for details we refer to the paper by Kramers and Wannier where this variational approach was suggested: H.A.Kramers and G.H.Wannier, *Phys. Rev.* **60**, 263 (1941).

Bibliographical Notes

For a general discussion of phase transitions see, for instance:
1. G.E.Uhlenbeck, *Brandeis Summer Institute Lectures in Theoretical Physics* 1962, Benjamin, New York (1963), Vol.3, p.1.
2. D.Ruelle, *Statistical Mechanics*, Benjamin, New York (1969), Chapter 5.
3. H.E.Stanley, *Introduction to Phase Transitions and Critical Phenomena*, Oxford University Press, Oxford (1971).
4. R.K.Pathria, *Statistical Mechanics*, Pergamon Press, Oxford (1972), Chapter 12.
5. E.M.Lifshitz and L.P.Pitaevskii, *Statistical Physics*, Pergamon Press, Oxford (1980), Chapter XIV.
6. J.M.Yeomans, *Statistical Mechanics of Phase Transitions*, Clarendon Press, Oxford (1992).

Section 9.1. See References 1 to 4 and also:
7. B.Kahn, *On the Theory of the Equation of State*, Thesis, University of Utrecht (1938).
8. B.Kahn and G.E.Uhlenbeck, *Physica* **5**, 399 (1938).
9. D.ter Haar, *Queen's Papers on Pure and Applied Mathematics*, No 11, 84 (1968).

Section 9.2. The thermodynamic theory of condensation was given by Becker and Döring:
10. R.Becker and W.Döring, *Ann. Physik* **24**, 719 (1935).
The statistical theory presented here is based on Wergeland's work:
11. H.Wergeland, *Trans. Norw. Acad. Sc., Oslo, Math.-Phys. Section*, No 11 (1943).
See also
12. D.ter Haar, *Proc. Cambridge Phil. Soc.* **49**, 130 (1953).
13. D.ter Haar, *Elements of Statistical Mechanics*, Rinehart, New York (1954), Chapter 9 and the references given there.

Section 9.3. Mayer's theory first appeared in the following papers:
14. J.E.Mayer, *J. Chem. Phys.* **5**, 67 (1937).
15. J.E.Mayer and P.G.Ackermann, *J. Chem. Phys.* **5**, 74 (1937).
16. J.E.Mayer and S.F.Harrison, *J. Chem. Phys.* **6**, 87 (1938).
16. S.F.Harrison and J.E.Mayer, *J. Chem. Phys.* **6**, 101 (1938).

See also Reference 4, § 12.2 and
17. J.E. and M.G.Mayer, *Statistical Mechanics*, Wiley, New York (1940), especially Chapters 13 and 14.
18. G.E.Uhlenbeck and G.W.Ford, *Lectures in Statistical Mechanics*, American Mathematical Society, Providence, RI (1963), Chapters II and III.
The discussion of the behaviour of cluster sizes when one approaches the gas-liquid transition is based on §§ 12.2, 9.2, and 9.4 of Reference 4.

Section 9.4. Yang and Lee's basic papers were published in 1952:
19. C.N.Yang and T.D.Lee, *Phys. Rev.* **87**, 404 (1952).

20. T.D.Lee and C.N.Yang, *Phys. Rev.* **87**, 410 (1952). See also Ref.4, § 12.3.

Various authors have studied the important question of the distribution of the zeros of the grand partition function; see, for instance,

21. P.C.Hemmer and E.Hiis Hauge, *Phys. Rev.* **133A**, 1010 (1964).

22. P.C.Hemmer, E.Hiis Hauge, and J.O.Aasen, *J. Math. Phys.* **7**, 410 (1966).

23. J.Stephenson and R.Couzens, *Physica* **129A**, 201 (1984).

24. J.Stephenson, *J. Phys.* **A20**, 4513 (1987).

Section 9.5. See References 3 to 6 and for an early review of experimental data also

25. F.C.Nix and W.Shockley, *Rev. Mod. Phys.* **10**, 1 (1938).

Section 9.6. The mean-field approximation was originally suggested by Gorsky:

26. W.Gorsky, *Z. Physik* **50**, 64 (1928).

It was developed by Bragg and Williams in the following papers:

27 W.L.Bragg and E.J.Williams, *Proc. Roy. Soc. (London)* **A145**, 699 (1934).

28 W.L.Bragg and E.J.Williams, *Proc. Roy. Soc. (London)* **A151**, 540 (1935).

29 W.L.Bragg and E.J.Williams, *Proc. Roy. Soc. (London)* **A152**, 231 (1935).

See also References 3, 4, and 6.

Section 9.7. The quasi-chemical method is due to Fowler and Guggenheim; see

30. R.H.Fowler and E.A.Guggenheim, *Proc. Roy. Soc. (London)* **A174**, 189 (1940).

Section 9.8. Critical phenomena have been extensively discussed in recent years. For general discussions we can refer to the following books:

31 J.J.Binney, N.J.Dowrick, A.J.Fisher, and M.E.J.Newman, *The Theory of Critical Phenomena*, Oxford University Press (1992).

32 J.Zinn-Justin, *Quantum Field Theory and Critical Phenomena*, Oxford University Press (1989).

33. D.Chandler, *Introduction to Modern Statistical Mechanics*, Oxford University Press (1988).

See also References 3, 4, and 6.

The idea of scaling transformations is due to Kadanoff; see, for instance,

34. L.P.Kadanoff, *Physics* **2**, 263 (1966).

This idea led to the renormalisation group methods through the work of Wilson; see, for instance,

34. K.G.Wilson and J.B.Kogut, *Phys. Rept.* **12**, 75 (1974).

The description of the renormalisation of the two-dimensional Ising ferromagnet on a square lattice follows Reference 33, § 5.7; see also

35. H.J.Maris and L.P.Kadanoff, *Am. J. Phys.* **46**, 652 (1976).

Section 9.9 The transfer matrix approach to evaluating the partition function of an Ising ferromagnet was first suggested by Kramers and Wannier; see

36. H.A.Kramers and G.H.Wannier, *Phys. Rev.* **60**, 252, 263 (1941).
    See also

37. E.W.Montroll, *J. Chem. Phys.* **9**, 706 (1941).

38. E.N.Lassettre and J.P.Howe, *J. Chem. Phys.* **9**, 747, 801 (1941).

The exact calculation of the partition function of the two-dimensional Ising ferromagnet on a square lattice was first given by Onsager in 1944; see

39. L.Onsager, *Phys. Rev.* **65**, 117 (1944).
    See also

40. B.Kaufman, *Phys. Rev.* **76**, 1232 (1949).

41. R.M.F.Houtappel, *Physica* **16**, 425 (1950).
    and especially

42 R.J.Baxter, *Exactly Solved Models in Statistical Mechanics*, Academic Press, London (1982).

SOLUTIONS TO SELECTED PROBLEMS

We shall give here solutions to some of the problems in the book. We give neither solutions to all the problems nor necessarily all details of the solutions to those problems which are here included. The solutions given should therefore be considered to be hints rather than complete solutions. The numbering is by chapter and problem number: 5.11, for instance, indicates problem 11 of Chapter 5.

1.1. The number of atoms striking unit area per unit time is

$$\frac{n}{2\pi} \sqrt{\frac{2\pi kT}{m}} = \tfrac{1}{4} n\bar{c},$$

where n is the number of particles per unit volume and $\bar{c}$ is given by Eq.(1.107).

1.9. The transport equation in this case is

$$\frac{\partial f(\vartheta, \varphi)}{\partial t} = A \int [f(\vartheta', \varphi') - f(\vartheta, \varphi)] \frac{d^2\omega'}{4\pi} = A[n - f(\vartheta, \varphi)],$$

with the solution

$$f(\vartheta, \varphi) = n e^{-At}.$$

2.4. The terms on the right-hand side of the transport equation are now

$$A = 4\pi\eta (f_0 + u\chi_1 + v\chi_2), \qquad B = 4\pi\eta f_0,$$

and if we linearise the transport equation with respect to the χ_i, its solution is

$$\chi_1 = \frac{-\lambda}{c(1 + s^2)} (f_1 - sf_2),$$

$$\chi_2 = \frac{-\lambda}{c(1 + s^2)} (sf_1 + f_2),$$

where λ is the mean free path,

$$\lambda = \frac{c}{4\pi\eta},$$

s is the ratio of the mean free path to the radius of curvature of an electron moving with speed c in the magnetic field B,

$$s = \frac{eB\lambda}{mc},$$

and the f_i are given by the equations

$$f_1 = \frac{\partial f_0}{\partial x} - eE_x \frac{\partial f_0}{\partial \varepsilon},$$

$$f_2 = \frac{\partial f_0}{\partial y} + eE_y \frac{\partial f_0}{\partial \varepsilon},$$

with ε the electron energy.

2.5. The electrical conductivity σ is given by the equation

$$\sigma = -\frac{8\pi e^2}{3m^2} K_1,$$

where the integrals K_i are given by the relation

$$K_i = \int_0^\infty \lambda \varepsilon^i \frac{\partial f_0}{\partial \varepsilon} d\varepsilon,$$

which for a Maxwell distribution gives

$$K_i = -i! \, n \left(\frac{m}{2\pi}\right)^{3/2} \beta^{-n+\frac{1}{2}} \lambda,$$

where n is the number of electrons per unit volume. Hence we get

$$\sigma = \tfrac{4}{3} ne^2 \lambda \left(\frac{\beta}{2\pi m}\right)^{1/2} = \frac{8}{3\pi} \frac{ne^2 \lambda}{m\bar{c}},$$

where here $\bar{c}$ is the rooth mean square velocity given by Eq.(1.106).

2.6. The Hall coefficient is given by the equation

$$R = \frac{3m^2}{8\pi e^2 B} \frac{L_2}{L_1^2 + L_2^2},$$

where the L_i are given by

$$L_i = \int_0^\infty \frac{\varepsilon \lambda s^{i-1}}{1 + s^2} \frac{\partial f_0}{\partial \varepsilon} d\varepsilon.$$

2.7. The thermal conductivity coefficient is given by the equation

$$\kappa = \frac{8\pi}{3m^2 T} \frac{K_2^2 - K_1 K_3}{K_1},$$

where the K_i were defined in the solution to problem 2.5.

For the Wiedemann-Franz ratio we have

$$\frac{\kappa}{\sigma T} = -\frac{1}{e^2 T^2} \frac{K_2^2 - K_1 K_3}{K_1^2}.$$

In the case of a Maxwell distribution the expression for the thermal conductivity reduces to

$$\kappa = \frac{16\pi}{3\pi} \frac{k^2 n\lambda T}{m\bar{c}} = \frac{8}{3\pi} n\bar{c}\lambda k = \frac{16}{9\pi} c_v \bar{c}\lambda,$$

where c_v is the specific heat per unit volume, and the expressions for the Wiedemann-Franz ratio reduces to

$$\frac{\kappa}{\sigma T} = \frac{2k^2}{e^2}.$$

2.8. The partition function of one particle is given by Eq.(2.701):

$$Z_\mu = K \int e^{-\beta\varepsilon} dp_1 \cdots dp_s \, dq_1 \cdots dq_s.$$

(a) In this case we have $\varepsilon = sp$ and hence

$$Z_\mu = \frac{8\pi KV}{(\beta s)^3}.$$

(b) In this case we have

$$q_1 = r, \qquad q_2 = \vartheta, \qquad q_3 = \varphi,$$

and

$$Z_\mu = KV \left(\frac{2\pi m}{\beta}\right)^{3/2}.$$

(d) In this case we have

$$q_1 = x, \quad q_2 = y, \quad q_3 = z, \quad q_4 = \vartheta, \quad q_5 = \varphi, \quad q_6 = \psi,$$

and

$$Z_\mu = 64\pi^5 KVm^{3/2} AC^{1/2}(kT)^3.$$

(h) We now have

$$Z_\mu = 64\pi^5 KVm^{3/2} AC^{1/2}(kT)^3 \frac{\sinh \beta\mu B}{\beta\mu B}.$$

2.9. (a) In the case covered by (a) of the previous problem the specific heat per particle is given by the equation

$$c_v = 3k.$$

(b) The magnetisation M is given by the equation

$$M = \mu N \mathcal{L}(\beta \mu B),$$

where $\mathcal{L}$ is the Langevin function given by Eq.(2.412).
In the limit as $T \to \infty$ we find

$$M \to \frac{N\mu^2 B}{3kT}.$$

3.5. The radiation energy density in this case is given by the equation

$$\varrho(\omega) = \frac{\hbar}{\pi^2 c^3} \frac{n(\omega)^3 \omega^3}{e^{\beta \hbar \omega} + 1}.$$

3.6. We have

$$c_v = \frac{k\beta^2 (\varepsilon_1 - \varepsilon_2)^2 e^{-\beta(\varepsilon_1 + \varepsilon_2)}}{(e^{-\beta \varepsilon_1} + e^{-\beta \varepsilon_2})^2} = k \frac{x^2 e^{-x}}{(1 + e^{-x})^2},$$

where $x = \beta(\varepsilon_2 - \varepsilon_1)$.
The maximum occurs for $x \approx 2.40$, that is, $T \approx 0.42 (\varepsilon_2 - \varepsilon_1)/k$.
3.7. The specific heat is in this case

$$c_v = \frac{2k\beta^2 \varepsilon^2 e^{\beta \varepsilon}}{(e^{\beta \varepsilon} + 1)^2}.$$

3.11. The low-temperature behaviour of the specific heat is now:

$$c_v \propto T^{3/n}.$$

3.18. To first order in γ we find

$$c_v = k \frac{x^2 e^x}{(e^x - 1)^2} \left[1 + 2\gamma \frac{e^{2x}(x - 2) + 4xe^x + x + 2}{(e^x - 1)^2} \right],$$

where $x = \beta \hbar \omega$.

4.2. In the case of Gentile statistics we have

$$\frac{N_j}{Z_j} = \frac{1}{e^{-\alpha + \beta E_j} - 1} - \frac{d + 1}{e^{(d+1)(-\alpha + \beta E_j)} - 1}.$$

4.3. We have

$$\frac{c_A c_B}{c_{AB}} = \left(\frac{2\pi\mu}{\beta h^3}\right)^{3/2} e^{-\beta x} \frac{Z_A Z_B}{Z_{AB}},$$

where μ is the reduced mass,

$$\mu = \frac{m_A m_B}{m_{AB}}.$$

4.7. The equation of the curve is

$$P v^{5/3} = \text{const.}$$

4.8. For $T < T_0$ we have

$$n_0 = N \left[1 - \left(\frac{T}{T_0}\right)^{3/2}\right].$$

4.11. The equation of the adiabatic of a perfect non-relativistic monatomic quantum gas is

$$PV^{5/3} = \text{const.}$$

4.21. The number of electrons, n_c, per unit volume in the conduction band is given by the equation

$$n_c = \sqrt{2n_b} \left(\frac{2\pi m}{\beta h^2}\right)^{3/4} e^{-\frac{1}{2}\beta\Delta\varepsilon},$$

where n_b is the number of impurity levels per unit volume.

The Fermi level ε_F is given by the relation

$$\varepsilon_F = -\tfrac{1}{2}\Delta\varepsilon.$$

The distribution of the electrons in the conduction band over the various energies is given by the equation

$$f(\varepsilon)\, d\varepsilon = \frac{2}{\sqrt{\pi}} n_c \beta^{3/2} \sqrt{\varepsilon}\, d\varepsilon.$$

4.22. In the case when $n_b = 10^{24}$ m^{-3} and $\Delta\varepsilon = 10^{-3}$ eV we cannot treat the gas of conduction electrons classically in the temperature range

$$T_1 \leqslant T \leqslant T_2, \qquad T_1 \approx 1.5 \text{ K}, \qquad T_2 \approx 170 \text{ K}.$$

4.23. The Richardson current j_{Rich} is given by the expression

$$j_{\text{Rich}} = \frac{4\pi m e}{h^3} k^2 T^2 e^{-\beta\varphi},$$

where φ is the work function.

4.25. The difference between the different parts of this problem and Problems 2.4 to 2.7 lies in the zeroth approximation distribution function. We now have for f_0 the Fermi-Dirac rather than the Maxwell distribution:

$$f_0 = \frac{2m^3}{h^3} \frac{1}{1 + e^{-\alpha + \beta \varepsilon}},$$

and in practically all cases of interest its derivative can be approximated by a δ-function:

$$\frac{\partial f_0}{\partial \varepsilon} \approx -\frac{2m^3}{h^3} \delta(\varepsilon - \mu_0),$$

where μ_0 is the partial chemical potential—strictly speaking, its value at $T = 0$—given by the equation

$$\mu_0 = \frac{h^2}{2m} \left(\frac{3n}{8\pi}\right)^{2/3}.$$

We must note here that the expression for $\partial f_0 / \partial \varepsilon$ only gives the zeroth approximation; if one wants the corrections to this expressions for finite temperatures one must use an expansion like the one given by Eq.(4.830).

Substituting the above expression for $\partial f_0 / \partial \varepsilon$ into the expressions for the K_i and the L_i, which we gave in the solutions to problems 2.5 and 2.6, we have

$$K_i = -\frac{2m^3}{h^3} \lambda \mu_0^i,$$

and

$$L_i = -\frac{2m^3}{h^3} \frac{\lambda s_0^{i-1}}{1 + s_0^2} \mu_0,$$

where

$$s_0 = \frac{eB\lambda}{m\bar{c}},$$

with $\bar{c}$ now the velocity of the electrons at the Fermi surface,

$$\bar{c} = \sqrt{\frac{2\mu_0}{m}}.$$

We shall see presently that, in fact, we need in some cases the corrections to the expressions we have just given for the K_i and L_i. For the K_i, for instance, this means that we have to use the equation

$$K_i = -\frac{2m^3}{h^3} \lambda \mu_0^i - \frac{\pi^2}{3} i(i-1)\lambda \frac{m^3}{\beta^2 h^3} \mu_0^{i-2}.$$

(a) We now find for the electrical conductivity

$$\sigma = \frac{ne^2\lambda}{m\bar{c}}.$$

(b) The zeroth approximation for the thermal conductivity gives zero, but using the first term in the expansion for the K_i we find that κ is given by the equation

$$\kappa = \frac{\pi^2}{3} \frac{k^2 n \lambda}{m\bar{c}} T.$$

(c) For the isothermal Hall coefficient we find:

$$R = \frac{3m^2}{8\pi e^2 B} \frac{L_2}{L_1^2 + L_2^2}$$

$$= -\frac{1}{ne},$$

which shows that the Hall effect can be used to determine the carrier density, a fact which is used extensively in the study of semiconductors.

(d) The Wiedemann-Franz ratio is now given by

$$\frac{\kappa}{\sigma T} = \frac{\pi^2}{3} \frac{k^2}{e^2}.$$

(e) For the conductivity in a magnetic field we find

$$\sigma(B) = -\frac{8\pi e^2}{3m^2} \frac{L_1^2 + L_2^2}{L_1}$$

$$= \sigma(0).$$

If we take the next term in the expansion for the L_i into account there is a dependence on the magnetic field.

4.26. (i) The spin paramagnetic susceptibility χ per unit volume is at $T = 0$ given by the expression

$$\chi = \chi_0 = \frac{3n\mu_B^2}{2\varepsilon_F},$$

where ε_F is the Fermi energy (Fermi level).

(ii) At low, but finite temperatures the spin paramagnetic susceptibility per unit volume is give by the expression

$$\chi \approx \chi_0 \left[1 - \frac{\pi^2}{12} \left(\frac{kT}{\varepsilon_F} \right)^2 \right].$$

4.27. The diamagnetic susceptibility per unit volume is given by the expression

$$\chi \approx -\frac{n\mu_B^2}{2\varepsilon_F}.$$

5.1. If we write

$$J_n = \frac{(\mathcal{J} - \bar{\mathcal{J}})^n}{\bar{\mathcal{J}}^n},$$

we have

$$J_0 = 1, \qquad J_1 = 0.$$

The recursion formula is

$$J_n = \frac{2(n-1)}{sN}\,(J_{n-1} + J_{n_2}),$$

whence it follows that

$$J_2 = \frac{2}{sN}, \qquad J_3 = \frac{8}{(sN)^2}, \qquad J_4 = \frac{12}{(sN)^2}\left\{1 + \frac{4}{sN}\right\}.$$

5.2. In a canonical grand ensemble we have

$$\langle(A_k - \langle A_k\rangle)(A_l - \langle A_l\rangle)\rangle = \frac{1}{\beta^2}\frac{\partial^2 q}{\partial a_k \partial a_l} + \frac{1}{\beta}\left\langle\frac{\partial^2 \varepsilon}{\partial a_k \partial a_l}\right\rangle,$$

and

$$\langle(\varepsilon - \langle\varepsilon\rangle)(A_k - \langle A_k\rangle)\rangle = -\frac{1}{\beta}\frac{\partial^2 q}{\partial a_k \partial \beta} - \frac{1}{\beta}\left\langle\frac{\partial\varepsilon}{\partial a_k}\right\rangle$$

$$= \frac{1}{\beta}\frac{\partial\langle\varepsilon\rangle}{\partial a_k} - \frac{1}{\beta}\left\langle\frac{\partial\varepsilon}{\partial a_k}\right\rangle.$$

5.6. After introducing new variables,

$$x_i = \frac{p_i}{\sqrt{2m}}, \qquad\qquad i = 1,\ldots,N,$$

$$x_i = \sqrt{\tfrac{1}{2}m}\,\omega\, q_{i-N}, \qquad i = N+1,\ldots,2N,$$

we find

$$\Omega(\varepsilon) = \frac{2^N}{\omega^N}\,\mathcal{V}_{2N},$$

where $\mathcal{V}_{2N}$ is the volume of a $2N$-dimensional sphere of radius $\sqrt{\varepsilon}$.
Hence we find that the average energy of the system satisfies the equation

$$E = NkT.$$

5.9. We have the following relations:

$$\langle\Delta V^2\rangle = -T\left(\frac{\partial V}{\partial P}\right)_T, \qquad \langle\Delta V\,\Delta T\rangle = 0, \qquad \langle\Delta T^2\rangle = \frac{T^2}{C_v}.$$

5.10. Similarly we have

$$\langle\Delta P^2\rangle = -T\left(\frac{\partial P}{\partial V}\right)_S, \qquad \langle\Delta P\,\Delta S\rangle = 0, \qquad \langle\Delta S^2\rangle = C_p.$$

5.13. We have

$$\left\langle (\varepsilon - \langle \varepsilon \rangle)^3 \right\rangle = 2k^2 T^3 C_v + k^2 T^4 \frac{dC_v}{dT},$$

which in the case of a perfect gas gives

$$\left\langle (\varepsilon - \langle \varepsilon \rangle)^3 \right\rangle = 3Nk^3 T^3.$$

6.4. The average value $\langle G \rangle$ of $\widehat{G}$ is given by the expression

$$\langle G \rangle = \int \langle x' | \widehat{\varrho} | x \rangle \, dx \langle x | \widehat{G} | x' \rangle \, dx'.$$

If $\widehat{\mathcal{H}}(x)$ is that part of the Hamiltonian which describes the behaviour of the degrees of freedom we are interested in — those denoted by x — the equation of motion for $\widehat{\varrho}$ will be

$$i\hbar \langle x | \dot{\widehat{\varrho}} | x' \rangle = \langle x | \widehat{\mathcal{H}} \widehat{\varrho} - \widehat{\varrho} \widehat{\mathcal{H}} | x' \rangle.$$

6.7. In the macrocanonical ensemble the density matrix can be written in the form

$$\langle x' | \widehat{\varrho} | x \rangle = \sum_n e^{\beta(\psi - \widehat{\mathcal{H}})} \varphi_n^*(x') \varphi_n(x),$$

where the φ_n are a complete orthonormal set of functions, or

$$\langle x' | \widehat{\varrho} | x \rangle = \sum_n e^{\beta(\psi - E_n)} \varphi_n^*(x') \varphi_n(x),$$

if the φ_n are the complete set of orthonormal eigenfunctions of the Hamiltonian $\mathcal{H}$.

6.8. Corresponding to an adiabatic switching on of the interaction we write

$$\widehat{\mathcal{H}}' = \widehat{V} e^{-i\omega t + \varepsilon t}, \qquad \varepsilon \to +0.$$

The equation of motion for $\Delta \widehat{\varrho}$ is

$$i\hbar \dot{\Delta \widehat{\varrho}} = \left[\widehat{V}, \widehat{\varrho}(-\infty) \right]_- e^{-i\omega t + \varepsilon t} + \left[\widehat{\mathcal{H}}_0, \Delta \widehat{\varrho} \right]_-.$$

Changing to the interaction representation for $\Delta \widehat{\varrho}$, substituting this expression into the equation of motion, and going back to the original representation we find

$$\Delta \widehat{\varrho} = -\frac{i}{\hbar} \int_{-\infty}^{t} e^{i\widehat{\mathcal{H}}_0(\tau - t)} \left[\widehat{V}, \widehat{\varrho}(-\infty) \right]_- e^{-i\widehat{\mathcal{H}}_0(\tau - t)} e^{-i\omega t + \varepsilon t} \, d\tau,$$

from which the Kubo formula follows straightforwardly.

6.9. The electric conductivity tensor can be expressed in terms of the auto-correlation function of the electric current as follows:

$$\sigma_{kl} = \int_0^\infty d\tau \, e^{-i\omega\tau} \int_0^\beta \left\langle \hat{j}_{k\text{int}}(-i\hbar\lambda)\hat{j}_{l\text{int}}(\tau) \right\rangle_0 d\lambda.$$

6.11. The density matrix is given by the equation

$$\hat{\varrho} = \sum_n e^{\beta(\psi - E_n)} \varphi_n^*(x')\varphi_n(x),$$

where the φ_n are a complete set of orthonormal eigenfunctions of the Hamiltonian $\hat{\mathcal{H}}$ of the system corresponding to the eigenvalues E_n. The arguments x and x' stand for the N position coordinates r_i and r_i' ($i = 1, \ldots, N$).

We can rewrite the equation for $\hat{\varrho}$ in the form

$$\hat{\varrho} = \sum_n e^{\beta\psi - \beta\hat{\mathcal{H}}} \varphi_n^*(x')\varphi_n(x),$$

provided we stipulate that here $\hat{\mathcal{H}}$ acts only on the argument x and not on x'.

We must now distinguish between Boltzmann, Bose-Einstein, and Fermi-Dirac systems. This is done by selecting from the complete orthonormal set only those which satisfy the relevant commutation rules. We thereto introduce the functions

$$\Delta_{\text{Bo}} = \sum_n \varphi_n(x')\varphi_n(x),$$

$$\Delta_{\text{BE}} = \frac{1}{N!} \sum_P \sum_n \varphi_n(x')\varphi_n(x_P),$$

$$\Delta_{\text{FD}} = \frac{1}{N!} \sum_P \varepsilon_P \sum_n \varphi_n(x')\varphi_n(x_P),$$

where P stands for a permutation of the N particles and where ε_P was introduced in § 4.2.

Using the relation

$$\sum_n \varphi_n(x')\varphi(x) = \prod_{i=1}^N \delta(r_i' - r_i),$$

we have for the Δ

$$\Delta_{\text{Bo}} = \prod_i \delta(r_i' - r_i),$$

$$\Delta_{\text{BE}} = \frac{1}{N!} \sum_P \prod_i \delta(r_i' - r_{Pi}),$$

$$\Delta_{\text{FD}} = \frac{1}{N!} \sum_P \varepsilon_P \prod_i \delta(r_i' - r_{Pi}).$$

The Hamiltonian of the system is just the kinetic energy operator:

$$\widehat{\mathcal{H}} = -\frac{\hbar^2}{2m} \sum_i \nabla_i^2,$$

and using the Fourier representation,

$$\delta(\boldsymbol{r}' - \boldsymbol{r}) = \int e^{2\pi i(\boldsymbol{k}\cdot[\boldsymbol{r}'-\boldsymbol{r}])} \, d^3\boldsymbol{k},$$

for the δ-functions we find finally for the density matrix the following expressions:

$$\widehat{\varrho}_{\mathrm{Bo}} = \frac{e^{\beta\psi}}{v_0^N} e^{-(m/2\beta\hbar^2)\sum_i (\boldsymbol{r}_i'-\boldsymbol{r}_i)^2},$$

$$\widehat{\varrho}_{\mathrm{BE}} = \frac{e^{\beta\psi}}{v_0^N} \sum_{\mathrm{P}} e^{-(m/2\beta\hbar^2)\sum_i (\boldsymbol{r}_i'-\boldsymbol{r}_{\mathrm{P}i})^2},$$

$$\widehat{\varrho}_{\mathrm{FD}} = \frac{e^{\beta\psi}}{v_0^N} \sum_{\mathrm{P}} \varepsilon_{\mathrm{P}} e^{-(m/2\beta\hbar^2)\sum_i (\boldsymbol{r}_i'-\boldsymbol{r}_{\mathrm{P}i})^2},$$

where v_0 is given by Eq.(4.736).

6.12. The spin density matrix is given by the equation

$$\widehat{\varrho} = \tfrac{1}{2}\left[\widehat{\mathbf{1}} + (\boldsymbol{P}\cdot\widehat{\boldsymbol{\sigma}})\right]$$

$$= \frac{1}{2}\begin{pmatrix} 1+P_z & P_x - iP_y \\ P_x + iP_y & 1 - P_z \end{pmatrix}.$$

The equation of motion for $\boldsymbol{P}$ is

$$\dot{\boldsymbol{P}} = -\gamma\left[\boldsymbol{B}\wedge\boldsymbol{P}\right].$$

6.17. The quantity Ω satisfies the equation (compare Eq.(4.220))

$$\Omega = \frac{(N + N_0)!}{N_0! N!},$$

whence follows

$$\frac{S}{N\hbar\omega} = (N\hbar\omega + E)\ln(N\hbar\omega + E) - E\ln E - N\hbar\omega\ln(N\hbar\omega),$$

and

$$\beta\hbar\omega = \ln\frac{N\hbar\omega + E}{E},$$

which leads immediately to the Planck formula.

6.18. The Lagrangian multiplier α does not occur as there is no restriction on the number of photons or phonons.

6.20. The energy level density is given by the equation

$$\varrho(E, Z, N) \;=\; \frac{2}{3} \left(\frac{2\mu_0}{AQ^5} \right)^{1/4} e^{\pi \sqrt{AQ/\mu_0}},$$

where μ_0 is the energy at the top of the Fermi seas of the protons and the neutrons — which we have assumed to be the same — and $Q = E - E_0$ is the excess of the energy above the ground state energy.

The factor in front of the exponential is a slowly varying function of the energy E and could have been written simply as $\phi_1(E)$.

6.21. In this case the energy level density is given by

$$\varrho(E, Z, N) \;=\; \phi_2(E) \, e^{\sqrt{\frac{2}{3} f(\mu_0)Q}},$$

where $\phi_2(E)$ is a slowly varying function of E and where μ_0 and Q have the same meaning as in the previous problem; the expression found in the previous problem is a special case of the above equation.

6.41. The leading term of the low-temperature spcific heat of a degenerate Fermi gas is given by the equation

$$c_v \;=\; KTZ(\varepsilon_{\mathrm{F}}),$$

where K is a constant and where $Z(\varepsilon_{\mathrm{F}})$ is the single-particle energy level density at the Fermi energy.

6.44. In the case of a relativistic Boltzmann gas we have

$$\overline{\frac{p^2 c^2}{2\varepsilon}} \;=\; \tfrac{3}{2}kT,$$

where p and ε are the momentum and energy of a particle.

7.5. It follows from Eq.(B) that

$$Z_N^{(\mathrm{conf})} \;=\; V^N \left[\int \psi_{12} \frac{d^3 r_1}{V} \frac{d^3 r_2}{V} \right]^{\frac{1}{2}N(N-1)}$$

$$=\; V^N \left\{ 1 + \frac{\beta_1}{V} \right\}^{\frac{1}{2}N(N-1)},$$

and hence we get Eq.(D) in the thermodynamic limit.

For the next approximation one needs to take into account the three-particle correlations. This means that one must multiply by the correction factor

$$\frac{\overline{\psi_{12}\psi_{13}\psi_{23}}}{\overline{\psi_{12}}\,\overline{\psi_{13}}\,\overline{\psi_{23}}} \;=\; \frac{\overline{\psi_{12}\psi_{13}\psi_{23}}}{\left(\overline{\psi_{12}}\right)^3}$$

to the power $\binom{N}{3}$ which is the number of triplets in the system.

If, as usual, we put

$$\psi_{ij} = 1 + f_{ij},$$

the correction factor to $Z_N^{(\text{conf})}$ will be

$$\left[\frac{1 + 3\overline{f_{12}} + 3\overline{f_{12}}^2 + \overline{f_{12}f_{13}f_{23}}}{1 + 3\overline{f_{12}} + 3\overline{f_{12}}^2 + \left(\overline{f_{12}}\right)^3} \right]^{\frac{1}{6}N(N-1)(N-2)}.$$

We note that $\overline{f_{12}}$ is proportional to V^{-1} while $\overline{f_{12}f_{13}f_{23}}$ is proportional to V^{-3}. Therefore the configurational partition function per particle will be proportional to

$$\left\{ 1 + \overline{f_{12}f_{13}f_{23}} + \mathcal{O}(V^{-3}) \right\}^{\frac{1}{6}(N-1)(N-2)}.$$

In the thermodynamic limit we thus find

$$Z_N^{(\text{conf})} = V^N e^{\frac{1}{2}nN\beta_1 + \frac{1}{3}n^2N\beta_2},$$

where β_2 is the second irreducible cluster integral,

$$\beta_2 = \overline{f_{12}f_{13}f_{23}}.$$

8.11. We shall denote the left-hand side of Eq.(8.454) by Φ,

$$\Phi = \frac{\overline{n}}{1 - 2\overline{n}} = \frac{v}{(2\pi)^3} \frac{d^3q}{e^{\beta E_q} - 1}.$$

At low temperatures small q-values will dominate so that we can integrate over the whole of q-space and we have

$$\Phi = \frac{v}{(2\pi)^3} \int_0^{2\pi} d\varphi \int_0^\pi \sin\theta \, d\theta \int_0^\infty q^2 \, dq \sum_{r=1}^\infty e^{-2r\beta\sigma K(0)\eta(q)}.$$

Expanding $\eta(q)$ in powers of q and integrating we get a power series in β^{-1}, that is, in T. The result is

$$\Phi = \zeta\left(\tfrac{3}{2}\right)\kappa^{3/2} + \tfrac{3\pi}{4}\nu\zeta\left(\tfrac{5}{2}\right)\kappa^{5/2} + \pi^2\omega\nu^2\zeta\left(\tfrac{7}{2}\right)\kappa^{7/2} + \dots,$$

where $\zeta(n)$ is Riemann's ζ-function,

$$\kappa = \frac{3}{4\pi\nu\beta K(0)},$$

while ν and ω depend on the kind of lattice we are dealing with:

$$\nu = 1, \qquad \omega = \tfrac{33}{32}, \qquad \text{(sc)};$$
$$\nu = 2^{-4/3}, \qquad \omega = \tfrac{281}{288}, \qquad \text{(bcc)};$$
$$\nu = \sqrt{2}, \qquad \omega = \tfrac{15}{16}, \qquad \text{(fcc)}.$$

Solving for σ we arrive at the following power series at low temperatures in the approximation we have used throughout:

$$\sigma = 1 - a_0 T^{3/2} - a_1 T^{5/2} \ldots - b_0 T^3 - b_1 T^4 - \ldots .$$

We may mention that this series expansion is exact up to the term in $T^{5/2}$; however, the exact expansion does not contain a term in T^3.

8.12. The expansion of $\coth(\tfrac{1}{2}\beta E_q)$ in terms of t_0 and t_1 is the following:

$$\coth(\tfrac{1}{2}\beta E_q) = \frac{1}{t_0}\left[1 + (1 - t_0^2) \sum_n \left(\frac{-t_1}{t_0}\right)^n \right].$$

Expanding t_1 in powers of β we find

$$\Phi = \frac{1}{2t_0}\left[1 + \frac{\beta K(0)}{t_0} + \frac{z+1}{z}\frac{\beta^2 K^2(0)}{t_0^2} + \cdots \right].$$

From this equation we get the susceptibility:

$$\chi = \beta\mu_B^2\left[1 + \frac{\beta}{\beta_c} + \frac{z-1}{z}\left(\frac{\beta}{\beta_c}\right)^2 + \cdots \right],$$

which for $\beta \ll \beta_c$ can be written approximately in the Curie-Weiss form:

$$\chi = \frac{\mu_B^2}{k(T - T_c)},$$

with

$$T_c = \frac{1}{k\beta_c},$$

where β_c is given by Eq.(8.461).

8.13. The equations of motion for the Green functions $\ll \hat{b}_g; \hat{b}_f^\dagger \gg$ and $\ll \hat{b}_g^\dagger; \hat{b}_f^\dagger \gg$ are

$$E \ll \hat{b}_g; \hat{b}_f^\dagger \gg = \frac{\sigma}{2\pi}\delta_{fg} + [2\mu_B B - 2NN_z\mu_B^2 + 2\sigma K(0)] \ll \hat{b}_g; \hat{b}_f^\dagger \gg$$
$$+ \sum_p [-2\sigma I(g-p) + \sigma\mu_B^2(N_x + N_y)] \ll \hat{b}_p; \hat{b}_f^\dagger \gg$$
$$+ \sigma\mu_B^2(N_x - N_y) \sum_p \ll \hat{b}_p^\dagger; \hat{b}_f^\dagger \gg,$$

$$E \ll \hat{b}_g^\dagger; \hat{b}_f^\dagger \gg = -[2\mu_B B - 2NN_z\mu_B^2 + 2\sigma K(0)] \ll \hat{b}_g^\dagger; \hat{b}_f^\dagger \gg$$
$$- \sum_p [-2\sigma I(g-p) + \sigma\mu_B^2(N_x + N_y)] \ll \hat{b}_p^\dagger; \hat{b}_f^\dagger \gg$$
$$- \sigma\mu_B^2(N_x - N_y) \sum_p \ll \hat{b}_p; \hat{b}_f^\dagger \gg,$$

where we have used the decoupling approximations mentioned in the statement of this problem as well as the fact that the extra term in the Hamiltonian in terms of the $\hat{b}_f$ and the $\hat{b}_f^\dagger$ has the form

$$\tfrac{1}{2}N^2 N_z \mu_B^2 - 2N N_z \mu_B^2 \sum_f \hat{n}_f + (N_x + N_y)\mu_B^2 \sum_{f,g} \hat{b}_f^\dagger \hat{b}_g$$
$$+ N_z \mu_B^2 \sum_{f,g} \hat{n}_f \hat{n}_g + \tfrac{1}{2}\mu_B^2 (N_x - N_y) \sum_{f,g} \left(\hat{b}_f \hat{b}_g + \hat{b}_f^\dagger \hat{b}_g^\dagger \right).$$

We again Fourier transform with respect to the lattice sites and use Eqs.(8.446) and (8.447) as well as the equation

$$\ll \hat{b}_g^\dagger ; \hat{b}_f^\dagger \gg = \frac{1}{N} \sum_q \Gamma_q \, e^{i([g-f]\cdot q)}.$$

The result is

$$(E + \overline{E}_q)\, G_q = \frac{\sigma}{2\pi} + \sigma\mu_B^2 (N_x - N_y) N \delta_{q0} \Gamma_q,$$
$$(E + \overline{E}_q)\, \Gamma_q = -\sigma\mu_B^2 (N_x - N_y) N \delta_{q0} G_q,$$

with

$$\overline{E}_q = E_q + N\sigma\mu_B^2 \left[(N_x + N_y)\delta_{q0} - 2N_z \right].$$

We note that for $q \neq 0$ the quantity $\overline{E}_q$ differs from E_q in that B is replaced by $B - N_z(N\sigma\mu_B)$, that is, by the field including the demagnetisation.
Solving the equations for G_q and Γ_q we find

$$q \neq 0: \qquad \Gamma_q = 0, \qquad G_q = \frac{\sigma}{2\pi(E - \overline{E}_q)},$$
$$q = 0: \qquad G_0 = \frac{\sigma}{2\pi} \frac{E - E_0}{E^2 - E_{\text{res}}^2},$$

with

$$E_{\text{res}}^2 = 4\mu_B^2 \left[B + N\sigma\mu_B(N_x - N_z) \right] \left[B + N\sigma\mu_B(N_y - N_z) \right].$$

Using these results we find the following expression for χ_+:

$$\chi_+ = -4\pi\mu_B^2 \sum_f \ll \hat{b}_g ; \hat{b}_f^\dagger \gg = -\frac{4\pi\mu_B^2}{N} \sum_f \sum_q G_q \, e^{i([f-g]\cdot q)}$$

$$= \lim_{\varepsilon \to 0+} \left[-\frac{4\pi\mu_B^2}{N} \sum_{q\neq 0} \frac{\sigma}{2\pi} \frac{1}{\omega + i\varepsilon - E_q} \sum_f e^{i([f-g]\cdot q)} \right]$$

$$- \lim_{\varepsilon \to 0+} \left[\frac{4\pi\mu_B^2}{N} \sum_f \frac{\sigma}{2\pi} \frac{\omega + i\varepsilon - E_0}{(\omega + i\varepsilon)^2 - E_{\text{res}}^2} \right]$$

$$= \lim_{\varepsilon \to 0+} \left[-\frac{\sigma\mu_B^2}{E_{\text{res}}} (\omega + i\varepsilon - E_0) \left\{ \frac{1}{\omega + i\varepsilon - E_{\text{res}}} - \frac{1}{\omega + i\varepsilon + E_{\text{res}}} \right\} \right].$$

We now use Eq.(8.427) and the fact that both ω and E_{res} are positive so that $\delta(\omega + E_{\text{res}}) = 0$ and find finally for χ'' the expression

$$\chi'' = \frac{2\pi\mu_B^2(\omega + \overline{E}_0)}{E_{\text{res}}} \delta(\omega - E_{\text{res}}),$$

which shows that absorption takes place only at $\omega = E_{\text{res}}$. In the approximation used the linewidth is zero.

8.14. In terms of the $\hat{b}_f$ and the $\hat{b}_f^\dagger$ the Hamiltonian has the form

$$\hat{\mathcal{H}} = -\mu_B BN + 2\mu_B B \left[\sum_f \hat{n}_f + \sum_g \hat{n}_g \right] - 2I \sum_{f,g} \hat{b}_f^\dagger \hat{b}_g$$

$$- \tfrac{1}{2}zNI + 2zI \left[\sum_f \hat{n}_f + \sum_g \hat{n}_g \right] - 2I \sum_{f,g} \hat{n}_f \hat{n}_g.$$

Consider now the Green functions $\ll \hat{b}_f; \hat{b}_h^\dagger \gg$ and $\ll \hat{b}_g; \hat{b}_h^\dagger \gg$, where h may be either a f- or g-site. If l denotes either f or g the equations of motion of those Green functions will be

$$E \ll \hat{b}_l; \hat{b}_h^\dagger \gg = \frac{\delta_{lh}}{2\pi}(1 - 2\hat{n}_l) + (2\mu_B B + zI) \ll \hat{b}_l; \hat{b}_h^\dagger \gg$$

$$- 2I \sum_k \ll \hat{b}_k; \hat{b}_h^\dagger \gg + 4I \sum_k \ll \hat{n}_l \hat{b}_k - \hat{n}_k \hat{b}_l; \hat{b}_h^\dagger \gg,$$

where the sums over k are over the z nearest neighbours of the site l.

Using again the decoupling approximation (8.443) we proceed as before, Fourier transforming with respect to the lattice and bearing in mind that now $\langle n_f \rangle$ is not necessarily equal to $\langle n_g \rangle$. To fix the ideas we assume that the site h belongs to the f-sublattice so that $\delta_{gh} = 0$. Denoting $\ll \hat{b}_f; \hat{b}_h^\dagger \gg$ by G and $\ll \hat{b}_g; \hat{b}_h^\dagger \gg$ by Γ we find the following expressions for those two Green functions:

$$G = \frac{\sigma_f}{4\pi N} \sum_q e^{i q \cdot [f - h]} \frac{E - 2\mu_B B + 2zI\sigma_f}{(E - E_1)(E - E_2)},$$

$$\Gamma = \frac{zI\sigma_f\sigma_g}{2\pi N} \sum_q e^{i q \cdot [g - h]} \frac{\gamma_q}{(E - E_1)(E - E_2)},$$

where we have

$$\sigma_l = 1 - 2n_l,$$

and

$$\gamma_q = \frac{1}{z} \sum_\delta e^{i(\delta \cdot q)},$$

with $\delta = f - g$, f and g being nearest neighbours and the summation being over all z nearest neighbours, while E_1 and E_2 are the solutions of the equation

$$\left[E - 2\mu_B B + 2zI\sigma_f\right]\left[E - 2\mu_B B + 2zI\sigma_g\right] - 4z^2 I^2 \sigma_f \sigma_g \gamma_q^2 = 0.$$

For the sublattice magnetisations σ_l we find

$$\sigma_l = \frac{1}{1 + 2\Psi_l},$$

with

$$\Psi_l = \frac{1}{2N} \sum_q \left\{ f(1) + f(2) + \frac{2zI(\sigma_l - \sigma_{l'})}{E_1 - E_2} \left[f(1) - f(2)\right] \right\},$$

where $l' = g$, if $l = f$ and $l' = f$, if $l = g$ while

$$f(i) = \frac{1}{e^{\beta E_i} - 1}.$$

If there is no external magnetic field we can assume that $\sigma_f = -\sigma_g (= \sigma_0)$. We then find that

$$\Psi_f = \frac{1}{2} \left[\frac{1}{N} \sum_q \frac{2zI\sigma_0}{E_0} \coth\left(\tfrac{1}{2}\beta E_0\right) - 1 \right],$$

with

$$E_0 = 2I\sigma_0 \sqrt{1 - \gamma_q}.$$

If T_N is the Néel temperature we find that $\Psi_f \to \infty$ as $T \to T_N$, while

$$\sigma_0 \approx \frac{1}{\Psi_f + 2},$$

and

$$\Psi_f \approx \frac{1}{N} \sum_q \frac{kT}{2zI\sigma_0(1 - \gamma_q^2)}.$$

Considering the limit as $\sigma_0 \to 0$ we find the following expression for the Néel temperature:

$$T_N = \frac{zI}{2k} \left[\frac{1}{N} \sum_q \frac{1}{1 - \gamma_q} \right]^{-1}.$$

9.2(iii). We find that at $v = v_c$ we have

$$\frac{\partial^n P}{\partial v^n} = \mathcal{O}\left(N^{(n-2)/3}\right).$$

9.3. In terms of u_0 one has

$$\xi = u_0 - \tfrac{1}{2}u_0^2, \qquad \eta = u_0, \qquad \zeta = \frac{u_0}{1 - u_0}.$$

9.8. In the mean-field approximation the critical temperature of an Ising ferromagnet is given by the equation

$$T_c = \frac{\Phi_0}{4k}.$$

9.17. The eigenvalues of the one-dimensional Ising ferromagnet are given by Eq.(9.924) so that Eq.(9.921) is of the form

$$Z = e^{NK}\left\{\left(\cos\tfrac{1}{2}\theta + \sqrt{\eta - \sin^2\tfrac{1}{2}\theta}\right)^N + \left(\cos\tfrac{1}{2}\theta - \sqrt{\eta - \sin^2\tfrac{1}{2}\theta}\right)^N\right\},$$

where

$$\theta = 2iC \qquad \text{and} \qquad \eta = e^{-4K}.$$

Introducing a new variable, φ, through the equation

$$\cos\varphi = \frac{\cos\tfrac{1}{2}\theta}{\sqrt{1-\eta}}$$

the equation $Z = 0$ reduces to

$$\cos N\varphi = 0,$$

with the solutions

$$\varphi = \pm\frac{(k-\tfrac{1}{2})\pi}{N}, \qquad k = 1,2,\ldots,N.$$

9.18. We shall assume that the lattice gas consists of N_{at} atoms distributed over N lattice sites.

(i) The grand partition function $Z_{l.g.}$ of the lattice gas is given by the equation

$$Z_{l.g.} = \sum_{N_{at}=0}^{\infty} e^{\alpha N_{at}} \sideset{}{'}\sum_{N_{n.n.}} W(N_{at}, N_{n.n.}) e^{\beta\varepsilon N_{n.n.}},$$

where $N_{n.n.}$ is the number of nearest-neighbour pairs on the lattice where both sites are occupied by atoms, where $W(N_{at}, N_{n.n.})$ is the number of configurations with given values of N_{at} and $N_{n.n.}$, and where the second summation is over all values of $N_{n.n.}$ compatible with a given value of N_{at}.

Comparing the equation for $Z_{l.g.}$ with the expression for the partition function of an Ising ferromagnet in a magnetic field we see that the two are the same if we replace N_{at} by the number of up-spins, ε by J, and $y \equiv e^{\alpha}$ by $e^{-\beta z J + 2\beta\mu_B B}$.

(ii) In the mean-field approximation the specific volume v satisfies the equation

$$\frac{1}{v} = \frac{N_{at}}{N} = \tfrac{1}{2}(1 + R_0),$$

where R_0 is the equilibrium value of the order parameter, whereas the pressure P follows from the grand potential q which by the results of (i) corresponds to the free energy of the Ising ferromagnet:

$$\frac{Pv}{kT} = -\frac{F}{kT} = \beta E_1 - \beta E_0(1 - R^2)$$
$$+ \left[\ln 2 - \tfrac{1}{2}(1 + R)\ln(1 + R) - \tfrac{1}{2}(1 - R)\ln(1 - R)\right].$$

The isotherms, which now are equivalent to the P, R_0-curves, are found from the equations

$$P = \mu_B B - \tfrac{1}{4}\varepsilon(1 + R_0^2) - \tfrac{1}{2}kT \ln \tfrac{1}{4}(1 - R_0^2),$$
$$R_0 = \tanh \tfrac{1}{2}(z\varepsilon R_0 + \beta\mu_B B).$$

It is interesting to find the values of P, v, and T at the critical point, that is, the point where $R_0 = 0$ for $B = 0$. The result is

$$T_c = \frac{z\varepsilon}{2k}, \qquad v_c = 2, \qquad P_c = -\tfrac{1}{2}kT_c + kT_c \ln 2,$$

and for the combination $P_c v_c / kT_c$ we find the value 0.386 as against the value 0.375 for the van der Waals gas.

(iii) From the results of the previous problem and of (i) it follows that if we write

$$y = e^{-\beta z J + i\theta}$$

the zeros of the grand partition function of the lattice gas satisfy the relation

$$\cos\theta = -\eta + (1 - \eta)\cos\frac{(2k - 1)\pi}{N}, \qquad k = 1, 2, \ldots, N.$$

We see that, indeed, the zeros lie on a segment of a circle in the y-plane which in the limit as $N \to \infty$ is continuous with a gap where the circle cuts the real axis. This gap lies between $\theta = -\theta_0 = -\arccos(1 - 2\eta)$ and $\theta = +\theta_0$. The gap only closes up if $\eta = 0$, that is, when $T = 0$: the one-dimensional Ising ferromagnet does not show a phase transition at any finite temperature.

9.19. The functions χ and ψ of Eq.(E) are found to satisfy the equations

$$\chi^a = \sum_{\{\sigma,\sigma'\}} e^{K\zeta + \left(\frac{1}{2}K + H\right)(\xi + \xi')\left(\frac{1}{2}C + B\right)(\eta + \eta')},$$

and

$$\psi^a = \sum_{\sigma} e^{2H\xi + 2C\eta},$$

where

$$\xi' = \frac{1}{a} \sum \sigma_i' \sigma_{i+1}', \qquad \eta' = \frac{1}{a} \sum \sigma_i', \qquad \text{and} \qquad \zeta = \frac{1}{a} \sum \sigma_i \sigma_i'.$$

Evaluating these functions by the method used in §9.9 to find Eq.(9.921) we find a cubic equation for χ:

$$\chi^3 - 2\chi^2 \left[e^{2K+2H} \cosh 2(B+C) + e^{-K} \cosh(K+2H) \right]$$
$$+ 4\chi \sinh(K+2H) \left[e^{K+2H} \cosh(2B+C) + e^{2K} \cosh(K+2H) \right]$$
$$- 8e^K \sinh^3(K+2H) = 0,$$

the largest root of which gives χ while ψ has been found in §9.9 and is equal to

$$\psi = e^{2H} \cosh 2B + \sqrt{e^{4H} \sinh^2 2B + e^{-4H}}.$$

In zero field, so that $C = 0$, $B = 0$ is always a solution since both $\partial\psi/\partial B$ and $\partial\chi/\partial B$ are equal to zero when $B = 0$. Putting $B = 0$ the equation for χ redues to

$$\chi^2 - 4\chi \cosh K \cosh(K+2H) + 4\sinh^2(K+2H) = 0.$$

Introducing now a variable s through the equation

$$s\lambda \cosh 2H = \sinh(K+2H)$$

and using the relation

$$\cosh K \cosh(K+2H) = \sinh K \sinh(K+2H) + \cosh 2H,$$

we find that

$$\lambda = \max_s \frac{2}{(s - \sinh k)^2 + 1 - \sinh^2 K},$$

or

$$\lambda = \frac{2}{1 - \sinh^2 K}.$$

This solution is valid only at sufficiently high temperatures, since for low temperatures the expression for λ would become negative. This means that at low temperatures the relation $B = 0$ leads to a metastable solution. One must therefore look for a solution with $B \neq 0$; the resulting equations are much more complicated. The two regimes join at the transition temperature which in the present approximation is given by the equation

$$T_c = 1.21 \frac{J}{k},$$

which can be compared to values of $1.13J/k$ and $2J/k$ for the exact solution and the mean-field approximation, respectively.

INDEX

A priori phases 238

a priori probability 26, 60, 71, 88, 92, 201, 203, 209, 222, 232, 237, 238, 262, 361

absolute temperature 102, 159, 225

acceptor level 139

accessible state 93

action and angle variables 74

action variable 193

activity 183

adiabatic changes 192

adiabatic invariant 193

advanced Green function 302

Aharanov-Bohm effect 146

allowed band 138

anharmonicity 85

anisotropic harmonic oscillator 73

annihilation operators 132, 297, 305

anticommutator 300

antiferromagnetism 313, 316, 338

anyons 134

asymmetric top 55

attractive fixed point 360

averages 40, 105, 154, 181, 218

Avogrado number 6

Band structure 138

bare mass 289

barometer formula 36

Bernoulli numbers 129

binary alloys 335, 338, 339, 342

black-body radiation 244, 289

Bogolyubov Hamiltonian 299

Bohr-van Leeuwen theorem 56

Boltzmann constant 102

Boltzmann factor 361

Boltzmann gas 112

Boltzmann statistics 89, 114, 135

Boltzmann transport equation § 2.5

Boltzmann weight 92, 110

Boltzmann's H 143, 172, 182, 194, 207

Bose-Einstein distribution 99

Bose-Einstein statistics 89

Bose-Einstein weight 110

bosons 59, 93

Boyle-Gay-Lussac law 5

Bragg-Williams approximation 342

Brillouin function 84

Brillouin zone 308, 309

Brownian motion 302

Canonical coordinates 39

canonical ensemble 156

canonical equations of motion 39

canonical grand ensemble 180, § 6.4

causal Green function 302

centre of mass velocity 12

chaotic motion 205

chemical constant 113

chemical potential 103, 126

chemical reactions 179

classical limit 90

classical statistics 89, 114

cluster 270, 325

cluster expansion 284

cluster integral 324

coarse-grained density 207, 208, 236

coefficient of probability 153

coexisting phases 280, 317

collisions § 1.4

commutation relations 296, 300, 307

Made in the USA
Lexington, KY
17 July 2013